Dynamics of Membrane Assembly

NATO ASI Series

Advanced Science Institutes Series

A series presenting the results of activities sponsored by the NATO Science Committee, which aims at the dissemination of advanced scientific and technological knowledge, with a view to strengthening links between scientific communities.

The Series is published by an international board of publishers in conjunction with the NATO Scientific Affairs Division

A Life Sciences	Plenum Publishing Corporation
B Physics	London and New York
C Mathematical and Physical Sciences	Kluwer Academic Publishers Dordrecht, Boston and London
D Behavioural and Social Sciences	
E Applied Sciences	
F Computer and Systems Sciences	Springer-Verlag Berlin Heidelberg New York
G Ecological Sciences	London Paris Tokyo Hong Kong
H Cell Biology	Barcelona Budapest
I Global Environmental Change	

NATO-PCO DATABASE

The electronic index to the NATO ASI Series provides full bibliographical references (with keywords and/or abstracts) to more than 30 000 contributions from international scientists published in all sections of the NATO ASI Series. Access to the NATO-PCO DATABASE compiled by the NATO Publication Coordination Office is possible in two ways:

– via online FILE 128 (NATO-PCO DATABASE) hosted by ESRIN,
  Via Galileo Galilei, I-00044 Frascati, Italy.

– via CD-ROM "NATO-PCO DATABASE" with user-friendly retrieval software in English, French and German (© WTV GmbH and DATAWARE Technologies Inc. 1989).

The CD-ROM can be ordered through any member of the Board of Publishers or through NATO-PCO, Overijse, Belgium.

Series H: Cell Biology Vol. 63

Dynamics of Membrane Assembly

Edited by

Jos A. F. Op den Kamp

Centre for Biomembranes and Lipid Enzymology
Padualaan 8, 3584 CH Utrecht, The Netherlands

 Springer-Verlag Berlin Heidelberg GmbH

Proceedings of the NATO Advanced Study Institute on Dynamics of Membrane
Assembly, held at Cargèse (France) June 17–29, 1991

ISBN 978-3-662-02862-9   ISBN 978-3-662-02860-5 (eBook)
DOI 10.1007/978-3-662-02860-5

Library of Congress Cataloging-in-Publication Data
Dynamics of membrane assembly / edited by Jos A. F. op den Kamp.
  (NATO ASI series. Series H, Cell biology ; vol. 63)
"Published in cooperation with NATO Scientific Affairs Division." "Proceedings of the NATO Advanced Study
Institute on Dynamics of Membrane Assembly, held at Cargèse (France) June 17–29, 1991" – T.p. verso. In-
cludes bibliographical references and index.

1. Membranes (Biology) – Congresses. 2. Cell membranes – Formation-Congresses. I. Kamp, Jos A. F. op den
(Jos Arnoldus Franciscus), 1939– . II. North Atlantic Treaty Organization. Scientific Affairs Division. III. NATO
Advanced Study Institute on Dynamics of Membrane Assembly (1991 : Cargèse, France) IV. Series. QH601.D89
1992 574.87'5 – dc20

Originally published by Springer-Verlag Berlin Heidelberg New York in 1992.
Softcover reprint of the hardcover 1st edition 1992

Typesetting: Camera-ready by authors
31/3145-5 4 3 2 1 0 – Printed on acid-free paper

PREFACE

The meeting on "Dynamics of Membrane Assembly.", sponsored by NATO Scientific Affairs Division as an Advanced Study Institute and by FEBS as a Lecture Course was held in Cargèse, France, in June 1991. The program included introductory lectures, specialized up-to-date contributions and poster sessions. Emphasis was laid on the new developments in the field of membrane biogenesis, in particular on the biosynthesis of phospholipids and the application of modern genetic techniques in these studies; on the membrane insertion and translocation of proteins; on intracellular protein and membrane traffic; and on the mutual interactions between the various events occurring during membrane biogenesis.

Much progress in these research areas has been made in recent years and the ASI provided an excellent opportunity to illustrate this progress in comparison with previous meetings on a similar topic. Not only graduate students and postdocs took advantage from this program but also experienced scientists were given the opportunity to obtain a complete overview of recent progress and the remodelling of ideas and concepts.

The present publication presents most of the important contributions by teachers as well as students in the meeting. It is clearly illustrated here that our knowledge of lipid biosynthesis is increasing rapidly, be it that progress is most obvious in those organisms in which genetic techniques are readily applicable such as bacteria and yeast. The study of these processes in more complex organisms has not, as yet, reached this level. A similar conclusion can be drawn for the complicated processes of protein insertion and translocation.and the intracellular membrane transport. Nevertheless detailed information on the basic mechanistic characteristics of these events become available, in particular from studies on model and reconstituted systems. Integration of the information regarding biosynthesis of individual membrane constituents, the assembly into functional membranes, and the regulation of these steps is illustrated in various contributions throughout.

An excellent overview of the meeting and an in depth summary of the contents of this book is written by Dr. A. P. Pugsley. (see pages 1-10) I would like to thank him in particular for this contribution and for his enthusiasm which ensured, together with the contributions of all the other participants, the success of this meeting.

December, 1991

Jos A.F. Op den Kamp

CONTENTS

THE STRUCTURE, GENESIS AND DYNAMICS OF BIOLOGICAL MEMBRANES

Anthony P. Pugsley
Unité de Génétique Moléculaire
Institut Pasteur
25, rue du Dr. Roux
Paris 75724
France

Introduction

The technical and intellectual skills required to solve some of the complex problems that face modern scientists are highly valued by society as a whole and by scientists in different disciplines in particular. This does not mean, of course, that the relative values of the various disciplines are not hotly disputed, especially when scientists in one discipline feel that others are being favored with greater financial support, or where such support is perceived as being wasted to the detriment of research into other subjects. Within the narrower field of molecular cell biology, the relative importance of model systems, subcellular components, types of organisms and different approaches are frequently discussed. Of course, the aim of scientific research within a particular discipline is to advance our knowledge of the topic in its entirety, but this is not an ideal world and there are certainly insufficient funds and far too few research scientists to do more than skim the surface of the wealth of potential that the subject possesses. The choice of system or approach is, therefore, usually dictated by such things as familiarity with a particular organism or experimental approach or by other than the purely fundamental interest of a particular topic (medical or biotechnological applications, for example).

With such limited resources available, it is essential that scientists using different approaches to answer the same basic question maintain close contact. This is particularly important for students and junior research scientists who are just starting their careers; the broader their initial grasp of a particular subject the greater their potential to make significant progress as they develop. This is essentially the purpose of NATO-sponsored summer schools such as the one on which this book is based.

NATO ASI Series, Vol. H 63
Dynamics of Membrane Assembly
Edited by J. A. F. Op den Kamp
© Springer-Verlag Berlin Heidelberg 1992

What are the basic questions that we are trying to answer in the discipline of the membrane structure and biology, the topic of this book? We should start by trying to understand the structure of the individual membrane components, the ways in which they interact with each other and their relative importance in the various biological functions of membranes. It should be obvious to anyone with even the most rudimentary knowledge of biology that membranes are essential to life, but do they do more than provide a simple dialysis bag within which essential activities such as DNA replication, transcription and translation occur?

Lipid bilayers: dialysis bags or life belts?

Most biological membranes are lipid bilayers whose composition can vary in terms of both head group and fatty acyl chain structure between different organisms, between different membranes in the same cell and even between the same membrane from the same cell type grown under different conditions. To take just one example, the fatty acyl composition of membrane lipids is often dramatically altered in response to changing environmental conditions such as temperature. These changes are not fortuitous, but instead reflect that fact that the lipids existing in the cell before the environmental conditions changed function less well in the new environment. They must therefore be replaced by better-adapted lipids. The ways in which membrane lipid composition is regulated is one of the least studied areas of molecular biology, and yet the tools with which to probe deep into the complex control patterns that exist are now available, especially in the bacterium *Escherichia coli* and in the yeast *Saccharomyces cerevisiae* (see articles by W. Dowhan and by S. Henry *et al*).

The structures of many lipids are known in exquisite detail and we also know a great deal about how their structure can influence membrane organization and other physical properties. If we accept that cells equip themselves with a mixture of membrane lipids that are ideally suited to their lifestyle, we are well on the way to accepting that that have more than a purely mechanical role to play in the cell. Therefore, one would not expect enforced changes in lipid composition to be well tolerated since this should disrupt many physiological functions without necessarily affecting the permeability barrier created by the lipid bilayer. This particular question can now be addressed directly since the development of molecular genetic approaches has permitted scientists to selectively block the production of certain lipids by inactivating a gene coding for an enzyme that is essential for the biosynthesis of that lipid. As described in the article by Dowhan, *E coli* contains three major species of membrane lipids: phosphatidyl ethanolamine (PE), phosphatidyl glycerol (PG) and a glycolipid called lipopolysaccharide (LPS) that is found exclusively in the outer leaflet of the outer membrane. If *E. coli* has evolved a lipid composition optimally suited to its particular needs, then one would expect mutations that drastically alter the ratio of these lipids or eliminate one of them entirely to be poorly tolerated, if at all. This has long been known to be the case for LPS, the cellular concentration of which cannot be reduced appreciably without compromising cell viability. Moreover, Dowhan describes experiments showing that this is also the case for the acidic head group phospholipid PG, which accounts for up to 25% of total phospholipids. *E. coli* must have a certain critical level of PG to remain viable. Why is PG required? Dowhan's mutants have been used to show that protein translocation across the cytoplasmic membrane is reduced when this membrane contains suboptimal levels of PG (see article by Brundage and Wickner), and other studies show that acidic phospholipids are required for chromosome replication initiation. Dowhan considers it unlikely that PG has a catalytic role, and favors instead the idea that it creates a favorable environment for productive interactions between specific proteins required for these processes and the membrane. It has also been shown by deKruijff and others that precursors of exported proteins of *E. coli* can interact directly and efficiently with negatively-charged phospholipids in monolayers.

The results of similar studies with mutants containing reduced amounts of PE, which normally accounts for over 70% of phospholipids, were the most unexpected and surprising. Growth of *E. coli* stops when the level of PE drops to ca 30%, but addition of magnesium ions to the medium allows the cells to grow under conditions where only trace amounts of PE are present in the membrane. Under these circumstances, PG becomes the major phospholipid [up to 80%, the remainder comprising the normally minor cardiolipin (CL)], but the lipid:protein and lipid:LPS ratios in the membrane remain unchanged. Rather surprisingly, the PE-depleted cells do not synthesize novel phospholipids or amplify the production of any of the numerous minor phospholipids they normally contain. PE is one of the phospholipids that are capable of forming non-bilayer hexagonal phases in lipid suspensions. The biological significance of such structures, if they do indeed exist in nature, remains unresolved, but does the behavior of Dowhan's mutants prove that such structures are unimportant? The answer seems to be no because the mutants remain viable only as long as they retain the ability to produce CL and as long as the medium contains divalent cations, which induce CL to form non-bilayer structures similar to those formed by PE.

These results with the PE-depleted strain are not at all easy to interpret. For example, divalent cations also stabilize the outer membrane and may also neutralize the highly negatively charged membrane surface resulting from the increased production of PG and CL. Nevertheless, Dowhan's studies represent a first, vital step towards determining the roles of individual lipids in metabolic activities. For the present, it is impossible to determine whether the trace amounts of the depleted lipids that invariably remain in the mutant strains are required as cofactors, substrates or regulators. In addition, the biological significance of minor phospholipids also needs to be evaluated.

The yeast *S. cerevisiae* has a different, more complex lipid composition than *E. coli*. Despite this complexity and the added complications caused by lipid movement between membranes of different organelles, Henry and her colleagues have been able to isolate mutations that dramatically alter bulk lipid composition. The levels of two of the normally abundant lipids, phosphatidyl serine (PS; $\geq$9% of total phospholipids) and phosphatidyl choline (up to 56% of total phospholipids) can be drastically reduced by introducing mutations into the *OPI3* and *CHO1* genes respectively. These mutations tip the balance of phospholipid composition in favor of other phospholipids or lead to auxotrophy for certain phospholipid precursors; PC, for example is replaced by similar amounts of monomethylethanolamine in the *opi3* mutant, and *cho1* mutants must be supplemented with choline or ethanolamine. This certainly complicates attempts to determine the relative importance of the naturally-occurring phospholipids in cell physiology and metabolism, but the genetic approach is well worth persuing, and should soon lead to a better understanding of how phospholipid composition is regulated.

The idea that membrane proteins are surrounded by very specific types of lipids with which they interact (so-called boundary lipids) has long been discredited. Nevertheless, we know that certain membrane proteins require certain types of phospholipids for activity, and we therefore cannot totally discount the existence of specific lipid-protein interactions. This idea is taken up by Milton Schlesinger in his article on the lipid modifications of proteins. This is one of the most rapidly expanding areas of interest in membrane and protein biology. Although we now know a great deal about how fatty acyl groups are attached to proteins in eukaryotic cells and are beginning to identify the enzymes, cofactors and substrate recognition sites involved, we still have no clear ideas on what the function of the fatty acids really are. Schlesinger proposes that at least some protein-linked fatty acyl groups do not serve merely (if at all) as membrane anchors, but instead ensure that an appropriate fatty acid is located in the correct position adjacent to the transmembrane segments of the polypeptide to which they are attached. This is an interesting idea which should be given more attention.

As Schlesinger points out, different fatty acids in different proteins are likely to be there for quite different reasons. The PI glycolipid proteins (see articles by Overath *et al* and by Foltz *et al*) use their C-terminally attached glycolipid anchors to remain anchored to the cell surface. Rather surprisingly, the PI glycolipid anchor replaces an apparently equally hydrophobic peptide segment that exists in the precursor polypeptide. Why replace one hydrophobic structure by another? Among the ideas that have been raised are the possibility that the PI glycolipid anchor allows the protein to be released by the action of a PI-specific phospholipase or that it endows the protein with greater lateral mobility in the membrane. The membrane association of other lipidated proteins, such as those carrying a myristoyl group, is also critically dependent on the presence of the lipid moiety. Schlesinger's own elegant work has shown that palmitoyl groups attached to certain viral proteins are required for efficient viral assembly, suggesting that the palmitate may be involved in the restructuring of the lipid bilayer that may be necessary to allow the complex process of virus assembly to occur (see above). Another possibility that the fatty acyl groups are conformational determinants that ensure that segments of the polypeptide to which they are attached are brought close to the surface of the membrane without being forced to penetrate into it.

Lipid shuffling

An added complication (or interest) of lipid metabolism in eukaryotes is that most of it is restricted to one or a very few regions of the cell, necessitating the transport of lipids from one organelle to another, and thus from one part of the cell to another. Understanding lipid traffic is crucial to understanding how the different organelles and membranes maintain their individual lipid composition and hence their identity. This topic is dealt with in this volume by Jean Vance, who describes her experience with two specific pathways for intermembrane phospholipid transfer. Vance lists four plausible methods by which such transfer could occur: free diffusion of lipids through the cytosol, vesicle-mediate transport, carrier protein-mediated transport, and direct contact between different organelles. All of these modes of lipid transport could conceivably exist within a single cell, and each one has its own unique characteristics. For example, Vance points out that free diffusion of lipid monomers, if it exists, must be restricted to free fatty acids, phosphatidic acids and CDP-diacylglycerol because other lipids are only sparingly soluble in water. Vesicle mediated lipid transfer is appealing in terms of organelle biogenesis and protein traffic because proteins are also transferred from the endoplasmic reticulum (ER) to organelles along the secretory pathway and then on to the cell surface in vesicles. This bulk flow of proteins must surely necessitate the movement of large quantities of lipids, and yet Vance provides strong experimental evidence to argue that ER to plasma membrane traffic of lipids can occur when vesicle-mediate protein traffic is negligible. Carrier-mediated lipid transport would seem to be involved in this case, but these carriers are unlikely to return "unladen" to the ER. Another aspect of lipid transport that is not resolved is how lipids get from the ER to the Golgi apparatus. Indeed, the highly complicated pathways for the forward and retrograde transport of both proteins and lipids between organelles in the secretory pathway and the plasma membrane, and in particular the role of exocytosis and endocytosis need to be carefully reevaluated in the light of Vance's proposals.

In the other example of lipid traffic studied by Vance, PS is transported from the ER to mitochondria, where it is decarboxylated by an inner membrane enzyme whose active site faces the intermembrane space. Vance proposes that this traffic is mediated by occasional direct contact between the ER and the mitochondrion. This pathway again has some remarkable features; for example, newly-synthesized PS seems to be the preferred substrate. Soluble transfer proteins do not seem to be involved. The major question raised by Vance's proposal, which is based largely on studies conducted *in vitro* with mixtures of ER-derived microsomes and mitochondria, is whether the ER and mitochondria actually encounter each other sufficiently often for efficient transfer to occur. A possible clue that this might occur is the isolation by Vance and her colleagues of a subcellular fraction that seems to have a mixture of properties that are usually attributed individually to microsomes or to mitochondria (fraction X). More work needs to be carried out on this fraction to see whether is does indeed have the capacity to transfer PS from one membrane to another and how the interaction between the different membranes is maintained.

A rather unusual example of the importance of lipid transfer proteins, noted both by Henry *et al* and by Dowhan (who was directly involved in the work) is the demonstration that the phosphatidyl inositol/phosphatidyl choline (PI/PC) transfer protein is essential for the vesicle-mediated exit of secretory proteins from the Golgi apparatus in yeasts. Extragenic suppressors of mutations blocking the activity of PI/PC transfer protein are affected in PC metabolism, leading Henry to propose that charge imbalance caused by the accumulation of PC in the absence of PI transfer may disrupt vesicle formation at the surface of specific regions of the Golgi that pinch off to form secretory vesicles

Membrane proteins: deceptions and misconceptions

Nearly every membrane protein biologist I know dreams of one day being able to see a high resolution, three-dimensional map of his favorite protein. The technical problems involved in performing such high resolution analysis on membrane proteins are enormous. The crystal structures of only 6 integral membrane proteins are known, which is nowhere near a sufficiently large data base from which to construct models for other, unrelated proteins. Many membrane biologists therefore resort to much cruder techniques that nevertheless allow at least some details of the topology of membrane proteins to be determined. Gene fusion techniques were very successfully used by Ron Kaback, Colin Manoil and their colleagues to construct a topology map of the lactose transporter protein (LacY) of the *E. coli* cytoplasmic membrane. The structure of LacY seems to be typical of many plasma membrane proteins in that its many transmembrane segments (in this case 12) are composed of mainly hydrophobic or neutral residues which together have a strongly predicted tendency to form an alpha helix (see article by Ron Kaback). There are several models for the ways in which transmembrane segments of polytopic membrane proteins insert into membranes. Some such segments seem to be able to insert independently, while others insert in pairs has a hairpin structures. Still more complicated cooperative patterns of insertion are also possible, but it is assumed that the final overall topology of polytopic proteins represents the net sum of all of the individual insertion events. One would therefore expect the final topology to be very susceptible to changes in the number of potential transmembrane segments. For example, removing or adding potential transmembrane segments to such polytopic proteins as LacY should not only inactivate the protein, but might also cause the region following the inserted or deleted segment to appear in reverse orientation. The topology of such derivatives of LacY protein was determined by tagging its end with a readily identifiable reporter protein. Kaback and his colleagues found that as long as the first and the last four transmembrane segments of LacY were retained, many drastic rearrangements in the center of the polypeptide did not affect the orientation of the last transmembrane segment, which invariably terminated in the cytoplasm. It would be informative to determine the overall topology of some of the altered proteins with uneven numbers of hydrophobic domains to see, for example, whether one of them remains looped out of the membrane or, conversely, if a relatively long hydrophobic segment of the polypeptide that normally spans the membrane only once could loop back upon itself so that it now crosses the membrane twice instead.

Getting proteins into and across membranes

The problem of understanding how proteins insert into and are transported across membranes is considerably more complex than that of determining the structure and function of a protein once it is inserted in a membrane. The articles by Stirling and Schekman, Brundage and Wickner, Stader *et al*, Schatz, Fridén *et al* and Benedetti *et al* all provide details of biochemical and genetic approaches to studying this problem in several different systems (protein transport across the bacterial cytoplasmic membrane and outer membrane, into the lumen of the ER and into mitochondria). Although there may be some exceptions (see below), it is now widely believed that proteins do not normally penetrate into or cross membranes unassisted. Not only is energy required to catalyse the process, but, soluble-, peripheral membrane- and intergral- membrane proteins are directly or indirectly involved. What are the functions of these proteins? The soluble proteins are thought to interact with nascent polypeptide chains to prevent them folding in such a way that they cannot be subsequently transported through the target membrane. Indeed, nascent polypeptides that are waiting to be transported across membranes are often referred to as unfolded. This is clearly a misrepresentation since proteins are never in an unfolded state. Instead, their association with cytosolic proteins (chaperonins) causes them to fold in such a way that they can be released and threaded as <u>locally unfolded segments</u> through membranes. It is becoming increasingly clear that chaperonins are also required for the correct assembly of soluble protein complexes, and it has been proposed that they function as a sort of work bench around which nascent polypeptides are wrapped while waiting to be assembled into oligomers or to interact with specific target membranes. Something very similar probably happens each time a polypeptide is extruded on the trans side of a membrane, as in the case of proteins imported into mitochondria (where chaperonins are clearly involved in the import and subsequent oligomerization reactions), transport into the lumen of the ER and export into the bacterial periplasm.

Do certain soluble chaperonins specifically recognize proteins that are destined to interact with membranes, and if so, what features of these proteins do they recognize and do they help pilot proteins to the target membrane? These questions have only been partially resolved in bacteria, where the "secretory chaperonin" SecB protein clearly recognizes only a subclass of secretory proteins. Questions regarding the "signal" that SecB protein recognizes and its role in piloting remain unanswered. In Eukaryotes, however, the signal recognition particle (SRP) clearly plays a key role both as a "competence factor" to maintain secretory proteins in a translocation competent state and as a pilot via its interaction with the signal peptide on nascent polypeptides and via its interaction with a specific receptor on the ER membrane.

Although membrane components involved in protein translocation have been identified by sophisticated biochemical and genetic techniques, their roles in translocation remain mysterious. Attempts to determine the chronological order of events as proteins are transported into and across membranes (e.g, articles by Schatz and by Stader *et al*) are beginning to provide promising results, as are attempts to identify the membrane components with which nascent polypeptides interact (import/export receptors), and it should soon be possible to isolate protein complexes that are fully functional in translocation, to study the role of specific lipids in the membrane and to answer important questions regarding the degree of contact between fatty acyl chains and lipid head groups and the proteins in transit through the membrane.

Another important question which needs to be answered is how proteins become embedded in the membrane. Recent studies suggest that translocation complexes such as that described by Brundage and Wickner and by Schatz assemble from their various independent components only when required (assembly being triggered by the arrival at the membrane of the nascent secretory protein). Disassembly may then be triggered by the entry of a long hydrophobic segments of integral membrane proteins so that they can diffuse laterally and become permanently embedded in the membrane. An interesting alternative possibility recently proposed by Günter Blobel is that translocation channels remain permanently assembled but can be triggered to open laterally to release polypeptide segments into the membrane. Genetic and biochemical studies of protein-protein interactions should eventually enable us to visualize each of the individual steps that occur as proteins are transported through membranes and to develop more refined models for protein transport through lipid bilayers.

It is still not clear whether integral polytopic membrane proteins use the same machinery as fully translocated proteins to interact with and to penetrate into membranes. Genetic studies by Colin Stirling reported in this volume suggest that there might be differences in the requirements for export of integral and soluble proteins in yeast. Other recent studies from Jon Beckwith and his colleagues suggest that at least some integral cytoplasmic membrane proteins of *E. coli* are assembled by a pathway that does not require the gene products that were identified in his laboratory as being required for the export of soluble and outer membrane proteins. If this is the case, then the assembly of LacY protein in the cytoplasmic membrane could conceivably occur spontaneously. If so, we have to ask how the protein becomes properly folded in the membrane. Kaback's studies showing that individual segments of LacY protein produced independently in the cell can assemble to form a functional lactose permease are highly interesting in this respect because not only must the two segments be inserted in the correct orientation, but individual transmembrane helices must locate and interact with their cognate partners in the membrane. This phenomenon is not unique to LacY, and recent studies by Don Engelman and others show that individual transmembrane alpha helices inserted independently into membranes can interact to form functional oligomers. We must therefore conclude that protein-protein interactions can occur within the matrix of a lipid bilayer and that such interactions are important if not critical for membrane protein assembly and function.

ROLE OF PHOSPHOLIPIDS IN CELL FUNCTION

William Dowhan
Department of Biochemistry and Molecular Biology
University of Texas Medical School
Houston, Texas 77096
U. S. A.

Phospholipids play a dual role in cells as both essential structural components of the cell membrane and as important metabolic intermediates and regulatory molecules. The wide diversity of phospholipid species found in both eukaryotic and prokaryotic cells suggests a broader spectrum of functions for phospholipids than are currently appreciated. Because of this pleiotropic requirement for normal cell function, it has often been difficult to assign a specific role to a particular phospholipid or group of phospholipids. In most cell types it is difficult to effect systematic and extensive alteration of the normal phospholipid composition by alterations of the growth conditions. Therefore, many earlier studies relied on *in vitro* biochemical studies without the necessary *in vivo* confirmation of physiological significance. Utilization of classical genetic approaches has established the essential role of specific phospholipids for cell viability but have not provided a precise understanding at the molecular level for their requirement (Raetz and Dowhan, 1990).

Several important questions relating to phospholipids remain unanswered. What is the importance of membrane phospholipid composition to membrane integrity and membrane mediated processes? Which cellular processes require specific phospholipid species for proper function? Are phospholipids metabolic precursors to important nonlipid products? Do phospholipids act as metabolic regulatory molecules? How do phospholipids move between intracellular organelles? What is the relationship between intracellular phospholipid and protein trafficking? Methods for answering these types of questions will be described in the context of studying the biochemical and physiological properties of recently isolated mutants of both

NATO ASI Series, Vol. H 63
Dynamics of Membrane Assembly
Edited by J. A. F. Op den Kamp
© Springer-Verlag Berlin Heidelberg 1992

Escherichia coli (Heacock and Dowhan, 1987 and 1989; DeChavigny *et al.*, 1991) and yeast (Aitken *et al.*, 1990; Bankaitis *et al.*, 1989 and 1990; Cleves *et al.*, 1991).

Role of Phospholipids in *Escherichia coli*

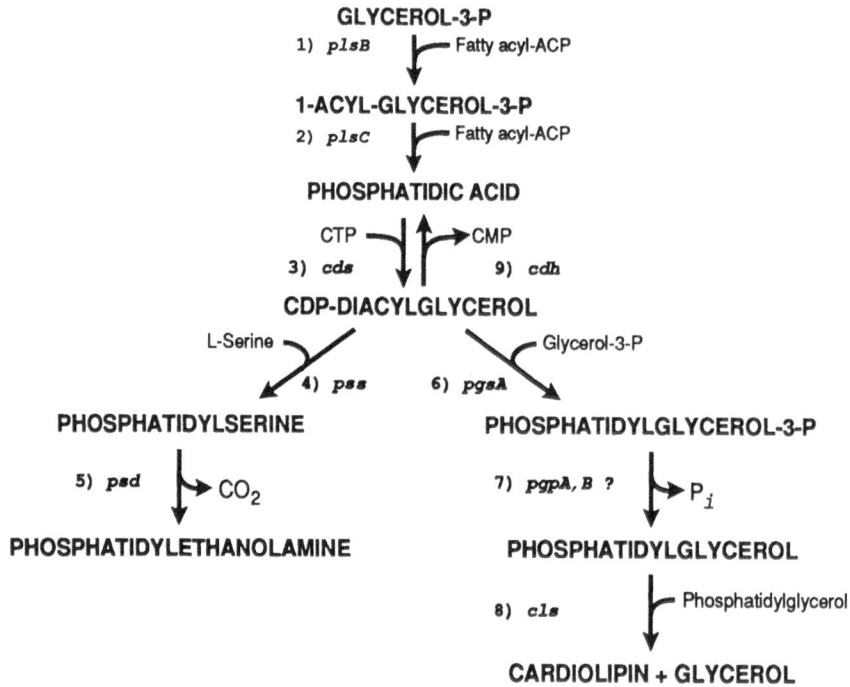

Figure 1.   Pathway of phospholipid metabolism in *E. coli* and the associated genes (Raetz and Dowhan, 1990).   The name of each gene is listed with the respective step catalyzed by the following enzymatic activities: 1) Glycero-*P* acyltransferase; 2) Acylglycero-*P* acyltransferase (Coleman, 1990); 3) CDP-diacylglycerol synthase; 4) Phosphatidylserine synthase; 5) Phosphatidylserine decarboxylase; 6) Phosphatidylglycero-*P* synthase; 7) Phosphatidylglycero-*P* phosphatases; 8) Cardiolipin synthase; 9) CDP-diacylglycerol hydrolase.

Overview of phospholipid metabolism--Figure 1 outlines the pathway for the biosynthesis of the major phospholipids of *E. coli* along with the names of those gene loci which have been identified and characterized. The zwitterionic phospholipid phosphatidylethanolamine (PE) accounts for 70-80% of the total glycerophospholipid while the acidic phospholipids phosphatidylglycerol (PG, 15-20%) and cardiolipin (CL, 5% or less) make up the remainder of the major phospholipid species. The remaining intermediates account for less than 5% of the total phospholipid. Phospholipid biosynthesis is confined to the inner cytoplasmic membrane but the products of biosynthesis are distributed between the inner and outer membranes. Although it is unclear how phospholipid metabolism is regulated, both genetic and biochemical evidence indicate Step 1 is responsible for regulating the ratio of total phospholipid to protein and Steps 4 and 6 determine the ratio of zwitterionic to acidic phospholipid in the cell membrane (Raetz and Dowhan, 1990).

The phospholipid composition of *E. coli* is invariant under a broad spectrum of growth conditions and is even unaltered by increasing any of the enzymes in Figure 1 by 10- to 20-fold using plasmid-borne copies of the respective structural gene. On the other hand mutations which result in a reduction in the effective catalytic efficiency of a particular step brings about the expected alteration in phospholipid composition. Therefore, it is not surprising that conditional lethal mutations have been isolated in Steps 1-3 (McIntyre *et al.*, 1977; Ganong and Raetz, 1982; Coleman, 1990). In addition conditional lethal mutations have been isolated in Steps 4 and 5 (Ohta and Shibuya, 1977; Raetz *et al.*, 1979; Hawrot and Kennedy, 1978) which would appear to support an essential role for PE; however, recent experiments discussed below have established that addition of divalent metal ions to the growth medium can completely suppress the requirement of *E. coli* for PE (DeChavigny *et al.*, 1991). Initial mutations is Steps 6-8 had indicated that acidic phospholipids may not be essential (Miyazaki *et al.*, 1985; Icho and Raetz, 1983; Nishijima and Raetz, 1979). In fact null alleles of the *cls* locus grow normally indicating that CL is not essential (Nishijima *et al.*, 1988). However, as described below PG has been established to be essential (Heacock and Dowhan, 1987), and no growth conditions have yet been found to suppress this requirement.

Since Steps 4 and 6 determine the balance between zwitterionic and acidic phospholipids, genetic manipulation of the *pss* and *pgsA* genes have been the target of attempts to alter phospholipid composition. Through the use of point mutants in these genes drastic alterations in phospholipid composition are possible, but these mutants cannot be used to make systematic changes in composition in order to study the effects on cell function. A molecular genetic approach is described below which has resulted in a strain of *E. coli* in which the acidic phospholipid composition can be systematically varied over a wide range (Heacock and Dowhan, 1989). This strain has made possible the correlation of acidic phospholipid content with changes in several cellular processes to begin to define the role of PG in cell function.

Table I. Properties of strains with different *pgsA* alleles.

Strain	Genotype	Sp. Act 30°C[a]	Viability	PE	PG	CL	PA
				\multicolumn{4}{c}{%}

Strain	Genotype	Sp. Act 30°C[a]	Viability	PE	PG	CL	PA
R477	*pgsA+*	1.0	+	76	22	2	--
R477-10	*pgsA10*	0.02	+	82	17	1	--
HD3122	*pgsA3 lpp*	0.01	+	92	2	1	5
HD3031	*pgsA3*	0.04	-	74	5	20	1
HD38	*pgsA30*	0.01	-	91	4	2	1
HD1266	*pgsA30 lpp*	0.01	-	93	2	1	4
HDL1001[b]	*lacOP-pgsA1*	0.02	-	93	1	1	3
HDL1001[c]	*lacOP-pgsA1*	0.4	+	83	13	3	1
HDL11[b]	*lacOP-pgsA1 lpp*	0.02	+	91	2	1	1
TΦ30[b]	*lacOP-pgsA1 cls*	0.02	-	88	4	1	2

[a]PGP synthase, units/mg protein.
[b]Grown in the absence of 1 mM IPTG.
[c]Grown in the presence of 1 mM IPTG.

Absolute requirement for phosphatidylglycerol--Table I summarizes the growth phenotype and biochemical properties of several mutants in the *pgsA* locus. This gene encodes the phosphatidylglycero-P synthase which catalyzes the committed step to synthesis of the acidic phospholipids of *E. coli*. Point mutants such as the *pgsA10* (Nishijima and Raetz, 1979) and *pgsA3* (Miyazaki *et al.*, 1985) alleles nearly eliminate the detectable *in vitro* activity of this gene product and bring about significant reductions in the steady-state acidic phospholipid content. With regard to enzymatic activity either the *in vitro* activity of these mutants does not reflect their *in vivo* activity or there are multiple modes for phosphatidylglycero-*P* synthesis. On the other hand the phospholipid composition of the *pgsA3* mutant brings into question whether acidic phospholipids are essential.

In order to directly address these questions a null allele (*pgsA30*) of the *pgsA* gene was constructed. This allele was made (Heacock and Dowhan, 1987) by introducing the cloned gene into a plasmid which was temperature sensitive for replication and therefore maintained the gene in cells grown at 30°C but not in cells grown at 42°C. A null allele of the gene was also constructed by disrupting the coding region with a kanamycin drug marker. This disrupted allele was then exchanged for the wild type chromosomal allele in the presence of the plasmid-borne copy of the gene at 30°C. If the *pgsA* locus is not essential then such cells should grow at both 30°C and 42°C; however, if the locus is required, then cells will not be viable after the plasmid copy of the gene is cured from the cell by growth at 42°C.

Three unrelated genetic backgrounds (represented by strain HD38) were not viable after introduction of the *pgsA30* allele and curing the cell of the plasmid copy of the gene (Heacock and Dowhan, 1987). Asai *et al.* (1989) determined that genetic backgrounds which were defective in the major outer membrane lipoprotein (*lpp* gene product) suppressed the normally lethal phenotype brought about by the *pgsA3* allele (strain HD3031); however this mutation could not suppress the effect of the null allele (strain HD1266, Heacock and Dowhan (1987)). The difference between the *pgsA3* and *pgsA30* alleles is that the former mutation still allows for synthesis of a partially active gene product which is capable of some low level of synthesis of acidic phospholipid. The residue activity in strains carrying the *pgsA30* allele is a result of dilution of preexisting enzyme to some threshold level by growth of the strain through several generations after loss of the plasmid. The suppressive effect of the lack of the major outer membrane lipoprotein may be related to the requirement of PG as a substrate in the posttranslational modification of this protein (Wu *et al.*, 1983). At the 1% level of total phospholipid, the stoichiometry between the lipoprotein and PG is on the order of one (Nikaido and Vaara, 1987).

Therefore, these experiments have established the *pgsA* gene as essential in coding the major, if not only, phosphatidylglycero-*P* synthase. In addition PG is essential to the organism for viability for some function other than as a precursor to cardiolipin or the *lpp* gene product since neither of these molecules appear to be essential. However, the function of PG still remained unknown.

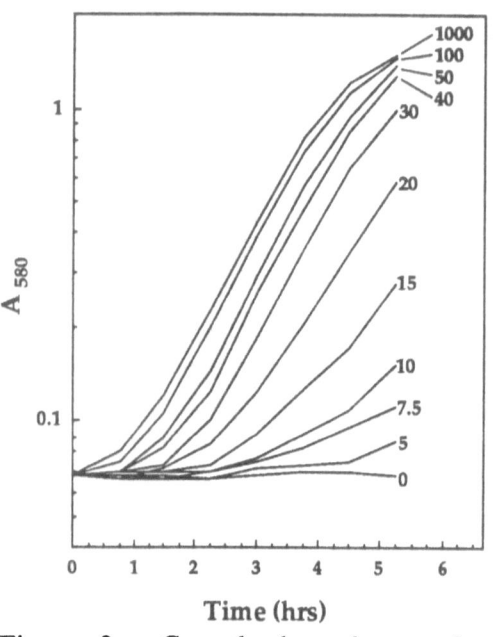

Figure 2. Growth dependence of strain HDL1001 on IPTG (μM) as measured by the increase of absorbance (A) of the cultures (Heacock and Dowhan, 1989).

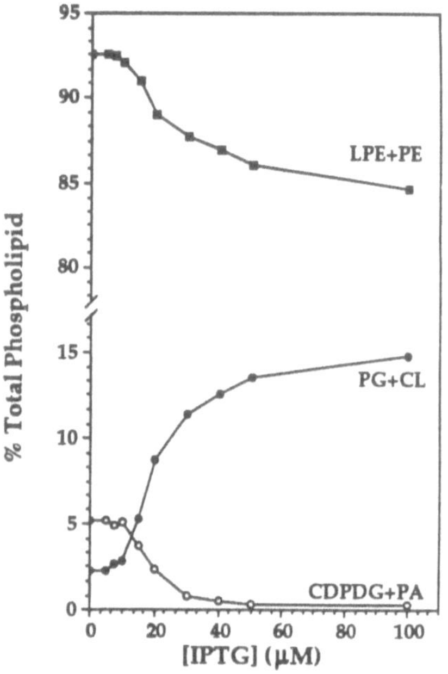

Figure 3. Dependence of phospholipid composition of strain HDL1001 on IPTP. (L, lyso; CDP-DG, CDP-diglyceride) (Heacock and Dowhan, 1989).

Regulation of acidic phospholipid content--The ability to systematically regulate the level of a specific phospholipid would be useful in defining the importance of phospholipid head group composition and the role of specific phospholipids. A solution to this problem has been the construction of a strain (HDL1001) in which the only functional copy of the *pgsA* gene is under the regulation of the *lacOP* promoter (Heacock and Dowhan, 1989). This strain is dependent

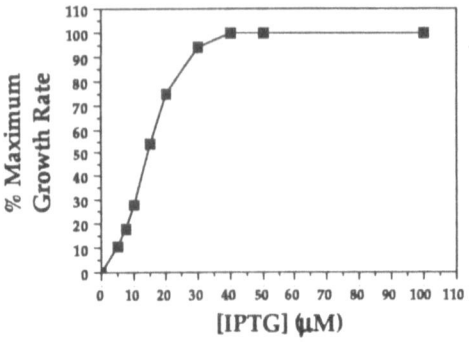

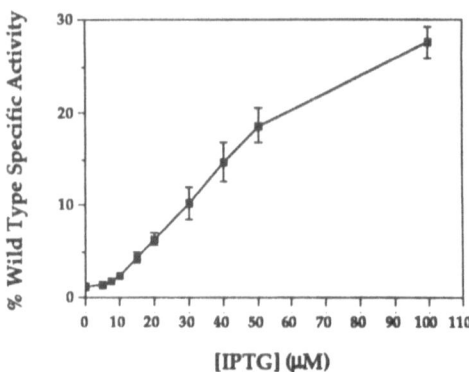

Figure 4. Growth rate of strain HDL1000 as a function of IPTG (Heacock and Dowhan, 1989).

Figure 5. Dependence of the specific activity of the synthase of strain HDL1001 on IPTG (Heacock and Dowhan, 1989).

on IPTG (isopropylthiogalactoside) for growth (Fig. 2), and growth arrest brought about by removal of IPTG is fully reversible by readdition of IPTG for up to several hours after arrest. The PG content (Fig. 3), growth rate (Fig. 4), and level of gene product (Fig. 5) are all proportionally affected within a similar concentration range of IPTG in the growth medium. The reversibility of growth arrest indicates that membrane integrity is not compromised by lack of acidic phospholipids. The fact that phospholipid composition does not begin to change until the specific activity of the synthase is reduced to 10-15% of wild type levels supports the general consensus that phospholipid biosynthetic enzymes are in great excess over the minimum level necessary to maintain normal function. The properties of this strain are much like strains carrying the *pgsA3* allele in that the requirement of IPTG is suppressed by a mutation in the *lpp* locus (strain HDL11). This is because there is some read through of the *lac* promoter in the absence of IPTG. Although there is a close correlation between growth rate and level of acidic phospholipid (Fig. 3 and 4) for strain HDL1001, it is unlikely that the absolute level of PG is limiting because the growth rate of strain HDL11 is independent of IPTG levels, and introduction of the *cls* mutation (strain TΦ30) does not suppress growth arrest even though the limiting level of PG is higher. New synthesis of PG

presumably occurs on the inner surface of the inner membrane where the posttranslational modification of the lipoprotein occurs (Wu *et al.*, 1983). The suppressor effect of the *lpp* mutation on growth rate reduction may be due to an increase in the availability of newly synthesized PG. Shibuya *et al.* (1985) have suggested that the active site for the cardiolipin synthase is on the outer surface of the inner membrane which would not be accessible to newly synthesized PG; therefore, lack of cardiolipin synthesis would not be expected to suppress mutations which limit the synthesis of PG (strain TΦ30). Membrane derived oligosaccharide (MDO) biosynthesis also utilizes PG as a substrate on the outer surface of the inner membrane (Bohin and Kennedy, 1984); addition of arbutin (an analogue of a precursor to MDO which stimulates PG turnover) to the growth medium also has no negative effect on the growth properties of strain HDL1001 or HD38.

What processes might be limited by the availability of newly synthesized PG which might in turn limit growth? As is discussed elsewhere in this symposium report, acidic phospholipids play a central role in the organization of the protein translocation machinery on the inner surface of the inner membrane. Specifically membrane association and activation of the SecA protein necessary for translocation of several outer membrane proteins requires acidic phospholipids (Lill *et al.*, 1990; Kusters *et al.*, 1991). Since strains HDL11 and HDL1001 are indistinguishable in their dependence on acidic phospholipids for efficient protein translocation, it is unlikely that inefficient protein translocation is limiting for growth. The other likely candidate may be DNA synthesis. Yung and Kornberg (1988) have shown a dependence on acidic phospholipids for the *in vitro* reconstitution of the DnaA-dependent reconstitution of initiation of DNA replication at the *oriC* locus of *E. coli*. There are several analogies between DnaA and SecA function. Both proteins are cytoplasmic and exhibit latent ATPase activity which is activated upon formation of a multicomponent complex in association with acidic phospholipids (or membranes containing acidic phospholipids). In both cases it is unlikely that phospholipids play a regulatory role since effects on these processes only occur when acidic phospholipid levels are artificially lowered; however, acidic phospholipids appear to play a passive role as an organizing locus on the membrane surface. There may be other examples in *E. coli* as well as in eukaryotic cells where the minor

phospholipid components such as the acidic lipids act as organization points for membrane associated processes.

Strains HDL1001 and HDL11 have proven to be useful reagents in the study of protein translocation. The ability to systematically set the phospholipid composition of the cell membrane should be useful in the study of a broad spectrum of membrane associated processes such as solute transport, electron transport, ATP synthesis, cell division, membrane fusion, *etc*.

Role of phosphatidylethanolamine--Under normal growth conditions both the *pss* and *psd* genes are essential for growth, but the growth phenotype of these mutants can be partially suppressed by adding mmolar concentrations of divalent metal ions to the growth medium (Ohta and Shibuya, 1977; Raetz *et al*., 1979; Hawrot and Kennedy, 1978). Mutants (*pss1*) temperature sensitive for growth and PS synthase activity (see Figure 1) grown in rich medium at the restrictive temperature arrest after the PE content drops to about half normal levels (Table II). Addition of 20 mM $MgCl_2$ to the growth medium suppresses the arrest of growth at 42°C, but the PE content still remains at 30-40% (Shibuya *et al*., 1985). Is the arrest of growth without Mg^{2+} due to reduced PE or elevated PG and CL? Is the remaining PE synthesized by an alternate pathway or a partially functional *pss* gene product? What is the mechanism of suppression by divalent metal ion? Can cells survive without PE? In order to answer these questions a null allele (*pss93*) of the *pss* gene was constructed (DeChavigny *et al*., 1991) using an analogous approach as outlined for the construction of the *pgsA30* allele.

Strain AH930 (*pss93*) grows normally at 30°C without supplementation of the growth medium with divalent metal ion provided the strain carries a functional plasmid borne copy of the *pss* gene (Table II). Curing the cell of the plasmid by growth at 42°C results in cell arrest after the PE level is reduced by 50% as was reported for the *pss1* allele. However, addition of $Ca^{2+} > Mg^{2+} > Sr^{2+}$ in the mmolar range to strain AH930 not only suppresses the growth phenotype in the order of effectiveness shown, but allows the cells to grow indefinitely with virtually no PE (Table II); therefore, the *pss* gene product is responsible for the catalysis of the committed step to the synthesis of PE. Monovalent cations (0.5 M) and osmotic stabilizers do not substitute

Table II. Properties of strains with different *pss* alleles

pss Allele	Growth Conditions	Viability	PE	PG	CL
				%	
pss+	42°C	+	70	24	6
pss1	42°C	-	32	48	20
pss1	42°C, Mg^{2+}	+	39	41	20
pss93	42°C	-	35	23	42
pss93	42°C, Mg^{2+}	+	<0.01	83	17

for divalent metal ions.   Cell arrest brought about by removing Mg^{2+} is fully reversible after the readdition of Mg^{2+} indicating that the cell membrane is relatively stable and still acting as a cell barrier.   None of the minor phospholipids or any novel phospholipid is elevated, and PE is completely replaced by increased levels of PG and CL since the protein to phospholipid ratio of strain AH930 is the same as wild type strains.   The fatty acid composition of the phospholipid fraction is very similar to wild type strains.

What might be the basis at the molecular level for the requirement of divalent metal in the absence of PE?   The nonspecific nature of the substitution would suggest a structural role for PE rather than a regulatory or essential precursor role.   The primary requirement for divalent metal ion would appear to be outside of the inner membrane and may be focused at the outer membrane.   Both Ca^{2+} and Sr^{2+} are excluded from *E. coli* by specific antiporter systems which maintain the intracellular concentration of these ions in the sub-μmolar range against a large extracellular concentration (Gangola and Rosen, 1987); intracellular Mg^{2+} concentration is near 100 mM (Chang *et al.*, 1986).   The only source of divalent metal ion for the exterior of the cell is the growth medium.   One possibility would be that this zwitterionic lipid normally acts as the membrane matrix and that divalent metal ion in association with the membrane surface of strain AH930 would act to damp the high concentration of negative charge in the absence of PE. Another role related to charge distribution would be the stability of the outer membrane which appears to require Ca^{2+}.   Normally the outer leaflet of the outer membrane is composed of a monolayer of lipid A and the inner leaflet of a monolayer of PE, PG and CL (Nikaido and Vaara, 1987). The increased negative charge density of the inner surface of this membrane in strains lacking PE may contribute to instability which is corrected by high concentrations of divalent metal ion in the growth medium.

Another specific role for PE which could be substituted by divalent metal ions would be a requirement for nonbilayer forming lipids. PE forms inverted hexagonal phases under most ionic conditions. The only other phospholipid in *E. coli* which forms such nonbilayer structures is cardiolipin induced by $Ca^{2+} > Mg^{2+} > Sr^{2+}$ in the indicated order of effectiveness (Vasilenka *et al.*, 1982). This possibility is supported by the following observations. The null allele of the *cls* gene reduces cardiolipin content from the normal 2-5% to < 0.1 % with little effect on the growth properties (Nishijima *et al.*, 1988). As indicated above, cells grown in the presence of divalent metal ions grow normally with no PE. Eliminating both the *pss* and *cls* genes is a lethal event which is not suppressed by divalent cations suggesting a requirement for either PE or CL but not both of these phospholipids. A complication with this argument is that the *pss* gene product appears to be responsible for synthesizing the residual CL in a *cls* null mutant (Nishijima *et al.*, 1988) which suggests that the trace levels of CL in *cls* strains may be essential. Experiments are currently underway to differentiate between a requirement for CL and/or PE as structural components of the membrane (which would require large amounts) from a role as substrate, cofactor or regulatory molecules (which would require only trace amounts). This can be accomplished by placing the *cls* and *pss* genes under exogenous regulation as described for the *pgsA* gene followed by studying the cell physiology and biochemistry in response to phospholipid head group composition.

Future prospectives--The above outlined approaches can be generalized to other steps of phospholipid metabolism in *E. coli*. Certainly placing the *pss* gene under exogenous regulation will make possible the study of the complete spectrum of phospholipid composition in relation to cell function. Placing the *plsB* gene under similar regulation will allow the study of the importance of the protein to phospholipid ratio of the cell membrane. The regulation of the concentration of CDP-diacylglycerol through control of the expression of the *cds* gene will shed light on the mechanism by which phospholipid composition is regulated at this critical branch point. Since all of these gene fusions will result in some low level of read through, it should be possible to isolate by pass or suppressor mutations for each of these steps analogous to the *lpp* suppression of the *lacOP-pgsA* fusion. Analysis of such mutations will lead to a better understanding of the function of phospholipids.

Relationship Between Intracellular Phospholipid and Protein Movement

In *E. coli* it is clear that acidic phospholipids play an important role in the process of movement of proteins through the inner membrane. Similar phospholipid involvement most likely occurs in the more complex process of movement of proteins by the vesicle-dependent organelle-localized secretory pathway of eukaryotic cells. Until recently the processes of synthesis and intracellular movement of proteins and phospholipids have been treated separately. There is a large and detailed literature which describes intracellular trafficking in relation to proteins, but the understanding of the intracellular movement of phospholipids and the molecular details of the joint involvement of these two groups of macromolecules is obscure. In yeast as well as in higher eukaryotic cells, the synthesis of phospholipids is compartmentalized (Carman and Henry, 1989). The biosynthetic pathway for the synthesis of the major phospholipids of yeast is shown in Fig. 6. Accept for one enzyme, the complete *de novo* synthesis of PE, PC (phosphatidylcholine) and PI (phosphatidylinositol) is carried out by endoplasmic reticulum-localized enzymes (via CDP-diacylglycerol and the methylation pathway); the sole exception is the phosphatidylserine (PS) decarboxylase which is associated with the inner mitochondrial membrane. Therefore, PS and PE must be shuttled between the mitochondria and the endoplasmic reticulum in order to effect net PE synthesis. CL and PG are made exclusively in the mitochondria where they remain. The terminal steps of the salvage pathways (via diacylglycerol) for PE and PC biosynthesis also appear to be localized to the endoplasmic reticulum although there have been

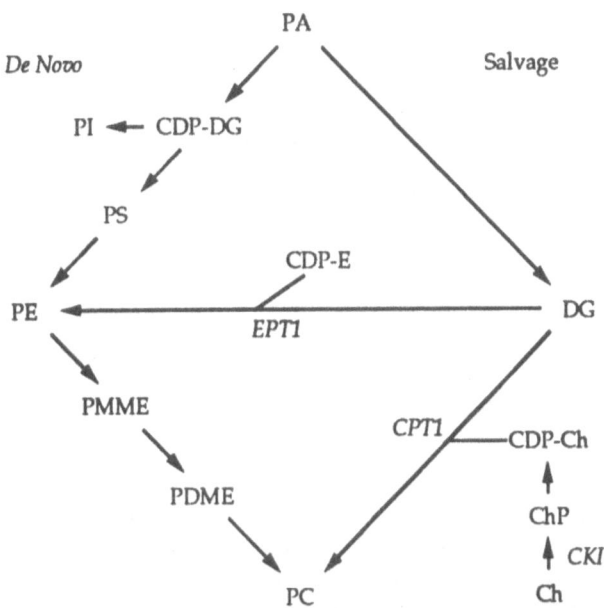

Figure 6. Pathway for the synthesis of phospholipids in yeast. Intermediates not mentioned in the text. PA, phosphatidic acid; PMME and PDME, phosphatidylmonomethylethanolamine and phosphatidyl-dimethylethanolamine; Ch, choline; Ch-P, choline phosphate; CDP-Ch, CDP-choline; CDP-E, CDP-ethanolamine. The steps catalyzed by the *CKI*, *EPT1* and *CPT1* gene products, respectively, are indicated.

reports of the salvage pathway also existing in the Golgi apparatus of liver cells (Meer, 1989). The remaining organelles appear to lack any significant phospholipid biosynthetic activity although there is considerable remodeling of existing phospholipids in higher eukaryotic cells. Since phospholipid biosynthesis is confined to primarily two organelles, how are phospholipids distributed to the various membrane systems for biogenesis and cellular regulation? How is phospholipid and protein synthesis and movement coordinated?

Phospholipid transfer proteins--Recent experiments employing a combined molecular genetic and biochemical approach for the study of phospholipid trafficking and protein secretion in yeast have yielded new insight into these questions. Phospholipid movement between cellular membranes could occur by diffusion of monomers, carrier protein facilitated transfer, vesicle mediated transport, or membrane

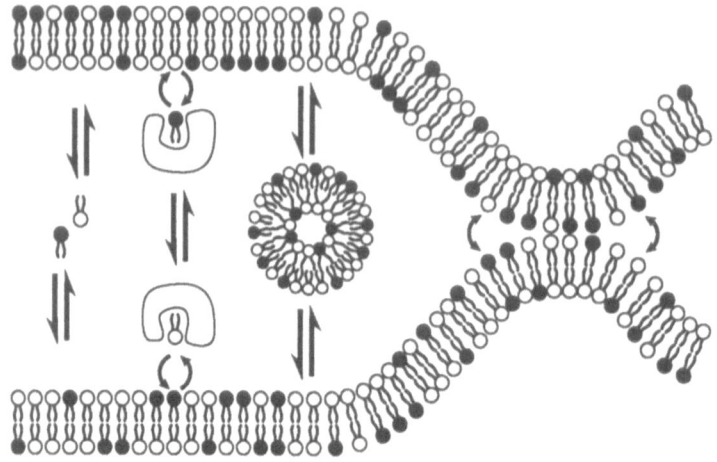

Figure 7. Potential mechanisms for intracellular movement between membranes. From left to right is depicted monomer diffusion through the cytoplasm, carrier protein (transfer protein) facilitated exchange, phospholipid mediated transport, and exchange by membrane fusion.

fusion (Figure 7). Although none of these possibilities have been excluded, protein mediated movement of individual phospholipid molecules between membranes finds considerable favor due to the presence of phospholipid transfer or exchange proteins in the cytoplasm of all somatic cells, yeast and some bacteria (Helmkamp, 1986). The most extensively characterized proteins fall into three specificity groups: exclusively phosphatidylcholine transfer protein (PC-TP); phosphatidylinositol and phosphatidylcholine transfer protein (PI-TP) with a higher affinity for the former; and non-specific lipid transfer protein (ns-LT) which catalyzes exchange of all phospholipids and cholesterol. These proteins are isolated with one mole or less of noncovalently associated lipid and catalyze exchange, but not net transfer, *in vitro* by acting as carriers through the aqueous phase rather than by facilitating collision or contact between membrane surfaces. Based on their ubiquitous distribution and their *in vitro* properties these proteins have been attractive candidates for the past 20 years as likely components of the phospholipid trafficking machinery. However, there was little *in vivo* evidence for their possible physiological role until recently.

In order to establish the requirement for phospholipid transfer proteins, a molecular genetic approach was take by Aitken *et al.* (1990) to isolate a mutant defective in the PI-TP of yeast. The PI-TP was purified to homogeneity and was verified to also effect the transfer of PC between microsomal membranes and artificial phospholipid vesicles. The partial amino terminal sequence of the purified protein was determined and used to design an oligonucleotide which would hybridize with the 5' end of the coding region for the gene (*PIT1*) encoding this protein. A yeast genomic library was screened using this oligonucleotide probe which yielded several clones which proved to contain the *PIT1* gene as evidenced by DNA and protein sequence comparison and expression of elevated PI- and PC-TP activities when introduced into yeast on multicopy number plasmids. The heterozygous diploid strain (*PIT1/pit1::LEU2*) was constructed by gene interruption of one of the two *PIT1* loci of a diploid wild type strain of yeast. After sporulation and analysis of the haploid spores, it was established that the *PIT1* gene is essential to yeast since no spores carrying the *pit1::LEU2* locus (mutant for PI-TP expression) were isolated. Therefore, for the first time a phospholipid transfer protein was demonstrated to have an essential *in vivo* function.

Vesicle-mediated protein secretion--Vesicle-mediated protein secretion in yeast follows a pathway quite homologous with the intracellular protein trafficking scheme of somatic cells (Clary *et al.*, 1990; Wilson *et al.*, 1989; Dunphy *et al.*, 1986). Some of the protein factors are sufficiently similar in structure and function to be interchangeable. Steps in the flow of proteins (Figure 8) from the endoplasmic reticulum to the Golgi apparatus to the plasma membrane have been defined genetically by a series of mutants temperature sensitive for the secretion of invertase in yeast (Novick *et al.*, 1980). These mutants have been used to clone the respective genes by complementation and to begin the process of identifying the factors involved in protein trafficking. Bankaitis *et al.* (1989) focused on *sec14* mutants which, when grown at the restrictive temperature, are characterized by an extensive proliferation of intracellular Golgi-related membranes and a defect in the exit of fully glycosylated, mature invertase from the Golgi apparatus. Complementation of the

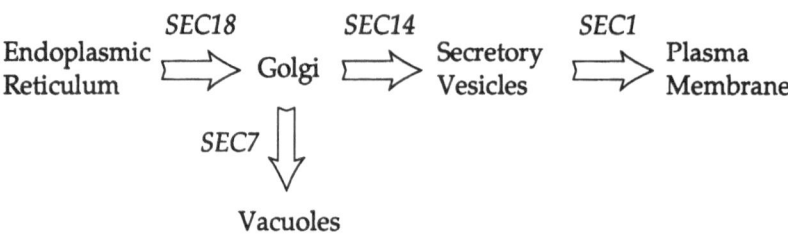

Figure 8. Intracellular movement of proteins along the vesicle-mediated secretory pathway in yeast. Movement between the various organelles is via defined protein laden phospholipid vesicles. One of several known mutations (Novick *et al.*, 1980) which genetically define each step of the pathway is shown.

temperature sensitive phenotype of such mutants with a plasmid-borne genomic library of yeast DNA lead to the cloning of the *SEC14* locus. Antibody directed against the *SEC14* gene product cross reacted with a predominantly cytoplasmic protein, but 20 to 40% of the cross reacting material was localized to a particulate fraction. The *SEC14* gene was shown by gene interrupted to be essential; however several non-allelic, suppressor mutations were isolated which apparently by passed the requirement for a functional *SEC14* gene product.

Comparison of the DNA sequences independently determined for the *PIT1* and *SEC14* genes lead to the surprising observation that these two genes were allelic (Bankaitis *et al.*, 1990); therefore, a defect in a phospholipid transfer protein results in a Golgi specific malfunction in vesicle mediated protein secretion in yeast. Further experimentation demonstrated a relationship between the *in vivo* phenotype of *sec14* temperature sensitive mutants and the *in vitro* function of the transfer protein isolated from such mutant extracts (Figure 9). PI-TP activity was present in the cytoplasm of the mutant strain when assayed at the permissive temperature of 25°C but could not be detected when the assay was carried out at the nonpermissive temperature of 37°C. The loss of transfer protein activity at 37°C was reversible (like the loss of viability) when the extracts pretreated at 37°C were assayed at 25°C; a plasmid-borne copy of the *PIT1/SEC14* gene not only corrected the growth phenotype but restored PI-TP transfer activity. A yet unexplained observation is the lack of any PC-TP activity in the mutant

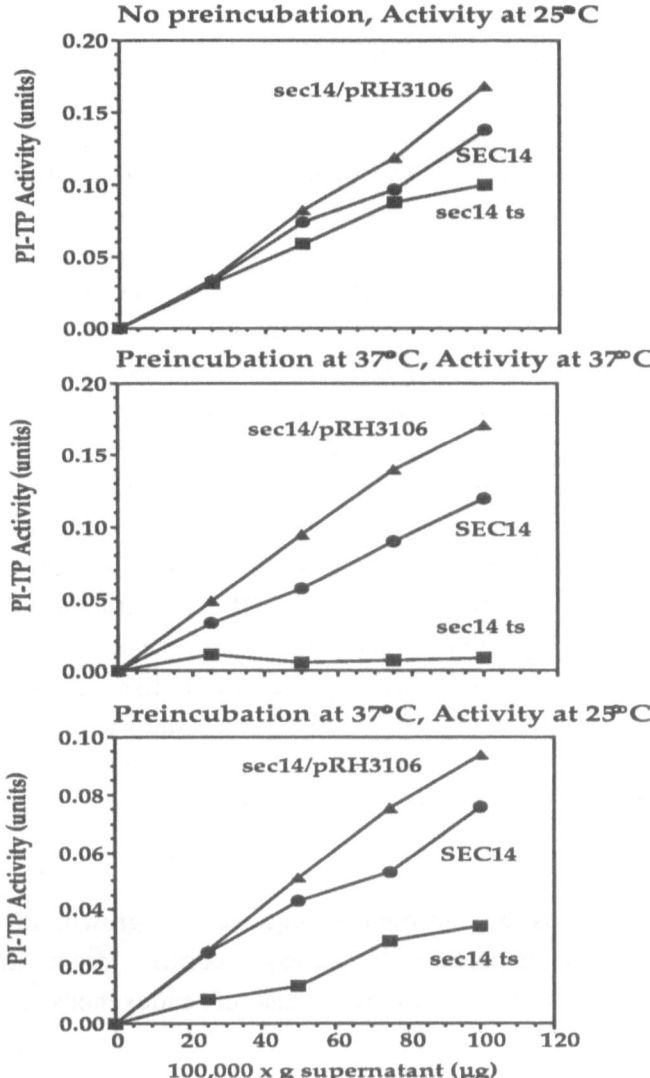

Figure 9. PI-TP activity in the supernatent fraction of yeast strains grown at the 25°C. *SEC14*, wild type; *sec14ts*, temperature sensitive mutant; *sec14ts*/pRH3106, mutant with a plasmid copy of the wild type *SEC14* gene.

extracts even though the plasmid copy of the gene restores such activity. Therefore, for the first time a specific *in vivo* process has been associated with a phospholipid transfer protein. This observation has focused attention on a possible link between protein and phospholipid trafficking.

Role of phospholipid metabolism in protein trafficking--How might a phospholipid transfer protein be linked to the process of intracellular movement of proteins? The beginnings of an answer to this question comes from an analysis of a series of recessive suppressors of both the temperature sensitive and null alleles of the *PIT1/SEC14* gene (Cleves *et al.*, 1991). The correction brought about by these suppressors does not result in the appearance of a new PI-TP activity or a new secretory pathway for invertase. Because these suppressors are mutations at a second site (i.e., nonallelic with *PIT1/SEC14*) and are recessive to a wild type allele of the suppressor locus, loss of function in the *PIT1/SEC14* gene can be corrected by a loss of function in one of several other genes. The identification of some of these suppressor loci by cloning and sequencing the suppressor loci has lead to the identification of phospholipid metabolism as an important factor in protein trafficking. Two of the suppressors have been identified as loss of function mutations in the choline salvage pathway for PC biosynthesis (Figure 6), namely, loss of function mutations in either the *CKI* or *CPT1* locus (Cleves *et al.*, 1991).

Because of the *de novo* pathway for synthesis of PC, apparently the salvage pathway is not absolutely required for growth of yeast but has been viewed as a means to salvage choline from the growth medium. Since yeast has no know means of endogenous synthesis of free choline, the fact that growth of a *pit1/sec14* mutant in choline-free medium does not suppress the growth phenotype suggests that the mechanism of by passing the need for a PI-TP is more complicated than simply lack of synthesis of PC by the salvage pathway. An additional phenotype of *cpt1* mutants is that they lack the normal repression in the initial rate of *de novo* PC biosynthesis relative to PI biosynthesis by inositol added to the growth medium (Morash *et al.*, 1990). This observation suggests that the mechanism of suppression may be related to a change in the initial rate of synthesis of PC versus PI leading to a change in the composition of the newly synthesized phospholipid pool. Since much of the newly synthesized phospholipid in the endoplasmic reticulum is probably destined for the Golgi apparatus to support vesicle mediated transport of secreted proteins, an alteration in the composition of this pool might result in a change in the phospholipid composition of the Golgi apparatus. Such a change might lead to a correction in an aberrant Golgi phospholipid composition which resulted from the lack of phospholipid transfer activity responsible for adjusting composition for proper Golgi function. On the other hand, if there were a mechanism for generating free choline in yeast, the subcellular localization of the salvage pathway to the Golgi apparatus, as has been suggested for somatic cells (Meer, 1989), could directly compensate for an altered phospholipid composition resulting from lack of a PL-TP.

Suppression of the lack of a functional *PIT1/SEC14* gene product brought about by lose of function mutations which result in changes in phospholipid metabolism is consistent with the PI-TP playing a secondary role in vesicle mediated protein trafficking rather than a direct role as a component of the trafficking machinery. These results also support the idea that Golgi membrane phospholipid composition is critical to one or possibly several processes which are localized to this organelle. The initial objective of making mutations in a phospholipid transfer protein has lead to a clearer understanding of the interrelationship between protein trafficking, phospholipid metabolism and regulation, and phospholipid trafficking. Although many questions still remain unanswered, the above results have shown that a complete understanding of the vesicle mediated protein secretion process will require a more extensive investigation of the role of phospholipids in this process.

Acknowledgement:

This work was supported in part by United States Public Health Service Grants GM 20487 and GM 35143 from the National Institutes of General Medical Sciences.

REFERENCES:

Aitken JF, van Heusden PH, Temkin M, Dowhan, W (1990) The gene encoding the phosphatidylinositol transfer protein is essential for cell growth. J Biol Chem 265:4711-4717

Asai Y, Katayose Y, Hikita C, Ohta A, Shibuya I (1989) Suppression of the lethal effect of acidic-phospholipid deficiency by defective formation of the major outer membrane lipoprotein in *Escherichia coli*. J Bacteriol 171:6867-6869

Bankaitis VA, Aitken JR, Cleves AE, Dowhan W. (1990) An essential role for a phospholipid transfer protein in yeast Golgi function. Nature 347:561-562

Bankaitis VA, Malehorn DE, Emr SD, Greene R (1989) The *Saccharoymyces cerevisiae* SEC14 gene encodes a cytosolic factor that is required for transport of secretory proteins from the yeast Golgi complex. J Cell Biol 108:1271-1281

Bohin J-P, Kennedy EP (1984) Regulation of the synthesis of membrane-derived oligosaccharides in *Escherichia coli*. J Biol Chem 259: 8388-8393

Carman GM, Henry SA (1989) Phospholipid biosynthesis in yeast. Annu Rev Biochem 58:635-669

Chang C-F, Shuman H, Somlyo AP (1986) Electron probe analysis, X-ray mapping, and electron energy-loss spectroscopy of calcium, magnesium, and monovalent ions in log-phase and in dividing *Escherichia coli* B cells. J Bacteriol 167:935-939

Clary DO, Griff IC, Rothman JE (1990) SNAPs, a family of NSF attachment proteins involved in intracellular membrane fusion in animals and yeast. Cell 61:709-721

Cleves AE, McGee TP, Whitters EA, Champion KM, Aitken JR, Dowhan W, Goebl M, Bankaitis VA (1991) Mutations in the CDP-Choline Pathway for Phospholipid Biosynthesis Bypass the Requirement for an Essential Phospholipid Transfer Protein. Cell 64:789-800

Coleman J (1990) Characterization of *Escherichia coli* cells deficient in 1-acyl-*sn*-glycerol-3-phosphate acyltransferase activity. J Biol Chem 265:17215-17221

DeChavigny A, Heacock PN, Dowhan W (1991) Sequence and Inactivation of the *pss* gene of *Escherichia coli*. J Biol Chem 266:5323-5332

Dowhan W, Heacock PN (1987) Construction of a lethal mutation in the synthesis of the major acidic phospholipids of *Escherichia coli*. J Biol Chem 262:13044-13049

Dunphy WG, Pfeffer SR, Clary DO, Wattenberg BW, Glick BS, Rothman JE (1986) Yeast and mammals utilize similar cytosolic components to drive protein transport through the Golgi complex. Proc Natl Acad Sci USA 83:1622-1626

Ganong BR, Raetz CRH (1983) pH-sensitive CDP-diglyceride synthetase mutants of *Escherichia coli*: Phenotypic suppression by mutations at a second site. J Bacteriol 153:731-738

Gangola P, Rosen BP (1987) Maintenance of intracellular calcium in *Escherichia coli*. J Biol Chem 262:12570-12574

Hawrot E, Kennedy EP (1978) Phospholipid composition and membrane function in phosphatidylserine decarboxylase mutants of *Escherichia coli*. J Biol Chem 253:8213-8220

Heacock PN, Dowhan W (1989) Alteration of the phospholipid composition of *Escherichia coli* through genetic manipulation. J Biol Chem 264:14972-14977

Helmkamp, Jr GM (1986) Phospholipid transfer proteins: Mechanism of action. J Bioenerg Biomemb 18:71-91

Icho T, Raetz CRH (1983) Multiple genes for membrane-bound phosphatase in *Escherichia coli* and their action on phospholipid precursors. J Bacteriol 153:722-730

Kusters R, Dowhan W, de Kruijff B (1991) Negatively charged phospholipids restore prePhoE translocation across phosphatidylglycerol-depleted *Escherichia coli* inner membranes. J Biol Chem 266:8659-8662

Lill R, Dowhan W, Wickner W (1990) The ATPase activity of SecA is regulated by acidic phospholipids, SecY, and the leader and mature domains of precursor proteins. Cell 60:271-280

McIntyre TM, Chamberlain BK, Webster RE, Bell RM (1977) Mutants of *Escherichia coli* defective in membrane phospholipid synthesis. J Biol Chem 252:4487-4493

Meer V (1989) Lipid traffic in animal cells. Ann Rev Cell Biol 5:247-275

Miyazaki C, Kuroda M, Ohta A, Shibuya I (1985) Genetic manipulation of membrane phospholipid composition in *Escherichia coli*: *pgsA* mutants defective in phosphatidylglycerol synthesis. Proc Natl Acad Sci USA 82:7530-7534

Morash RH, Hjelmstad RH, Bell RM (1990) Effect of mutations in choline and ethanolamine phosphotransferases of *S. cerevisiae* on phospholipid metabolism. FASEB J 4:A1926

Nikaido H and Vaara M (1987) Molecular architecture and assembly of cell parts: Outer membrane. In: Neidhardt, FC (ed) *Escherichia coli* and *Salmonella typhimurium*: Cellular and molecular biology. Vol. 1, Amer Soc Microbiol, Washington, DC, pp. 7-22

Nishijima S, Asami Y, Uetake N, Yamogoe S, Ohta A, Shibuya I (1988) Disruption of the *Escherichia coli cls* gene responsible for cardiolipin synthesis. J Bacteriol 170:775-780

Nishijima M, Raetz CRH (1979) Membrane lipid biogenesis in *Escherichia coli*: Identification of genetic loci for phosphatidyl-glycerophosphate synthetase and construction of mutants lacking phosphatidylglycerol. J Biol Chem 254:7837-7844

Novick P, Field C, Schekman R (1980) Identification of 23 complementation groups required for post-translational events in the yeast secretory pathway. Cell 21:205-215

Ohta A, Shibuya I (1977) Membrane phospholipid synthesis and phenotypic correlation of an *Escherichia coli pss* mutant. J Bacteriol 132:434-443

Raetz CRH, Dowhan W (1990) Biosynthesis and function of phospholipids in *Escherichia coli*. J Biol Chem 265:1235-1238

Raetz CRH, Kantor GD, Nishijima M, Newman KF (1979) Cardiolipin accumulation in the inner and outer membranes of *Escherichia coli* mutants defective in phosphatidylserine synthetase. J Bacteriol 139:544-551

Shibuya I, Yamagoe S, Miyazaki C, Matsuzaki H, Ohta A (1985) Biosynthesis of novel acidic phospholipid analogues in *Escherichia coli*. J Bacteriol 161:473-477

Vasilenko I, de Kruijff B, Verkleij AJ (1982) Polymorphic phase behavior of cardiolipin from bovine heart and from *Bacillus subtilis* as detected by ^{31}P-NMR and freeze-fracture. Biochim Biophys Acta 684:282-286

Wilson DW, Wilcox CA, Flynn GC, Chen E, Kuang WJ, Henzel WJ, Block MR, Ullrich A, Rothman JE (1989) A fusion protein required for vesicle-mediated transport in both mammalian cells and yeast. Nature 339:355-359

Wu HC, Tokunaga M, Tokunaga H, Hayashi S, Giam C-Z (1983) Posttranslational modification and processing of membrane lipoproteins in bacteria. J Cell Biochem 22:161-171

Yung BY-M, Kornberg A (1988) Membrane attachment activates DnaA protein, the initiation protein of chromosome replication in *Escherichia coli*. Proc Natl Acad Sci USA 85:7202-7205

BIOSYNTHESIS OF MEMBRANE PHOSPHOLIPIDS IN YEAST - GENETIC AND MOLECULAR STUDIES

Susan A. Henry[a], Sepp D. Kohlwein[b], John M. Lopes[a],
Michael J. White[a], Tina L. Gill[a], D. Michele Nikoloff[a],
John A. Ambroziak[a], Kimberly A. Hudak[a], Marci J. Swede[a]
and Deborah Allen[a]

[a]Department of Biological Sciences
Carnegie Mellon University
4400 Fifth Avenue
Pittsburgh, PA   15213   USA

Introduction

Baker's yeast, *Saccharomyces cerevisiae*, synthesizes its
membrane phospholipids by utilizing pathways largely common
to other eukaryotes (Figure 1). This simple unicellular
organism can easily be manipulated using well established
classical genetics and powerful molecular genetics. A large
collection of yeast mutants defective in various aspects of
biosynthesis and regulation of phospholipids has been
collected over the last several decades (Henry, 1982; Nikoloff
and Henry, 1991; Swede *et al.*, 1991). These mutants permit
the phospholipid content of the yeast membrane to be
manipulated to a substantial degree. Much progress has been
made recently in purifying the phospholipid biosynthetic
enzymes from yeast and in cloning the structural genes
encoding them (Carman and Henry, 1989). Several genes
involved in regulation of phospholipid biosynthesis have also
been cloned (Hoshizaki *et al.*, 1990; White *et al.*, 1991b).
Thus, in yeast, it is now possible to explore the regulation
and biosynthesis of phospholipids at a level of sophistication
not yet possible in any other eukaryotic organism.

[b]Department of Biochemistry, Graz University of Technology,
Petergasse 12, A 8010 Graz, Austria.

NATO ASI Series, Vol. H 63
Dynamics of Membrane Assembly
Edited by J. A. F. Op den Kamp
© Springer-Verlag Berlin Heidelberg 1992

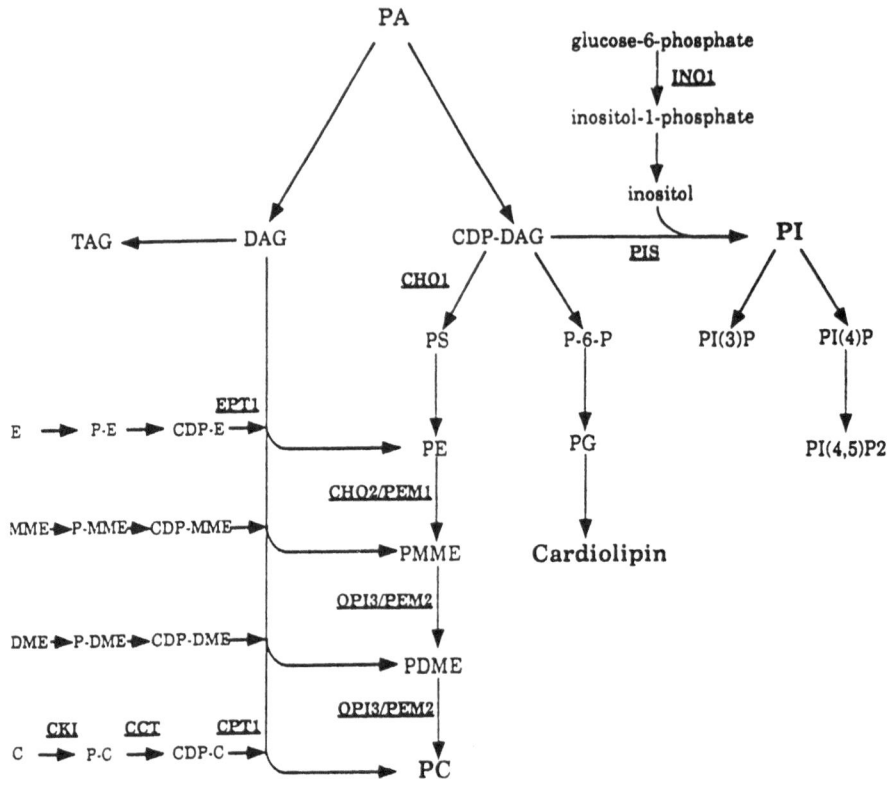

FIGURE 1 - PHOSPHOLIPID BIOSYNTHETIC PATHWAYS IN YEAST

LEGEND - The enzymatic activities for the synthesis of the following compounds have been established: PA, phosphatidic acid; DAG, diacylglycerol; CDP-DAG, cytidine diphosphate diacylglycerol; TAG, triacylglycerol; PI, phosphatidylinositol; PI(3)P, phosphatidylinositol-3-phosphate; PI(4)P, phosphatidylinositol-4-phosphate; PI(4,5)P2, phosphatidylinositol-4,5-phosphate; PGP, phosphatidylglycerol phosphate; PG, phosphatidylglycerol; PS, phosphatidylserine; PE, phosphatidylethanolamine; PMME, phosphatidylmonomethylethanolamine; PDME, phosphatidyldimethylethanolamine; PC, phosphatidylcholine; PE, phosphatidylethanolamine; CDP-E, cytidine phosphate ethanolamine. The structural genes that have been identified and their respective enzyme designations are indicated in italics.

Products Encoded by Phospholipid Structural Genes

INO1, Inositol-1-phosphate synthase; *PIS*, Phosphatidylinositol synthase; *CHO1*, Phosphatidylserine synthase; *CHO2/PEM1*, Phosphatidylethanolamine methyltransferase; *OPI3/PEM2*, Phospholipid methyltransferase; *EPT1 sn*-1,2-diacylglycerol ethanolamine phosphotransferase; *CKI*, Choline kinase; *CCT*, Cholinephosphate cytidyltransferase; *CPT1 sn*-1,2-diacylglycerol choline phosphotransferase.

Manipulation of Phospholipid Content in Yeast Mutants

Structural genes for which mutants have been described are indicated in Figure 1. Some of these mutants have manipulated further by substantially altered phospholipid compositions that can besupplementing the growth medium with phospholipid precursors (Table 1).

TABLE 1

PHOSPHOLIPID COMPOSITION OF SELECTED YEAST PHOSPHOLIPID MUTANTS

STRAIN	MEDIUM[a]	PI	PS	PE	PMME	PDME	PC	REF.[b]
					PERCENTAGE			
wild type	none	7.5	9.8	13.8	tr.	tr.	57.4	1
	I	24.8	5.2	16.0	tr.	tr.	46.0	1
cho2-Δ LEU2	I	27.3	5.2	49.6	tr.	tr.	6.8	2
cho2-1	I	24.8	4.1	51.0	tr.	tr.	9.1	2
cho2-1	I, MME	22.6	8.2	13.0	11.2	13.1	20.1	2
opi3-Δ	I	35.0	3.0	5.6	44.0	2.0	N	3
	I, MME	20.5	2.4	5.5	58.9	3.4	N	3
	I, DME	21.1	4.6	11.6	12.0	35.8	N	3
	I, C	24.7	5.1	12.2	8.9	1.8	39.4	3
chol-Δ	C	18.1	N	9.4	N	N	59.2	1
	I, C	27.4	N	7.1	N	N	57.0	1
chol-1	I, E	32.8	N	15.0	tr.	tr.	41.7	4
	I, C	29.8	N	4.8	tr.	tr.	57.8	4
ino2-2	I	36.5	4.6	28.3	1.3	9.5	11.9	5
ino4-26	I	36.4	4.6	18.5	2.5	7.8	17.8	5
opil-1	none	25.0	7.0	24.0	tr.	tr.	33.0	6
	I	28.0	6.0	22.0	tr.	tr.	32.0	6

LEGEND: Figures in the table represent percentage of lipid soluble phosphorus in each phospholipid. Percentages do not add up to 100% because not all phospholipids are reported. Unreported lipids include cardiolipin, phosphatidylglycerol, CDP-diacylglycerol and phosphatidic acid. N = not detected; tr. = trace quantity detected.

[a]Medium supplements: I, 75 μM inositol; C, choline; E, ethanolamine; MME, monomethylethanolamine; DME, dimethylethanolamine - all at 1 mM.

[b]References: 1. Kelley *et al.*, 1988. 2. Summers *et al.*, 1988. 3. McGraw and Henry, 1989. 4. Atkinson *et al.*, 1980. 5. Loewy and Henry, 1984. 6. Klig *et al.*, 1985.

One surprising outcome of growth studies on these mutants has been the discovery that yeast cells will tolerate very substantial changes in phospholipid content. For example, although *cho1* mutants do not synthesize phosphatidylserine (PS) they are able to grow if they are supplied with exogenous ethanolamine or choline. Ethanolamine and choline enter the phosphatidylcholine (PC) biosynthetic pathway (Figure 1) via the reactions described by Kennedy and Weiss (1956) bypassing PS as an intermediate in PC biosynthesis. However, *cho1* mutants supplied with choline are still defective in PS biosynthesis. The most stringent *cho1* mutants contain no detectable PS (Table 1). Surprisingly, *cho1* mutants do synthesize some PE even when supplied only with choline (Table 1). This residual PE synthesis is believed to be fueled by small amounts of ethanolamine liberated as a result of sphingolipid turnover (Atkinson, 1985).

Likewise, *opi3* mutants, defective in the final two methylations leading to the synthesis of PC (Figure 1), have no detectable PC in their membranes (Table 1) while accumulating elevated levels of PMME. However, *opi3* cells cultured under conditions that result in high PMME content are unable to grow at high temperatures, a defect that is eliminated if the mutants are supplied with exogenous choline or DME (McGraw and Henry, 1989). If supplied with choline or DME, *opi3* mutants have reduced PMME content (Table 1) and accumulate PDME (if DME is supplied) or PC (if choline is supplied).

The large collection of yeast phospholipid mutants provides an unparalleled opportunity for manipulating phospholipid composition *in vivo*. However, surprisingly few studies have been performed using these mutants to study membrane-associated functions or membrane-dependent processes.

Regulation of Phospholipid Biosynthesis in Yeast

In yeast, synthesis of PC is regulated in coordination with the synthesis of the inositol-containing lipids (Carman and Henry, 1989; White *et al.*, 1991a). The physiological purpose of this regulation is not known, but its effect is to balance synthesis of PC and phosphatidylinositol (PI) relative to each other. Since PI is a major negatively-charged phospholipid and PC is zwitterionic carrying a net neutral charge, this regulation may serve to balance net phospholipid charge (White et al., 1991a). Recently, it has been demonstrated that the gene encoding the yeast PI/PC transfer protein (Aitken et al., 1990) is identical to the *SEC14* gene (Bankaitis et al., 1989; 1990). The *sec14* mutants are blocked in secretory activity of the Golgi (Novick et al., 1980). Mutants that contain lesions in PC biosynthesis via the Kennedy (CDP-choline) pathway (i.e., *cki* and *cpt*; Figure 1) suppress the *sec14* phenotype suggesting that balancing PI and PC concentrations may be crucial within localized regions of the Golgi complex (Cleves et al., 1991).

Regulation of phospholipid biosynthesis occurs on a number of levels including direct enzyme inhibition by soluble phospholipid precursors (Kelley et al., 1988) and modulation of enzyme activity in response to the phospholipid composition of the membrane (Fischl et al., 1986; Hromy and Carman, 1986). Many of the enzymatic activities of phospholipid biosynthesis are also repressed in response to the presence of soluble precursors. Inositol-1-phosphate (I1P) synthase (Figure 1) and the enzymatic activities in the reaction series PA --> CDP-DG --> PS --> PE -->-->--> PC are regulated in response to the availability of the precursors inositol and choline (Carman and Henry, 1989). These activities are maximally derepressed when cells are grown in the absence of inositol and choline. The addition of choline by itself to the growth medium has little or no effect on specific activities of the coregulated enzymes. The addition of inositol causes partial repression and addition of inositol plus choline causes maximum repression of the entire set of coregulated enzymatic activities. The repression ratios of individual enzymes differ dramatically ranging from approximately two-fold for CDP-diacylglycerol synthase (Homann et al., 1985) to over 30-fold for I1P synthase (Culbertson et al., 1976; Hirsch and Henry, 1986). However, in mutants that have lesions in PC biosynthesis via the methylation pathway, the regulatory response to inositol is absent when PC biosynthesis is blocked. (This aspect of the regulation will be discussed in further detail in a subsequent section of this article.)

The coregulated enzymes are also controlled by a common set of regulatory genes. The *ino2* and *ino4* mutants are unable to express the I1P synthase subunit (Donahue and Henry, 1981) and have pleiotropic defects, including a deficiency in PC biosynthesis due to an inability to derepress the phospholipid N-methyltransferases (Loewy and Henry, 1984). Other regulatory mutants, isolated on the basis of an inositol overproduction (Opi⁻) phenotype (Greenberg *et al.*, 1982b), were subsequently found to have pleiotropic deficiencies in phospholipid regulation (Klig *et al.*, 1985). The *opi1* mutants express I1P synthase and PS synthase constitutively at a level two-fold higher than the wild type derepressed level (Greenberg *et al.*, 1982a; Klig *et al.*, 1985).

The *INO2* and *INO4* genes encode regulatory factors required for maximum expression (derepression) of all the coregulated enzymes (Hoshizaki *et al.*, 1990). The *OPI1* gene encodes a negative regulatory factor required for repression of these same activities (White *et al.*, 1991b). Recently, all three of these regulatory genes have been cloned. The *OPI1* gene contains an open reading frame encoding a predicted protein with properties characteristic of a DNA binding protein, including a leucine zipper and polyglutamine stretches (White *et al.*, 1991b). The *INO2* and *INO4* genes both contain open reading frames encoding predicted proteins that share similarity to a DNA binding motif that has been termed the helix-loop-helix motif (Hoshizaki *et al.*, 1990; M. Nikoloff, unpublished results).

Regulation of I1P synthase and PS synthase in response to soluble precursors and in response to the regulatory genes has been shown to occur at the level of transcription of the *INO1* and *CHO1* structural genes (Bailis *et al.*, 1987; Hirsch and Henry, 1986). The role of sequences in the promoter region of the *INO1* gene has been explored using portions of the promoter region fused to the *Escherichia coli lacZ* reporter gene (Table 2). A nine base pair (bp) repeated element (nonamer) has been found in the promoters of all the coregulated structural genes (Carman and Henry, 1989; Lopes et al., 1991). All gene fusions that support regulated expression of the reporter gene contain at least one copy of the nonamer. However, when the nonamer is inserted by itself in flanking DNA unrelated to the INO1 promoter, it fails to support expression of the gene fusion (Lopes et al., 1991).

TABLE 2

ß-GALACTOSIDASE ACTIVITY IN WILD TYPE AND cho2 STRAINS OF S. cerevisiae HARBORING INO1- AND INO1-CYC1-lacZ FUSIONS

STRAIN AND PLASMID	MEDIUM SUPPLEMENT			
	NONE	C[+a]	I[+b]	I[+]C[+ab]
wild type, INO1-lacZ	119.5[c]	93.5	13.0	1.0
wild type, INO1-CYC1 TATA-lacZ	144.0	129.3	1.5	0.0
cho2-1; INO1-lacZ	167.0	175.0	153.0	10.0

[a]1 mM choline   [b]75 μM inositol
[c]1 unit = (optical density @ 420 nm/min/mg of total protein) x 1000

ß-galactosidase assays conducted as described in Lopes et al. (1991).

An oligonucleotide containing the nine base pair element has been shown, in electrophoretic mobility shift assays, to support formation of a DNA:protein complex when incubated with yeast cell extracts. A similar complex forms in response to fragments of the INO1 promoter containing the nonamer element and can be competed by the oligonucleotide containing the nine bp sequence. Formation of the nonamer-dependent complex is unaffected by mutations at the ino2, ino4 and opi1 loci. However, additional DNA:protein complexes, that are not competed by the oligonucleotide containing the nine bp element, form on the region of the INO1 promoter spanning sequences from -259 to -154. These complexes are absent if extracts are prepared from strains harboring either *ino2* or *ino4* mutants (Lopes and Henry, 1991).

Role of Ongoing PC Biosynthesis in Regulation of Phospholipid Biosynthesis

The *cho1, cho2* and *opi3* mutants are defective not only in synthesis of PC (Figure 1), but also in regulation of I1P synthase in response to inositol. All of the *opi3* mutants and many of the *cho2* mutants were originally isolated on the basis of an Opi⁻ phenotype rather than failure to synthesize PC (Summers *et al.*, 1988; McGraw and Henry, 1989). In these mutants, PC biosynthesis is conditional and can be interrupted by removal of a soluble precursor that enters the pathway downstream of the metabolic lesion in question. Thus, *cho1* mutants have an Opi⁻ phenotype when starved for choline or ethanolamine (Letts and Henry, 1985). The *cho2* mutants also overproduce and excrete inositol only when grown under conditions that prevent PC biosynthesis. When *cho2* mutants are grown in medium containing MME, DME or choline, PC biosynthesis is restored and the Opi⁻ phenotype is eliminated (Summers *et al.*, 1988). In *cho2* mutants, this response correlates with elevated constitutive expression of I1P synthase subunit and *INO1* mRNA (Hirsch and Henry, 1986; Summers *et al.*, 1988). When *cho2* mutants are grown in the absence of MME, DME or choline, addition of inositol to the growth medium fails to repress *INO1* mRNA (in a wild type strain, addition of inositol produces at least 10-fold repression of I1P synthase). However, if MME or choline is present, addition of inositol leads to full repression of *INO1* mRNA in *cho2* mutants (Hirsch and Henry, 1986; Summers *et al.*, 1988). Similarly, *opi3* strains fail to repress the I1P synthase subunit and *INO1* mRNA in response to inositol unless DME or choline is present (McGraw and Henry, 1989).

PC is synthesized in the yeast cell either by methylation of endogenous PE or by incorporation of soluble precursors via the CDP-choline (Kennedy) pathway (Figure 1). Studies with *cho1, cho2* and *opi3* mutants indicate that restoration of PC or PDME biosynthesis by either route is sufficient to restore the transcriptional response to inositol. In each type of mutant (*cho1, cho2* or *opi3*) regulation of I1P synthase is restored only by those precursors that enter the PC pathway downstream of the respective metabolic lesion.

Overall, such studies indicate that yeast cannot regulate phospholipid biosynthesis in response to inositol if PC biosynthesis is blocked. However, when PC biosynthesis is restored, the presence of inositol produces a signal that is transmitted via the regulatory cascade that controls the *INO1* gene expression at the transcriptional level (Hirsch and Henry, 1986; McGraw and Henry, 1989).

New Approaches to the Isolation of Mutants Defective in Regulation of Phospholipid Biosynthesis

Mutants defective in the regulation of phospholipid biosynthesis have been isolated largely on the basis of phenotypes related to altered regulation of the *INO1* structural gene (Swede *et al.*, 1991). Mutants unable to derepress the *INO1* gene synthesize insufficient inositol to sustain growth and are, consequently, auxotrophic for inositol. Thus, the *ino2* and *ino4* mutants, which are defective in positive regulators of the *INO1* gene, were isolated as inositol auxotrophs (Culbertson and Henry, 1975). The *ino2* and *ino4* mutants also express repressed levels of *CHO1* transcript (Bailis *et al.*, 1987) and phospholipid N-methyltransferase activity (Loewy and Henry, 1984; Hoshizaki *et al.*, 1990). However, the *ino2* and *ino4* mutants are not choline auxotrophs because basal expression of enzymes in the pathway for PC biosynthesis via PE methylation (Figure 1) is sufficient to sustain growth. The *ino2* and *ino4* mutants do exhibit alterations in their phospholipid compositions including a reduction in PC and elevation of PE, PMME and PDME levels (Table 1).

Likewise, *opi1* mutants constitutively overexpress all of the coregulated enzymatic activities (Klig *et al.*, 1985). The *opi1* mutants express two to three times the I1P synthase activity and *INO1* mRNA levels of wild type derepressed cells (Greenberg *et al.*, 1982a; Hirsch and Henry, 1986) regardless of growth condition. The phospholipid compositions of *opi1* mutants do not differ dramatically from those of wild type strains except that the PI content of *opi1* cells remains high even when cells are grown in the absence of inositol (Table 1; Klig *et al.*, 1985).

Recently, we have begun to search for regulatory mutants by employing an *INO1* promoter-*lacZ* fusion as a reporter gene. A fully-regulated gene fusion containing *INO1* promoter sequences and a portion of the coding sequence fused inframe to the *lacZ* gene has been integrated in single copy at the *URA3* locus of a wild type strain. Wild type strains containing the fusion construct are dark blue when grown in the absence of inositol and choline in medium containing the chromogenic substrate, X-gal. These strains are white when grown on X-gal medium containing inositol and choline. The construction of an *INO1-lacZ* promoter fusion that is fully regulated when transformed into yeast cells has permitted us to devise sensitive mutant screens based upon colony color on medium containing X-gal. Using this phenotype, we screened for mutants unable to repress the fusion construct (Figure 2).

Many of the mutants identified in this screen were subsequently found to have an Opi⁻ phenotype. However, three mutants, all belonging to a single complementation group, (*cpe1* - <u>c</u>onstitutive <u>p</u>hospholipid <u>e</u>xpression), expressed the *INO1-lacZ* fusion constitutively but did not exhibit an Opi⁻ phenotype. The *cpe1* mutants are recessive, complement all existing classes of regulatory mutants, represent a single complementation group and are not linked to the *INO1-lacZ* fusion. One *cpe1* mutant has been analyzed in detail and was found to express the *INO1-lacZ* reporter, *INO1*, *CHO1*, *CHO2* and *OPI3* genes constitutively. The *ino2* and *ino4* mutations are epistatic to *cpe1* mutations suggesting that *cpe1* mutants represent a regulatory function distinct from previously-identified regulatory factors (J. Lopes and K. Hudak, unpublished data).

Mutant screen employing INO1-lacZ fusion

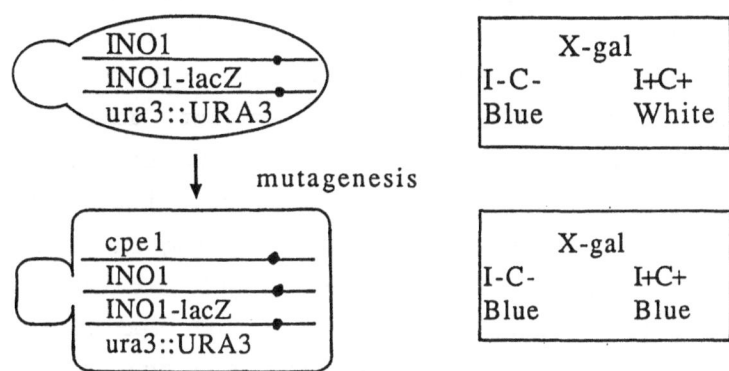

FIGURE 2

LEGEND - Schematic representation of a mutagenic strategy employed to isolate *cpe1* mutant. A strain harboring an *INO1-lacZ* fusion gene stably integrated at the *URA3* locus was mutagenized with EMS. While the wild type strain is able to repress expression of the fusion gene in response to inositol (1+) and choline (C+), *cpe1* mutant do not yield a blue colony phenotype on X-gal media (HC+).

We have also begun isolating suppressors of *ino2* and *ino4* mutants using *ino2* and *ino4* strains containing a single copy of the *INO1-lacZ* fusion integrated at the *URA3* locus. In each case, mutagenized cells are plated onto inositol-free medium to select for inositol prototropy. The inositol prototrophs are screened using the X-gal plate assay for colonies that exhibit a constitutive pattern of expression of the *INO1-lacZ* reporter (i.e., blue colonies under all growth conditions). In this analysis, we hope to identify mutants that define genes whose products interact with those of the *INO2* and *INO4* loci. This genetic analysis may also provide evidence concerning the nature of the interaction between *INO2* and *INO4* gene products. We have also integrated the *INO1-lacZ* fusion into *opi1* strains. On X-gal medium, these strains are blue under all growth conditions. We are searching for "bypass" suppressors using *opi1* null alleles as well as allele-specific suppressors of *opi1* point mutants that may define lesions in proteins that interact with the *OPI1* gene product. The mutagenized colonies are being screened for restoration of wild type *INO1* repression, indicated by a white colony phenotype on X-gal medium containing inositol and choline and blue colony color on medium lacking inositol.

We are also employing a *cho2* strain containing an *INO1-lacZ* fusion construct to screen for mutants defective in the PC-dependent response to inositol. The *cho2* strain containing the *INO1-lacZ* fusion fails to repress β-galactosidase activity in response to inositol (Table 2) resulting in blue colonies when grown on inositol-containing X-gal medium (Table 3). However, when choline is added to growth medium containing inositol, repression is restored (Table 2) and colonies grown on inositol- and choline-containing X-gal medium are white (Table 3). Using this parent strain, we are now searching for mutants that fail to respond to choline and thus, remain blue when choline is added to the growth medium.

In conjunction with our search for mutants that fail to respond to the PC-dependent signal, we have begun a comprehensive analysis of *INO1* regulation in mutants containing lesions in the Kennedy pathway for PC biosynthesis and in *sec14* mutants defective in the PI/PC phospholipid transfer protein. The *INO1-lacZ* fusion reporter gene is being systematically transformed into all of these strains. Colony phenotypes have been assessed on X-gal medium (Table 3) and β-galactosidase assays are being performed. Preliminary evidence (Table 3) suggests that a number of these mutations affect *INO1* regulation. Hopefully, analysis of *INO1* regulation in these strains will begin to elucidate the complex mechanisms controlling the proportional synthesis of inositol- and choline-containing phospholipids in yeast.

TABLE 3

REGULATION OF THE *INO1* GENE IN VARIOUS GENETIC BACKGROUNDS

	MEDIUM SUPPLEMENT			
STRAINS AND PLASMIDS	X[a]	XC[b]	XI[c]	XIC
wild type, *INO1-lacZ*	B	B	W	W
wild type, *INO1-CYC-lacZ*	B	B	W	W
cho2, INO1-lacZ	B	B	B	W
sec14, INO1-lacZ	B+	B-	B-	W
sec14, cki, INO1-lacZ	B+	B+	B-	B-/W
sec14, bsr3, INO1-lacZ	B+	B	W	W
sec14, bsd1, INO1-CYC-lacZ	B+	B+	B+	B+
#*cpt1, INO1-lacZ*	B+	B+	B-	W

[a]X = 0.004% X-gal   [b]C = 1 mM choline   [c]I = 75 μM inositol

*Strains courtesy of Dr. Vytas Bankaitis and Todd McGee.
#Strain courtesy of Drs. Robert Bell and Russell Hjelmstad.
References: Cleves *et al.*, 1991; Hjelmstad and Bell, 1987.

LEGEND: The strains were grown 1-2 days at 30°C on agar plates containing the indicated supplements. Strains having a blue phenotype that is visibly darker than wild type strains grown under derepressed conditions (lack of inositol and choline) are scored B+ (i.e., blue+). Strains having a blue phenotype similar to wild type derepressed are indicated with a B (blue). Strains that are a much lighter shade of blue than the wild type derepressed phenotype are rated B- and strains marked W appear white.

Acknowledgements

This work was supported by Public Health Service Research Grant GM19629 from the National Institutes of Health to SAH. JML is supported by NIH postdoctoral fellowship GM12099. DMN is the recipient of a predoctoral fellowship from the Western Pennsylvania Chapter of the American Heart Association. DA was a participant in a summer undergraduate research program underwritten by grants from the National Science Foundation and the Howard Hughes Medical Institute Undergraduate Biological Sciences Education Initiative. Work in SDK's laboratory was supported by the Fonds zur Förderung der Wissenschaftlichen Forschung in Österreich (Project No 7635). The authors are greatly indebted to S. Haslett for expert assistance in preparing the manuscript.

LITERATURE CITED

Aitken, J. R., G.P.H. van Heusden, M. Temkin and W. Dowhan. 1990. The Gene Encoding the Phosphatidylinositol Transfer Protein is Essential for Cell Growth. *J. Biol. Chem.*, **265**: 4711-4717.

Atkinson, K. D., B. Jensen, A. I. Kolat, E. M. Storm, S. A. Henry and S. Fogel. 1980. Yeast Mutants Auxotrophic for Choline or Ethanolamine. *J. Bacteriol.*, **141**: 558-564.

Atkinson, K. D. 1985. Two Recessive Suppressors of *Saccharomyces cerevisiae CHO1* that are Unlinked but Fall in the Same Complementation Group. *Genetics*, **111**: 1.

Bailis, A. M., M. A. Poole, G. M. Carman and S. A. Henry. 1987. The Membrane-Associate Enzyme Phosphatidylserine Synthase is Regulated at the Level of mRNA Abundance. *Mol. Cell. Biol.*, **7**: 167-176.

Bankaitis, V. A., D. E. Malehorn, S. D. Emr and R. Greene. 1989. The *Saccharomyces cerevisiae SEC14* Gene Encodes a Cytosolic Factor that is Required for Transport of Secretory Proteins form the Yeast Golgi Complex. *J. Cell Biol.*, **108**: 1271-1281.

Bankaitis, V. A., J. R. Aitken, A. E. Cleves and W. Dowhan. 1990. An Essential Role for a Phospholipid Transfer Protein in Yeast Golgi Function. *Nature*, **347**: 561-562.

Carman, G. M. and S. A. Henry. 1989. Phospholipid Biosynthesis in Yeast. **In:** *Ann. Rev. Biochem.*, Vol. 58, pp. 635-669.

Cleves, A. E., T. P. McGee, K. Champion, M. Goebl, W. Dowhan and V. A. Bankaitis. 1991. Mutations in the CDP-Choline Pathway for Phospholipid Biosynthesis Bypass the Requirement for an Essential Phospholipid Transfer Protein. *Cell*, **64**: 789-800.

Culbertson, M. R. and S. A. Henry. 1975. Inositol Requiring Mutants of *Saccharomyces cerevisiae*. *Genetics*, **80**: 23-40.

Culbertson, M. R., T. F. Donahue and S. A. Henry. 1976. Control of Inositol Biosynthesis in *Saccharomyces cerevisiae*. I. Properties of a Repressible Enzyme System in Extracts of Wildtype (*ino*[+]) Cells. *J. Bacteriol.*, **126**: 232-242.

Donahue, T. F. and S. A. Henry. 1981. Myoinositol-1-Phosphate Synthetase: Characteristics of the Enzyme and Identification of its Structural Gene in Yeast. *J. Biol. Chem.*, **256**: 7077-7085.

Fischl, A. S., M. J. Homann, M. A. Poole and G. M. Carman. 1986. Phosphatidylinositol Synthase from *Saccharomyces cerevisiae*. *J. Biol. Chem.*, **261**: 3178-3183.

Greenberg, M., P. Goldwasser and S. Henry. 1982a. Characterization of a Regulatory Mutant Constitutive for Inositol-1-Phosphate Synthetase. *Mol. Gen. Genetics*, **186**: 157-163.

Greenberg, M. L., B. Reiner and S. A. Henry. 1982b. Regulatory Mutations of Inositol Biosynthesis in Yeast: Isolation of Inositol Excreting Mutants. *Genetics*, **100**: 19-33.

S. A. Henry. 1982. The Membrane Lipids of Yeast: Biochemical and Genetic Studies. In: *Molecular Biology of Yeast*, Eds., J. Strathern, E. Jones, J. Broach, Cold Spring Harbor Press, Vol. 2, pp. 101-158.

Hirsch, J. P. and S. A. Henry. 1986. Expression of the *Saccharomyces cerevisiae* Inositol-1-Phosphate Synthetase (INO1) Gene is Regulated by Factors that Affect Phospholipid Synthesis. *Mol. Cell. Biol.*, **6**: 3320-3328.

Hjelmstad, R. H. and R. M. Bell. 1987. Mutants of *Saccharomyces cerevisiae* Defective in *sn-1,2* Diacylglycerol Cholinephosphotransferase. *J. Biol. Chem.*, **262**: 3909-3917.

Homann, M. J., S. A. Henry and G. M. Carman. 1985. Regulation of CDP-diacylglycerol Synthetase Activity in *Saccharomyces cerevisiae*. *J. Bacteriol.*, 163: 1265-1266.

Hoshizaki, D. K., J. E. Hill and S. A. Henry. 1990. The *Saccharomyces cerevisiae* INO4 Gene Encodes a Small, Highly Basic Protein Required for Derepression of Phospholipid Biosynthetic Enzymes. *J. Biol. Chem.*, **265**: 4736-4745.

Hromy, J. M. and G. M. Carman. 1986. Reconstitution of *Saccharomyces cerevisiae* Phosphatidylserine Synthase into Phospholipid Vesicles. *J. Biol. Chem.*, **261**: 15572-15576.

Kelley, M. J., A. M. Bailis, S. A. Henry and G. M. Carman. 1988. Regulation of Phospholipid Biosynthesis in *Saccharomyces cerevisiae* by Inositol. Inositol is an Inhibitor of Phosphatidylserine Synthase Activity. *J. Biol. Chem.*, **263**: 18078-18085.

Kennedy, E. P. and S. B. Weiss. 1956. The Function of Cytidine Coenzymes in the Biosynthesis of Phospholipids. *J. Biol. Chem.*, **222**: 193-214.

Klig, L. S., M. J. Homann, G. M. Carman and S. A. Henry. 1985. Coordinate Regulation of Phospholipid Biosynthesis in *Saccharomyces cerevisiae*: Pleiotropically Constitutive *opi1* Mutant. *J. Bacteriol.*, **162**: 1135-1141.

Letts, V. A. and S. A. Henry. 1985. Regulation of Phospholipid Synthesis in Phosphatidylserine Deficient (*cho1*) Mutants of *Saccharomyces cerevisiae*. *J. Bacteriol.*, **163**: 560-567.

Loewy, B. S. and S. A. Henry. 1984. The *INO2* and *INO4* Loci of *Saccharomyces cerevisiae* are Pleiotropic Regulatory Genes. *Mol. Cell. Biol.*, **4**: 2479-2485.

Lopes, J. M., J. P. Hirsch, P. A. Chorgo, K. L. Schulze and S. A. Henry. 1991. Analysis of Sequences in the *INO1* Promoter that are Involved in its Regulation by Phospholipid Precursors. *Nucl. Acids Res.*, **19**: 1687-1693.

Lopes, J. M. and S. A. Henry. 1991. Interaction of *trans* and *cis* Regulatory Elements in the *INO1* Promoter of *Saccharomyces cerevisiae*. *Nucl. Acids Res.* (In press).

McGraw, P. and S. A. Henry. 1989. Mutations in the *Saccharomyces cerevisiae opi3* Gene: Effects on Phospholipid Methylation, Growth and Cross-Pathway Regulation of Inositol Synthesis. *Genetics*, **122**: 317-330.

Nikoloff, D. M. and S. A. Henry. 1991. Genetic Analysis of Yeast Phospholipid Biosynthesis. In: *Ann. Rev. Gen.* (In press).

Novick, P., C. Field and R. Schekman. 1980. Identification of 23 Complementation Groups Required for Post-translational Events in the Yeast Secretory Pathway. *Cell*, **21**: 205-215.

Summers, E. F., V. A. Letts, P. McGraw and S. A. Henry. 1988. *Saccharomyces cerevisiae cho2* Mutants are Deficient in Phospholipid Methylation and Cross-Pathway Regulation of Inositol Synthesis. *Genetics*, **120**: 909-922.

Swede, M. J., K. A. Hudak, J. M. Lopes and S. A. Henry. 1991. *Strategies for Generating Phospholipid Synthesis Mutants in Yeast.* **In**: Methods in Enzymology: Phospholipid Biosynthesis, Eds. D. E. Vance and E. A. Dennis. (In press).

White, M. J., J. M. Lopes and S. A. Henry. 1991a. Inositol Metabolism in Yeast. *Adv. in Micro. Physiol.*, **32**: 1-51.

White, M. J., J. P. Hirsch and S. A. Henry. 1991b. The *OPI1* Gene of *Saccharomyces cerevisiae*, a Negative Regulator of Phospholipid Biosynthesis, Encodes a Protein Containing Polyglutamine Tracts and a Leucine Zipper. *J. Biol. Chem.*, **266**: 863-872.

GENETIC CONTROL OF MEMBRANE ASSEMBLY IN YEAST$

Erwin Lamping, Sepp D. Kohlwein, Susan A.Henry[1] and Fritz Paltauf *
Institut für Biochemie und Lebensmittelchemie
Technische Universität Graz
Petersgasse 12/II
A 8010 Graz, Austria

The synthesis of the membrane-forming phospholipids is a highly regulated process in yeast. Figure 1 summarizes the biosynthetic pathways of the major phospholipids in yeast. A number of enzymes in the *de novo* pathway for the synthesis of the major phospholipid, phosphatidylcholine (PtdCho), in *Saccharomyces cerevisiae* are subject to coordinate regulation [Carman & Henry, 1989]. These enzymes include CDP-diacylglycerol synthase [Homann et al., 1985], phosphatidylserine synthase [Klig et al., 1985; Poole et al., 1986] and the phospholipid-*N*-methyltransferases [Carson et al., 1984; Klig et al., 1985; Summers et al., 1988; Waechter & Lester, 1973; Yamashita et al., 1982]. Interestingly, the cytosolic enzyme, inositol-1-phosphate synthase, is also regulated in coordination with the other, membrane associated enzymes [Hirsch & Henry, 1985]. All of these enzymes exhibit a common pattern of regulation [Carman & Henry, 1989, Henry et al., 1984] : they are fully derepressed in the absence of inositol and choline, they are partially repressed in the presence of inositol and they are fully repressed in the presence of inositol plus choline. Choline alone without inositol has no repressing effect.

$ Research in the authors' laboratories was supported by research grants from the Fonds zur Förderung der wissenschaftlichen Forschung in Österreich (Project 7635) and grant GM 19629 from the NIH

[1]   Department of Biological Sciences
Carnegie-Mellon University
4400 Fifth Avenue
Pittsburgh, PA 15213
U.S.A.

*   to whom all correspondence should be addressed

The same enzymes are also under the control of a common set of regulatory genes, *OPI1, INO2* or *INO4*. *opi1* mutants fail to repress and *ino2* and *ino4* mutants fail to derepress expression of genes encoding phospholipid synthesizing enzymes. Strong evidence has emerged that ongoing synthesis of phosphatidylcholine is required for this regulation to occur [White et al., 1991]. If PtdCho synthesis is interrupted by a lesion in the *de novo* biosynthetic pathway, proper regulation of expression of genes encoding lipid synthesizing enzymes in response to soluble lipid precursors is abolished. In these mutants addition of soluble lipid precursors entering the PtdCho biosynthetic cascade beyond the metabolic lesion via the "Kennedy"-pathway may restore synthesis of PtdCho and the regulatory signal.

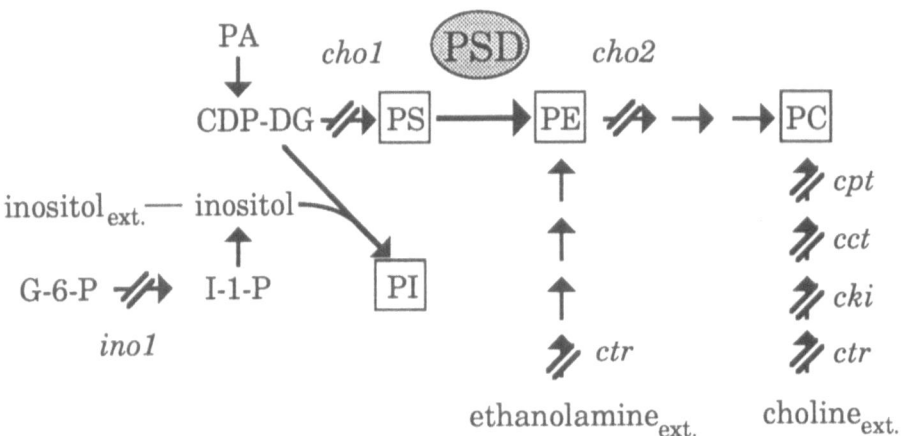

Figure 1. Phospholipid biosynthetic steps in *Saccharomyces cerevisiae*. PSD designates the step catalyzed by phosphatidylserine decarboxylase; PA : phosphatidic acid, CDP-DG : CDP-diacylglycerol, PS : phosphatidylserine, PE : phosphatidylethanolamine, PC : phosphatidylcholine, PI : phosphatidylinositol, G-6-P : glucose-6-phosphate, I-1-P : inositol-1-phosphate. ext. = external soluble precursors. Names in *italic* describe structural gene mutants : *ino1* : inositol-1-phosphate synthase; *cho1* : phosphatidylserine synthase; *cho2* : phosphatidylethanolamine-*N*-methyltransferase; *ctr* : choline transporter, *cki* : choline kinase, *cct* : choline phosphate:CTP cytidylyl-transferase, *cpt* : choline phosphotransferase.

Phosphatidylserine decarboxylase (PSD) catalyses a central step in phospholipid synthesis in yeast (Figure 1). The enzyme is located in the inner mitochondrial membrane [Zinser et al., 1991], whereas the preceding step, catalyzed by phosphatidylserine synthase, is located in a distinct light microsomal membrane fraction [Zinser et al., 1991]. The subsequent three-step methylation of its product, phosphatidylethanolamine, to the major cellular phospholipid, phosphatidylcholine, is located in the endoplasmic reticulum. This poses an interesting question as to the inter-membrane regulation of different metabolic steps in the PtdCho biosynthetic cascade. Localization of enzymes in different membrane compartments may well play an important role in the control of substrate flow through the biosynthetic pathways. Elucidation of the factors controlling PSD activity should thus provide some hints on the mechanisms involved in controlling cellular lipid synthesis.

In this report we describe the effect of soluble lipid precursors, inositol and choline, on the specific activity of phosphatidylserine decarboxylase in wild-type and in *opi1* and *ino2* or *ino4* regulatory mutants, respectively. We have further explored the effect of impaired synthesis of phosphatidylcholine on PSD specific activity under different supplementation conditions in *cho1* mutants, defective in the synthesis if phosphatidylserine, and in *cho2* mutants, defective in the methylation of phosphatidylethanolamine to phosphatidylmonomethylethanolamine. Results from these experiments strongly suggest that PSD is also regulated by the mechanisms of general control of phospholipid synthesis. Based on this observation, we have initiated a screening for mutants defective in the PSD reaction. By using a reporter gene construct integrated into the yeast genome, consisting of the yeast *INO1* promoter fused to the bacterial *lacZ* gene [Lopes et al., 1991], mutants were selected with altered regulation of *INO1* expression in response to soluble lipid precursors. One mutant exhibiting partially restored regulation of *INO1* expression upon the addition of inositol and ethanolamine to the growth medium, was further characterized.

PSD SPECIFIC ACTIVITY IN WILD-TYPE AND MUTANT CELLS

PSD specific activity was analyzed in total membrane preparations from wild-type *ade5* [Culbertson & Henry, 1975], and from the structural gene mutants *cho1* ([Bailis et al., 1987], defective in the synthesis of phosphatidylserine), *cho2* ([Summers et al., 1988], defective in the first methylation step of phosphatidylethanolamine to phosphatidylmonomethylethanolamine) and in the regulatory mutants *opi1* ([Greenberg et al., 1982], defective in a negative regulatory gene) and *ino2* or *ino4* (defective in positive regulatory genes [Culbertson & Henry, 1975]), respectively. The *in vitro* PSD assay was based on a procedure described by Carson et al. [1984], with modifications. Triton X-100 (0.12%) was used as the detergent to solubilize the substrate, PtdSer. Substrate concentration was 0.5 mM and membrane protein concentration was between 60 to 150 µg/300 µL. At 30°C the reaction was linear with respect to time up to 10 minutes. Total membranes were prepared from cells grown in the presence or absence of inositol and choline in the medium. In order to obtain comparable results from the different mutants, extreme care was taken to use

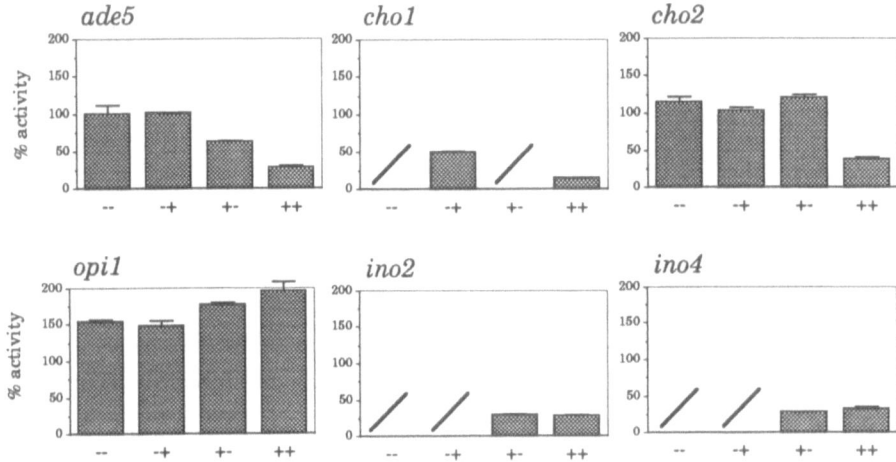

Figure 2. PSD specific activity in wild-type (*ade5*) and mutant cells: *cho1* is defective in PtdSer synthase, *cho2* is defective in PtdEtn-*N*-methyltransferase, *opi1* is a negative regulatory gene mutant, and *ino2* and *ino4* are positive regulatory gene mutants. — : no supplementation; –+ : supplementation with 1 mM choline; +– : supplementation with 100 µM inositol; ++ : supplementation with 100 µM inositol plus 1 mM choline. Slashes indicate "not determined" due to auxotrophic requirements of the mutants.

identical amounts of membrane preparations isolated from cells harvested at the same state of growth. For comparative studies, cells were harvested at the transition from growth to stationary phase, when PSD specific activity was highest (data not shown). Routinely, assays were performed in triplicate from at least two independent preparations. Standard deviation was below 10%.

Results of *in vitro* analyses of PSD in wild-type and lipid synthesis mutants are summarized in figure 2. PSD specific activity was reduced to about 58% in wild-type if inositol was present in the medium, compared to unsupplemented cells. Addition of choline alone had no effect on PSD specific activity. Combined addition of inositol and choline, however, reduced PSD specific activity to about 28% of wild-type activity. *opi1* mutants showed high levels of PSD activity, which was even stimulated by the addition of choline, indicating that the regulatory signal responding to the presence of inositol and choline is interrupted in the mutant. Similarly, *ino2* and *ino4* mutants, auxotrophic for inositol due to their inability to derepress the *INO1* gene encoding inositol-1-phosphate synthase, failed to partially derepress PSD activity in the absence of choline in the growth medium. The level of PSD specific activity in the regulatory mutants indicates that PSD expression may also be controlled by the negative regulatory gene *OPI1* and the positive regulatory genes *INO2* and *INO4*.

In order to further explore the reaction catalyzed by PSD in conjunction with the synthesis of its substrate PtdSer and further metabolic conversion of its product, PtdEtn, we have analyzed PSD specific activity in *cho1* and *cho2* mutants, defective in PtdSer synthase and PtdEtn-*N*-methyltransferase, respectively (figure 1). In *cho1* mutants, grown in the presence of ethanolamine or choline, PSD specific activity was reduced to about 50% of the wild-type activity. This reduced level of activity is probably due to the high frequency of petite-mutants generated in the *cho1* genetic background and/or to specific lipid requirements of PSD. Further addition of inositol led to a fully repressed level of activity comparable to wild-type cells (28% activity of derepressed level). In *cho2* mutants, addition of inositol had no effect on PSD specific activity. Combined addition of inositol plus choline again reduced PSD specific activity to the fully repressed level observed in the wild-type. These data further confirm previous observations that an intact PtdCho biosynthetic pathway is required for proper regulation by the soluble precursors to occur [White et al., 1991]. The mitochondrial enzyme PSD also responds to the same regulatory circuit. In accordance with results obtained from studies with other genes encoding enzymes involved in lipid synthesis we suggest that regulation of PSD may, at least in part, occur at the level of transcription.

ISOLATION OF MUTANTS DEFECTIVE IN PSD ACTIVITY

Wild-type strain JH334 [Lopes et al., 1991] harboring the *INO1* promoter fused to the bacterial *lacZ* gene encoding β-galactosidase was subjected to EMS-mutagenesis following standard procedures. In wild-type cells β-galactosidase is expressed in the absence of inositol in the medium, and can be monitored on plates by the conversion of the chromogenic substrate X-gal (5-bromo-4-chloro-3-indolyl-β-*D*-galactopyranoside) into a blue dye. Wild-type cells turn white upon addition of inositol to X-gal containing plates. In *cho2* mutants harboring the *INO1-lacZ* fusion gene repression of *INO1* controlled expression of β-galactosidase by inositol would be observed if PtdCho synthesis is restored upon addition of monomethyl- or dimethylethanolamine or choline to the inositol containing plates, thus entering the *de novo* pathway beyond the metabolic lesion. For a *psd* mutant – if viable – one would expect restoration of *INO1* repression by a combined addition of inositol plus ethanolamine to the medium. In a first screening, 120 mutants which did not respond to the addition of inositol (blue on X-gal + 100 µM inositol) were selected. Seven of the mutants responded to an addition of 1 mM ethanolamine to the inositol containing plates and turned light blue/white under these conditions. One particular mutant strain showed reduced incorporation of labeled serine into phospholipids and was further subjected to biochemical and genetic analysis. The primary isolate was characterized by a reduction of PSD specific activity to about 40% of wild-type activity. Genetic analysis, however, reveiled that this phenotype was the result of at least two mutations. Whereas one type of mutation did not have any detectable phenotype, the second mutation, designated as *psd*, had a strong *INO1* expression phenotype and a somewhat (10-20%) lower PSD specific activity. Table I summarizes *INO1*-promoter controlled expression of β-galactosidase activities in the wild-type strain JH334 harboring the *INO1-lacZ* fusion in comparison to the *psd* mutant under different supplementation conditions. In the *INO1-lacZ* genetic background,

<u>Table I</u>

β-galactosidase relative specific activities (%)

	no supplementation	100 µM inositol	1mM ethanolamine 100 µM inositol	1mM choline 100 µM inositol
wild type	100	12	5.5	3.7
psd	100	37	21	17

<u>Table I.</u> β-galactosidase specific activities in wild-type JH334 and *psd* mutant harboring the *INO1*-lacZ fusion gene in response to supplementation of cells with soluble lipid precursors.

strains harboring the *psd* mutation express β-galactosidase activity at a different level compared to the wild type under the respective supplementation conditions (Table I). Repression of β-galactosidase specific activity in the mutant is weaker (reduced to 1/3) in response to inositol, compared to wild-type. The additional repressing effect of choline in the presence of inositol is also significantly reduced. These data suggest that the metabolic signal required for proper regulation of *INO1* expression in response to the soluble lipid precursors is impaired in the *psd* mutant. In order to characterize the underlying defect, complementation analysis was performed to known structural and regulatory mutants defective in lipid synthesis. *psd* mutants were able to complement mutations affecting the *de novo* biosynthetic pathway for PtdCho synthesis, indicating that this mutation represents a new genetic locus unlinked to any of the known structural or regulatory genes involved in this pathway.

Table II summarizes specific activities of several phospholipid biosynthetic enzymes in the *psd* mutant and in wild-type. Phosphatidylserine decarboxylase specific activity was reduced about 10-20% compared to wild-type in unsupplemented cells. However, repression by inositol plus choline was significantly less pronounced in the mutant. The response of phosphatidylserine synthase specific activity to supplementation was identical compared to PSD. Choline kinase, on the other hand, was expressed at a significantly lower level (50%–30%) in the *psd* mutant compared to wild-type. In contrast,

Table II

relative specific activity (%)

	no supplementation		100 µM inositol		1mM ethanolamine 100 µM inositol		1mM choline 100 µM inositol	
	w303	psd	w303	psd	w303	psd	w303	psd
PSD	100	85	58	50	47	59	27	37
PIS	100	127	125	148	112	149	120	141
PSS	100	83	58	51	51	53	26	38
CKI	100	46	46	25	51	12	27	10

Table II. Relative specific activities of phosphatidylserine decarboxylase (PSD), phosphatidylinositol synthase (PIS), phosphatidylserine synthase (PSS) and choline kinase (CKI) in wild-type (w303) and the *psd* mutant.

phosphatidylinositol synthase was rather stimulated under all supplementation conditions in the *psd* mutant. As it appeared that repression of several phospholipid biosynthetic steps in response to an addition of choline to cell cultures growing in the presence of inositol was absent or significantly reduced, we have analyzed the uptake of choline into *psd* mutants and wild-type. In fact, *psd* mutants took up choline at a rate of about 16% of wild-type cells (data not shown); in *ctr* (choline transporter) mutants, which were analyzed as a control, uptake of choline was completely lacking. Complementation analysis, however, unveiled that *psd* is unlinked to *ctr* (choline transporter), *cki* (choline kinase) and *bsr2* (presumably allelic to *cpt1* (choline phospho transferase)). It has recently become apparent [Cleves et al., 1991] that mutations in the uptake and incorporation of choline into phosphatidylcholine are able to phenotypically suppress a defect in the essential gene *SEC14*, encoding a lipid transfer protein specific for phosphatidylinositol and phosphatidylcholine [Aitken et al., 1990]. Although a defect in the uptake/incorporation of choline was observed in the *psd* mutant, no suppression of *sec14ts* was detectable. This rather indirect evidence together with the complementation analyses suggest that the *psd* mutation might be unrelated to the structural genes encoding enzymes involved in the salvage pathway for PtdCho synthesis. It appears that the psd mutation represents a regulatory mutation ultimately affecting the generation of a proper signal in response to the intact synthesis of PtdCho in the cells. It may also, to some extent, affect genes involved in the salvage pathway. The observation of a pleiotropic defect in the *in vitro* activities of several lipid biosynthetic activities in the *psd* mutant indicates a global regulatory role of the wild-type gene in coordinated synthesis of phospholipids in yeast.

ACKNOWLEDGEMENTS

We wish to thank Dr. Brendel (Frankfurt, FRG) for providing a *ctr* mutant and Dr. Bankaitis (Urbana, IL) for providing *ctr*, *cki* and *brs2* mutants. This work was supported by the Fonds zur Förderung der wissenschaftlichen Forschung in Österreich (Project 7635) to S.D.K. and a grant from the NIH (GM19629) to S.A.H.

REFERENCES

Aitken JF, van Heusden GPH, Temkin M and Dowhan W (1990) The gene encoding the phosphatidylinositol transfer protein is essential for cell growth. J. Biol. Chem. 265: 4711–4717

Bailis AM, Poole MA, Carman GM and Henry SA (1987) The membrane-associated enzyme phosphatidylserine synthase is regulated at the level of mRNA abundance. Mol. Cell. Biol. 7: 67-176

Carman GM and Henry SA (1989) Phospholipid biosynthesis in yeast. Annu.Rev.Biochem. 58: 635-669

Carson MA, Emala M, Hogsten P and Waechter CJ (1984). Coordinate regulation of phosphatidylserine decarboxylase activity and phospholipid N-methylation in yeast. J. Biol. Chem. 259: 6267-6273

Cleves AE, McGee TP, Whitters EA, Champion KM, Aitken JR, Dowhan W, Goebl M and Bankaits VA (1991) Mutations in the CDP-choline pathway for phospholipid biosynthesis bypass the requirement for an essential phospholipid transfer protein. Cell 64: 789-800

Culbertson MR and Henry SA (1975) Inositol requiring mutants of *Saccharomyces cerevisiae*. Genetics 80: 23-40

Greenberg M, Goldwasser P and Henry SA (1982) Regulatory mutations of inositol biosynthesis in yeast: Isolation of inositol excreting mutants. Genetics 100: 19-33

Henry SA, Klig LS and Loewy BS (1984) The genetic regulation of biosynthetic pathways in yeast: Amino acid and phospholipid synthesis. Annu. Rev. Gen. 18: 207-231

Hirsch JP and Henry SA (1986) Expression of the *Saccharomyces cerevisiae* inositol-1-phosphate synthase *(INO1)* gene is regulated by factors that affect phospholipid synthesis. Mol. Cell. Biol. 6: 3320-3328

Homann MJ, Henry SA and Carman GM (1985) Regulation of CDP-diacylglycerol synthase activity in *Saccharomyces cerevisiae*. J. Bacteriol. 163: 1265-1266

Klig LS, Homann MJ, Carman GM and Henry SA (1985) Coordinate regulation of phospholipid biosynthesis in *Saccharomyces cerevisiae*: pleiotropically constitutive *opi1* mutant. J. Bacteriol. 162: 1135-1141

Lopes JM, Hirsch JP, Chorgo P, Schulze K and Henry SA (1991) Identification of sequences in the *INO1* promoter that are involved in its regulation by phospholipid precursors. Nucl. Acids Res. 19: 1687-1693

Poole MA, Homann MJ, Bae-Lee MS and Carman GM (1986) Regulation of phosphatidylserine synthase from *Saccharomyces cerevisiae* by phospholipid precursors. J. Bacteriol. 168: 668-672

Summers EF, Letts VA, McGraw P and Henry SA (1988) *Saccharomyces cerevisiae cho2* mutants are deficient in phospholipid methylation and cross-pathway regulation of inositol synthesis. Genetics 120: 909-922

Waechter CJ and Lester RJ (1973) Differential regulation of the *N*-methyltransferases responsible for phosphatidylcholine synthesis in *Saccharomyces cerevisiae*. Arch. Biochem. Biophys. 158: 401-410

White MJ, Lopes JM and Henry SA (1991) Inositol Metabolism in Yeast. Adv. Microbiol.Physiol. 32: 1-51

Yamashita S, Oshima A, Nikawa J and Hosaka K (1982) Regulation of the phosphatidylethanolamine methylation pathway in *Saccharomyces cerevisiae*. Eur. J. Biochem. 104: 611-616

Zinser E, Sperka-Gottlieb CDM, Fasch E-V, Kohlwein SD, Paltauf F and Daum D (1990) Phospholipid Synthesis and Lipid Composition of Subcellular Membranes in the Unicellular Eukaryote, *Saccharomyces cerevisiae*. J.Bacteriol. 173: 2026–2034

Control of Phosphatidylcholine Biosynthesis and Its Role in Lipoprotein Assembly

Dennis E. Vance, Susanne Lingrell, Daren G. Fast and Henkjan J. Verkade
Lipid and Lipoprotein Research Group and Department of Biochemistry
University of Alberta
Edmonton, Alberta T6G 2S2 Canada

The regulation of the biosynthesis of phosphatidylcholine (PC) has been studied intensely during the past decade and significant progress has been made in unraveling control mechanisms for regulation of PC biosynthesis (Vance, 1990; Tijburg et al., 1989). More recently our laboratory has also studied the importance of PC biosynthesis in the assembly of lipoproteins in liver cells. The purpose of the present chapter is to review the recent progress in these two areas of research with emphasis on the more recent contributions from the authors' laboratory.

Regulation of Phosphatidylcholine Biosynthesis

Over the past decade four major mechanisms have been identified for the regulation of PC biosynthesis. A summary of the current status of these mechanisms follows.

Regulation by Supply of Fatty Acids

The most striking results in various cells lines have been obtained by treatment of cells in culture with exogenous fatty acids (Vance, 1990). For example, treatment of HeLa cells with 0.35 mM sodium oleate results in a 5- to 10-fold stimulation of PC

NATO ASI Series, Vol. H 63
Dynamics of Membrane Assembly
Edited by J. A. F. Op den Kamp
© Springer-Verlag Berlin Heidelberg 1992

biosynthesis. This is apparently due to translocation of CTP:phosphocholine cytidylyltransferase (CT) from cytosol where it is inactive to membranes where it is active. In this instance the supply of CDP-choline made via the CT reaction limits the rate of PC biosynthesis in the HeLa cells. The membrane to which CT translocates is probably the endoplasmic reticulum (ER) (Vance and Vance, 1988; Terce et al., 1988). If the oleate is removed by addition of albumin to the cell medium, the rate of PC biosynthesis returns to normal and CT translocates from the membranes to the cytosol. The specificity for the translocation of CT has been studied (Cornell and Vance, 1987). A negative charge is not required since oleoyl alcohol is also a potent mediator of CT binding to membranes. The mechanism by which fatty acids induce CT translocation is still unclear. One possibility is that simply a dilution of the concentration of PC in the membranes somehow attracts the enzyme to the membrane where CT is activated. Another possibility is that the fatty acids influence PC biosynthesis by changing cellular diacylglycerol levels. Diacylglycerol will promote CT binding to membranes in vitro. Thus, diacylglycerol is a substrate for the cholinephosphotransferase reaction but may also affect PC biosynthesis by modulating the binding of CT to membranes. One major problem, however, is the lack of evidence that free fatty acids have a physiologically relevant role in modulation of PC biosynthesis (Vance, 1990).

Cyclic AMP and Okadaic Acid Inhibition of Phosphatidylcholine Biosynthesis

The second mechanism involves the inhibition of PC biosynthesis in hepatocytes by cAMP-analogues, glucagon and okadaic acid (Vance, 1990). There is good agreement in a variety of cell types that short term incubations (0.5-3h) with cAMP analogues cause an inhibition of PC biosynthesis. We showed a loss of CT activity from microsomes when cells were incubated with chlorophenylthio-cyclic AMP (CPT-cAMP) (Pelech et al., 1981), but failed to see an effect on CT translocation into cytosol of hepatocytes treated with glucagon (Pelech et al., 1984). After the successful purification of CT (Feldman and Weinhold, 1987), it became possible to determine if CT were an in vitro substrate for cAMP-dependent protein kinase (Sanghera and Vance, 1989). We were able to detect up to 0.2 moles of $^{32}P_i$ incorporated per mole of CT on serine residues and this coincided with a decreased binding of CT to membranes. Whether this occurred in intact hepatocytes remained an open question which was answered after an antibody was obtained to a synthetic peptide derived from the cDNA sequence for CT (Kalmar et al., 1990). This peptide when linked to KLH (keyhole limpet hemocyanin) generated an antibody in rabbits that quantitatively immunoprecipitated CT from hepatocyte cytosol (Jamil et al., 1991). Subsequent studies showed that CT was a protein phosphorylated on approximately 7 peptides and that the state of phosphorylation was unchanged when hepatocytes were incubated with CPT-cAMP. This result strongly suggested that the inhibition of PC biosynthesis in hepatocytes incubated with cAMP analogues was not due to a direct effect of the cAMP-dependent kinase on CT.

The emphasis of the research then shifted to other potential mechanisms that might alter PC biosynthesis. It is well known that cAMP causes an inhibition of fatty acid biosynthesis and it was conceivable that this would result in a decreased supply of diacylglycerol which could limit PC biosynthesis via CDP-choline:1,2-diacylglycerol cholinephosphotransferase. This hypothesis was tested and the results showed a nearly 50% reduction in diacylglycerol in cAMP-treated hepatocytes compared to control incubations. Correlation studies showed that the cAMP effect could be reversed by the addition of oleic acid which caused an increase in the concentration of diacylglycerol and the rate of PC biosynthesis. In these experiments, the correlation between the rate of PC biosynthesis and diacylglycerol levels in hepatocytes was r^2 = 0.93 (Jamil et al., 1991). The data strongly suggest that the rate of PC biosynthesis is inhibited in hepatocytes incubated with cAMP analogues because of an inhibition of fatty acid biosynthesis and consequent reduction of diacylglycerol levels.

In an alternative approach to understand the role of CT phosphorylation in regulation of PC biosynthesis, we have utilized okadaic acid, a potent inhibitor of protein phosphatases 1 and 2A (Cohen and Cohen, 1989; Cohen et al., 1990). In an initial study we showed that addition of okadaic acid to liver postmitochondrial supernatants inhibited the translocation of CT from cytosol to microsomes (Hatch et al., 1990). Subsequent studies showed an inhibition of PC biosynthesis in hepatocytes treated with okadaic acid and a reduced CT activity associated with the membrane fraction (Hatch et al., 1991a). The mechanism for the okadaic acid effect was explored further. Significant differences in the level of phosphorylation of CT were observed in hepatocytes treated with okadaic acid (Hatch et al., 1991b). However, knowing the linkage between diacylglycerol and the cAMP inhibition of PC biosynthesis, we also investigated the levels of diacylglycerol in okadaic acid treated hepatocytes. Various time course studies suggested that okadaic acid, which causes an inhibition of acetyl-CoA carboxylase and fatty acid synthesis (Cohen et al., 1990), inhibits PC biosynthesis by causing a reduction in the supply of diacylglycerol rather than by an effect on CT phosphorylation (Hatch et al., 1991b). The function of phosphorylation of CT therefore remains unknown and requires further investigation.

The present evidence supports the hypothesis that cAMP and okadaic acid inhibit fatty acid biosynthesis which lowers the level of diacylglycerol. The supply of this substrate for the cholinephosphotransferase reaction thus limits the rate of PC biosynthesis.

Phosphatidylcholine Feedback Inhibits Its Own Biosynthesis

Feedback inhibition of a pathway by the final biosynthetic product is well known in many metabolic pathways but not often observed in phospholipid biosynthetic pathways (Vance, 1990). Recent studies from our laboratory have provided convincing evidence for a potent feedback control mechanism in PC biosynthesis (Yao et al., 1990; Jamil et al., 1990). We had been investigating the role of PC biosynthesis in the secretion of lipoproteins in hepatocytes and found that cells depleted of choline were defective in the secretion of very low density lipoproteins whereas all other hepatic proteins as well as HDL were secreted normally (Vance, 1990). In the course of these studies we were curious what effect choline deficiency had on CT since we knew that the rate of PC biosynthesis was decreased by 70 to 80 %. We found a 2-fold increase in CT on the membranes of the choline deficient cells (Yao et al. 1990). We explored several ideas to explain the increased CT binding to cellular membranes and eliminated concentration of fatty acids or a change in the state of phosphorylation of CT. However, we noted a very strong correlation between the distribution of CT and the concentration of PC in the cell membranes. When choline was added to the deficient cells, the levels of PC rose over a 2 hour period by 15% and this coincided with a release of CT from membranes to the cytosol (r^2 = 0.98) (Jamil et al., 1990). The translocation could also have been due to a rise in the aqueous precursors of PC (e.g., choline, phosphocholine, CDP-choline). However, when PC levels were increased by the addition of methionine which promoted PC biosynthesis by the methylation of phosphatidylethanolamine, the release of CT into the cytosol was also observed. Similarly, addition of low levels of lyso-PC to the cells resulted in uptake and acylation to form PC and this again correlated with release of CT into the cytosol (Jamil et al., 1990). Taken together, these data show for the first time that PC is a potent feedback inhibitor of its own biosynthesis.

Treatment of various types of cells with phospholipase C results in enhanced translocation of CT from cytosol to membranes (Sleight and Kent, 1983) and specifically the ER in Krebs II cells (Tercé et al., 1988). They have speculated that the translocation of CT is due to a deficiency of PC in the cellular membranes but direct evidence for this hypothesis has not yet been presented. In light of the results from the choline deficiency experiments, a decrease in the levels of PC after phospholipase C digestion would be an attractive hypothesis.

Diacylglycerol as a Modulator of CT Translocation and PC Biosynthesis

As described above, the supply of diacylycerol as a substrate for the cholinephosphotransferase reaction can directly control the rate of PC biosynthesis. On the other hand, in vitro studies have shown that an increase in the concentration of diacylglycerol can enhance the binding of CT to membranes (Cornell and Vance, 1987). Recent evidence now suggests that, at least in HeLa cells, the concentration of diacylglycerol can regulate the binding of CT to membranes where it is activated and functions as the rate-limiting step of PC biosynthesis in these cells (Utal et al., 1991).

Diacylglycerol has been implicated as a mediator of CT translocation as a result of studies on the mechanism by which PC biosynthesis is stimulated by the tumor promoter, tetradecanoyl phorbol acetate (TPA). Incubation of HeLa cells with this compound causes a 3- to 4-fold increase in the rate of PC biosynthesis and a translocation of CT from cytosol to membranes (Vance, 1990; Utal et al., 1991). Over the years several studies investigated the mechanism by which TPA causes CT translocation and eliminated changes in the levels of fatty acids and changes in the state of phosphorylation of CT (Vance, 1990; Watkins and Kent, 1990; Utal et al., 1991). We considered the idea that a change in the level of diacylglycerol might be responsible for the translocation of CT and the increased rate of PC biosynthesis. Various different approaches suggested a strong correlation between the concentration of diacylglycerol, CT translocation and PC biosynthesis (Utal et al., 1991). The working model is that TPA activates protein kinase C which in turn activates within minutes a phospholipase C that degrades PC on the plasma membrane. The diacylglycerol generated migrates to the ER where it enhances the binding and activation of CT. Thus, the rate of PC biosynthesis is accelerated. A direct phosphorylation of CT by protein kinase C does not appear to be involved since pure CT is not a substrate for protein kinase C in vitro (Jamil and Vance, unpublished results). Moreover, the state of phosphorylation of CT in HeLa cells is unchanged after treatment with TPA (Watkins and Kent, 1990; Utal et al., 1991). In addition, HeLa cells, downregulated for protein kinase C by incubation with TPA for 24 h, showed a stimulation of PC biosynthesis and CT translocation when incubated with the soluble diacylglycerol, dioctanoylglycerol.

Thus, it can be concluded that diacylglycerol can modulate PC biosynthesis by two different mechanisms: 1) as a substrate for the cholinephosphotransferase reaction as demonstrated in the studies with cAMP and okadaic acid; 2) as a mediator of CT translocation and activation in HeLa cells treated with TPA.

Phosphatidylcholine Biosynthesis and Lipoprotein Secretion from Hepatocytes

PC is the major phospholipid in lipoproteins and together with sphingomyelin and lyso-PC account for more than 90% of the phospholipids in human and rat plasma lipoproteins (Vance and Vance, 1985). The major lipoprotein secreted by rat hepatocytes is very low density lipoprotein (VLDL). This is a large particle (diameter 30 to 70 nm) that contains triacylglycerol and cholesteryl ester in its core which is surrounded by a monolayer of phospholipids that contains one of two large apoproteins (apo B100 or apo B48), apoprotein E and apoproteins C. Apo B100 has a molecular weight of 513,000 and is a single, hydrophobic polypeptide, glycosylated on multiple sites (Young, 1990). The molecular weight of apo B48 is 48% that of apo B100. Rat liver secretes VLDL particles that contain either apo B100 or apo B48, but not both on the same particle. In contrast, human liver secretes only apo B100 containing particles. Apo B48, but not Apo B100, is secreted by intestinal cells in humans and rats. Apo E has a molecular weight of approximately 33,000 and several copies are associated with each VLDL particle. The apo C proteins have a molecular weight of approximately 10,000. The number of apo E and apo C copies associated with the lipoprotein varies due to exchange between VLDL and high density lipoproteins (HDL). In contrast, neither apo B48 nor B100 exchanges between lipoprotein particles.

HDL is the other major lipoprotein secreted by rat hepatocytes. It has apo E and apo A1 (molecular weight = 28,000) as the major constituents and apo Cs as minor components.

The scheme proposed for VLDL assembly and secretion in 1976 (Alexander et al.) still appears to be correct. The apoproteins are made on rough ER and the lipids are made on the smooth and rough ER. Glycosylation of the apoproteins occurs in the ER. The components are somehow assembled into nascent VLDL particles and transferred to the Golgi where the glycosylation pattern is modified. After transit through the Golgi the VLDL particles are packaged into secretory vesicles for secretion into the blood stream. Although the general outline for VLDL secretion is well accepted, the details of the assembly process are still sketchy.

Phosphatidylcholine Biosynthesis is Required for VLDL Secretion

Because of the relatively high content of PC in lipoproteins we postulated several years ago that the biosynthesis of PC may have a critical role in lipoprotein assembly. In order to test this idea we needed a selective method for inhibition of PC biosynthesis. Since there were and are no specific drugs to block PC biosynthesis, we decided to use the choline deficient rat as a model. By removing choline from the diet of young rats (50 g) for 3 days, the rate of PC biosynthesis in isolated liver cells is decreased to 30% of control values (Yao and Vance, 1988). Thus, although a complete block of PC biosynthesis was not achieved in these experiments, the rate was decreased significantly and allowed us to measure the effect on lipoprotein secretion from the hepatocytes.

In a typical experiment hepatocytes from 3 day choline deficient rats were prepared and plated on to Petri dishes (Yao and Vance, 1988). Five hours or more after plating, the hepatocytes had adhered to the dishes and the culture medium was changed and the experiment performed. The cells were either maintained in the choline (and methionine) deficient state or supplemented with either choline or methionine. The result were very clear. Secretion of triacylglycerol and PC was inhibited by approximately 70% in the choline deficient hepatocytes compared to cells incubated with either choline or methionine (Yao and Vance, 1988). The secretion of apoproteins associated with VLDL was also decreased as judged by labeling studies with [^{3}H]leucine. Thus, cells supplemented with either choline or methionine and cells without supplementation were incubated with labeled leucine for 12 hours. By ultracentrifugation the VLDL, low density lipoproteins (LDL), HDL and bottom fraction (proteins not associated with lipid that move to the bottom of the salt gradients and have a density > 1.18) were isolated and concentrated by binding to fumed silica (Cab-O-Sil). The lipids were extracted with chloroform/methanol and the proteins solubilized by a special buffer that contained 2% sodium dodecyl sulfate and 6 M urea and analyzed by polyacrylamide gel electrophoresis (Yao and Vance, 1988). Fluorography of the dried gel showed that the secretion of apo B100, apo B48 and apo E associated with VLDL was markedly inhibited (by approximately 70%) from choline deficient compared to choline supplemented hepatocytes. In addition, methionine supplemented cells also returned VLDL secretion back to normal rates. Surprisingly, there was no effect on the secretion of HDL or any of the other proteins secreted by hepatocytes (Yao and Vance, 1988).

These studies allowed us to make the following conclusions. 1) PC biosynthesis is required for the normal secretion of VLDL from cultured rat hepatocytes. 2) The requirement for PC can be provided by the CDP-choline pathway or via the conversion of phosphatidylethanolamine to PC by the methyltransferase. 3) The effect on secretion of VLDL is very specific since the secretion of all other proteins as well as HDL appeared to be unaffected. Apparently PC biosynthesis is not required for the operation of all other secretory processes in liver cells.

Three subsequent studies allowed us to reach additional conclusions. We tested the specificity of the head group of PC for VLDL secretion by supplementation of choline deficient hepatocytes with structurally related compounds: dimethylethanolamine, monomethyl-ethanolamine and ethanolamine (Yao and Vance, 1989). All of these compounds are transported into the hepatocytes and the corresponding phospholipid is made. Thus, after 4 hours of supplementation of choline deficient cells with dimethylethanolamine, the concentration of phosphatidyldimethylethanolamine was 30 nmol per mg protein, about 50% of the concentration of PC in these cells. In normal cells phosphatidyldimethylethanolamine is present at much less than 1 nmol per mg cell protein. Supplementation with ethanolamine and monomethylethanolamine did not cure the defect in secretion of VLDL from the choline deficient hepatocytes but even caused further inhibition (Yao and Vance, 1989). Dimethylethanolamine appeared to restore partially the secretion of apo B48-containing particle but had no effect on the secretion of apo B100 particles. We, therefore, concluded that the biosynthesis of phospholipids structurally related to PC would not substitute for PC in the secretion of VLDL. Why there is such a striking and specific requirement for PC biosynthesis is still not clear.

PC can also be made in hepatocytes from lyso-PC added to the medium of the cells. The lyso lipid is transported into the cell and its major fate is reacylation to PC. Supplementation of hepatocytes with lyso-PC (100µM) was as effective as choline in restoring the secretion of VLDL from choline deficient hepatocytes, as judged by the secretion of PC, triacylglycerol and apoproteins associated with VLDL (Robinson et al., 1989). Thus, PC biosynthesis via reacylation of lysoPC, methylation of phosphatidylethanolamine or the CDP-choline pathway will restore VLDL secretion to normal amounts in choline deficient hepatoctyes.

In the most recent study we were curious if the decreased secretion of VLDL observed in the cultured hepatocytes was reflected by reduced levels of VLDL in the plasma of choline deficient rats. We found that indeed the level of VLDL in these animals was decreased by approximately 70% compared to choline-supplemented animals (Yao and Vance, 1990). This finding agrees with numerous earlier studies showing a decrease in triacylglycerol in the plasma of choline deficient rats. In contrast to VLDL and as expected from the hepatocyte studies, the plasma HDL was present at normal levels in the choline deficient rats.

What is the Mechanism by which PC Biosynthesis is Required for VLDL Secretion?

The above mentioned studies provided convincing evidence that the active synthesis of PC was required for the normal secretion of VLDL. Where in the assembly process is PC biosynthesis required? Two pieces of evidence helped us to formulate our current working hypothesis. 1) The defect in secretion is specific for the apo B associated lipoproteins. 2) Sufficient PC is apparently available for the normal secretion of all other proteins and HDL suggesting that biosynthesis of PC might be specifically required for <u>assembly</u> of VLDL.

In contrast, the assembly of HDL could utilize a passive process, for example, exchange of PC between the membrane and HDL apoproteins.

Our present hypothesis is that PC biosynthesis is required for budding of the apo B-containing VLDL particles from the endoplasmic reticulum (ER) membrane into the lumen of the ER. This would be facilitated by biosynthesis of PC on the ER membrane in which the hydrophobic apo B would be intercalated upon translation (Figure 1). At the same time triacylglycerol and cholesterol ester synthesis would occur on the ER and fill the core of the pre-VLDL particle as it buds from the membrane into the lumen of the ER. In choline deficiency the biosynthesis of PC would be decreased and therefore limit the formation of VLDL. The apoB on particles not properly assembled would eventually be degraded by cellular proteases.

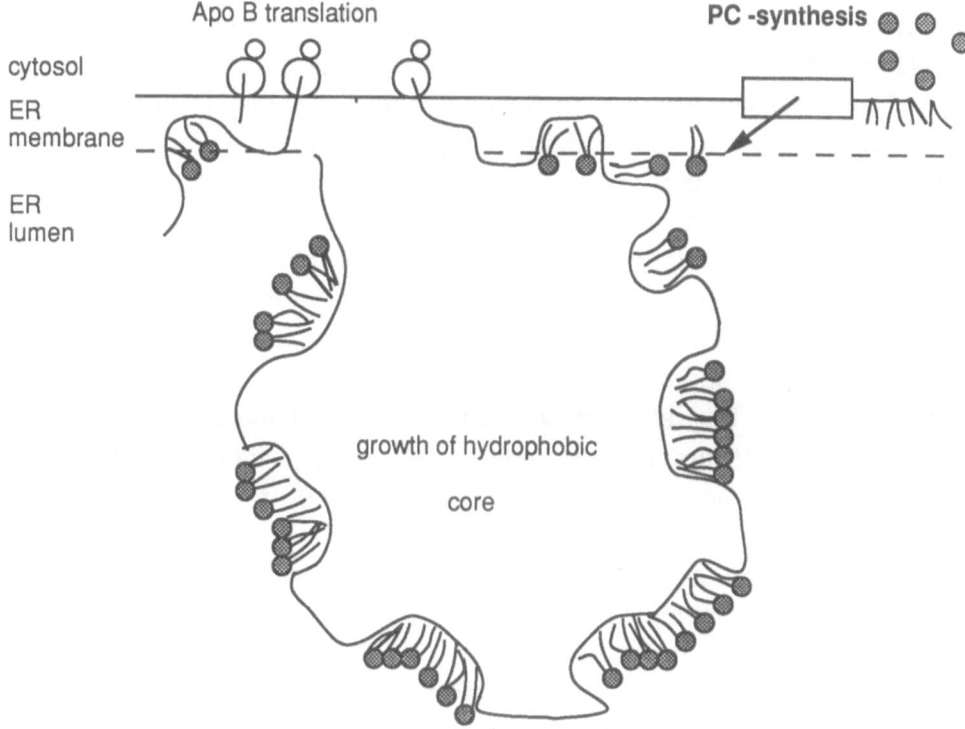

Figure 1   Proposed model by which PC biosynthesis may be required for VLDL assembly.

The above hypothesis suggested two different experimental approaches. If the proposed mechanism were correct, then we might expect to observe an accumulation of apo B100 and B48 in the ER of choline deficient rat liver and and a reduced transport of these two apo Bs transported to the Golgi. Thus, rough ER, smooth ER and Golgi were isolated from choline-deficient and choline-supplemented rat livers (Vance and Vance, 1988). The proteins from these fractions were analyzed by sodium dodecyl sulfate electrophoresis on polyacrylamide gels and transferred to polyvinylidene difluoride microporous membranes. The apo Bs and apo E were detected by immunoblotting with a rabbit antibody raised against VLDL. The results showed a similar amount of apoproteins in rough ER and an accumulation of apoproteins in the smooth ER of the choline deficient relative to choline supplemented animals. In contrast, less of the apo Bs were detected in the Golgi of the choline deficient compared to choline supplemented animals (Fast and Vance, unpublished observations).

These results are still preliminary since the transfer of the apo B100 from the gel to the membrane is not yet quantitative. A second potential pitfall is that sometimes the Golgi in particular can be contaminated by endosomes which contain VLDL and LDL particles derived from the plasma (Hornick et al., 1985). By a slight modification of our subcellular fractionation procedure, we have eliminated endosomal contamination of our ER and Golgi.

In the other approach to the mechanism by which PC biosynthesis is required for VLDL secretion, we reasoned that if there were a defect in the budding of the apo B into the ER lumen, there should be a difference in the movement of radiolabeled apo B from the ER to the Golgi. The success of an experiment to test this idea rests on two important technical achievements: 1) the ability to immunoprecipitate quantitatively apo B from ER and Golgi. 2) a successful separation of ER and Golgi from cultured hepatocytes.

We showed that our immunoprecipitation was quantitative by the following experimental approach. Hepatocytes were prepared and plated onto Petri dishes. One dish of cells was harvested, homogenized and ^{125}I-VLDL was added. The membrane associated proteins were solubilized by addition of a sodium dodecyl sulfate-containing buffer. The solubilized proteins were incubated with various concentrations of antibody to VLDL and precipitated by incubation with Protein A

Sepharose. The proteins were eluted in a sodium dodecyl sulfate buffer and analyzed by electrophoresis. The apoproteins were eluted from the gel and radioactivity was determined. The results showed that 100 to 150 µl of antibody would quantitatively precipitate the radiolabeled apo B added to the homogenate indicating we had recovered 100% of the apo B in the homogenate. Similar experiments with cell fractions enriched in Golgi or ER also indicated quantitative recovery of added apo B.

The separation of ER and Golgi has proved to be a more difficult task. When purified ER and Golgi (Vance and Vance, 1988) were separated on a sucrose gradient as described by Boren et al. (1990), a clear separation of the ER (assayed by NADPH cytochrome C reductase) from the Golgi (assayed by galactosyl transferase activity, an enzyme localized to trans Golgi) was achieved. However, when 10,000 X g supernatants from hepatocyte homogenates were separated by sucrose gradient centrifugation, there was significant contamination of the Golgi by cytochrome C reductase activity, suggesting an incomplete separation of the ER from the Golgi. Moreover, we have not yet assayed a marker for cis Golgi and therefore do not know where this fraction would be found on the sucrose gradients. More work is required to establish the separation of the ER and Golgi fractions from rat hepatocytes.

Nevertheless, several preliminary experiments have been conducted. Hepatocytes ± choline were incubated for 10 min with [^{3}H]leucine and chased for 0 to 120 min. At various times the cells were harvested, fractionation of ER and Golgi performed on sucrose gradients and the apo Bs immunoprecipitated and radioactivity determined. The radioactivity in apo B48 and B100 was similar in the ER from hepatocytes ± choline. In contrast, only 50% as much radioactivity was recovered in apo B48 and apo B100 in the Golgi fractions of choline deficient compared to choline supplemented cells. Similarly, 50% less apo Bs were recovered in the medium from choline deficient compared to supplemented hepatocytes. These results agree with the hypothesis that there is a defect in assembly of apo B into VLDL in the ER. Further experiments are required to confirm these results.

In conclusion, the results from both the pulse-chase studies and the immunoblot analyses are consistent with the hypothesis presented in Fig. 1. It will be a challenge to continue these studies and provide more detailed information on the mechanism by which PC is involved in the assembly of VLDL in the hepatocyte.

Acknowledgements - Our research was supported by a grant from the Heart and Stroke Foundation of Alberta. D.E.V. is a Medical Scientist of the Alberta Heritage Foundation for Medical Research. H.J.V. was supported by a Fellowship from the Alberta Heritage Foundation for Medical Research.

References

Alexander, C.A., Hamilton, R.L. and Havel, R.J. (1976) Subcellular localization of B apoproteins of plasma lipoproteins in rat liver. J. Cell Biol. 69:241-263.

Borén, J., Wettesten, M., Sjöberg, A., Thorlin, T., Bondjers, G., Wiklund, O. and Olofsson, S.-O. (1990) The assembly and secretion of apoB 100 containing lipoproteins in Hep G2 cells. Evidence for different sites for protein synthesis and lipoprotein assembly. J. Biol. Chem. 265:10556-10564.

Cohen, P. and Cohen, P.T.W. (1989) Protein phosphatases come of age. J. Biol. Chem. 265:21435-21438.

Cohen, P., Holmes, C.F.B. and Tsukitani, Y. (1990) Okadaic acid: a new probe for the study of cellular regulation. Trends Biochem. Sci. 15:98-102.

Cornell, R. and Vance, D.E. (1987) Translocation of CTP:phosphocholine cytidylyltransferase from cytosol to membranes in HeLa cells: stimulation by fatty acid, fatty alcohol, mono- and diacylglycerol. Biochim. Biophys. Acta. 919:26-36.

Feldman, D.A. and Weinhold, P.A. (1987) CTP:phosphorylcholine cytidylyltransferase from rat liver: isolation and characterization of the catalytic subunit. J. Biol. Chem. 262:9075-9081.

Hatch, G.M., Lam, T.-S., Tsukitani, Y. and Vance, D.E. (1990) Effect of NaF and okadaic acid on the subcellular distribution of

CTP:phosphocholine cytidylyltransferase activity in rat liver. Biochim. Biophys. Acta 1042: 374-379.

Hatch, G.M., Jamil, H. Utal, A.K. and Vance, D.E. (1991b) Feedforward regulation of phosphatidylcholine biosynthesis: Okadaic acid inhibits phosphatidylcholine biosynthesis by reducing 1,2-diacylglycerol content in isolated rat hepatocytes. submitted.

Hatch, G.M., Tsukitani, Y. and Vance, D.E. (1991a) The protein phosphatase inhibitor, okadaic acid, inhibits phosphatidylcholine biosynthesis in isolated rat hepatocytes. Biochim. Biophys. Acta 1081:25-32.

Hornick, C.A., Hamilton, R.L., Spaziani, E., Enders, G.H. and Havel, R.J. (1985) Isolation and characterization of multivesicular bodies from rat hepatocytes: an organelle distinct from secretory vesicles of the Golgi apparatus. J. Cell Biol. 100:1558-1569.

Jamil, H., Utal, A.K. and Vance, D.E. (1991) Cyclic AMP-induced inhibition of phosphatidylcholine biosynthesis is caused by a decrease in cellular diacylglycerol levels in cultured rat hepatocytes. Submitted.

Jamil, H., Yao, Z. and Vance, D.E. (1990) Feedback regulation of CTP:phosphocholine cytidylyltransferase translocation between cytosol and endoplasmic reticulum by phosphatidylcholine. J. Biol. Chem. 265:4332-4339.

Kalmar, G.B., Kay, R.J., Lachance, A., Aebersold, R. and Cornell, R.B. (1990) Cloning and expression of rat liver CTP:phosphocholine cytidylyltransferase: An amphipathic protein that controls phosphatidylcholine synthesis. Proc. Natl. Acad. Sci. USA 87:6029-6033.

Pelech, S.L., Pritchard, P.H., Sommerman, E.F., Percival-Smith, A. and Vance, D.E. (1984) Glucagon inhibits phosphatidylcholine biosynthesis via the CDP-choline and transmethylation pathways in cultured rat hepatocytes. Can. J. Biochem. and Cell Biol. 62:196-202.

Pelech, S.L., Pritchard, P.H. and Vance, D.E. (1981) cAMP analogues inhibit phosphatidylcholine biosynthesis in cultured rat hepatocytes. J. Biol. Chem. 256:8283-8286.

Pelech, S.L., Pritchard, P.H. and Vance, D.E. (1982) Prolonged effects of cyclic AMP analogues of phosphatidylcholine biosynthesis in cultured rat hepatocytes. Biochim. Biophys. Acta 713:260-269.

Robinson, B.S., Yao, Z., Baisted, D.J. and Vance, D.E. (1989) Lysophosphatidylcholine metabolism and lipoprotein secretion by cultured rat hepatocytes deficient in choline. Biochem. J. 260:207-214.

Sanghera, J.S. and Vance, D.E. (1989) CTP:phosphocholine cytidylyltransferae is a substrate for cAMP-dependent protein kinase in vitro. J. Biol. Chem. 264:1215-1223.

Sleight, R. and Kent, C. (1983) Regulation of phosphatidylcholine biosynthesis in mammalian cells:III. Effects of alterations in the phospholipid compositions of chinese hamster ovary and LM cells on the activity and distribution of CTP:phosphocholine cytidylyltransferase. J. Biol. Chem. 258:836-839.

Tercé, F., Record, M., Ribbes, G., Chap, H. and Douste-Blazy, L. (1988) Intracellular processing of cytidylyltransferase in Krebs II cells during stimulation of phosphatidylcholine synthesis. Evidence that a plasma membrane modification promotes enzyme translocation specifically to the endoplasmic reticulum. J. Biol. Chem. 263:3142-3149.

Tijburg, L.B.M., Geelen, M.J.H. and van Golde L.M.G. (1989) Regulation of the biosynthesis of triacylglycerol, phosphatidylcholine and phosphatidylethanolamine in the liver. Biochim. Biophys. Acta 1004:1-19.

Utal, A.K., Jamil, H. and Vance, D.E. (1991) Diacylglycerol signals the translocation of CTP:phosphocholine cytidylyltransferase in HeLa cells treated with 12-0-tetradecanoylphorbol 13-acetate. Submitted.

Vance, D.E. (1990) Phosphatidylcholine metabolism: masochistic enzymology, metabolic regulation, and lipoprotein assembly. Biochem. Cell Biol. 68:1151-1165.

Vance, J.E. and Vance, D.E. (1985) The role of phosphatidylcholine biosynthesis in the secretion of lipoproteins from hepatocytes. Can. J. Biochem. Cell Biol. 63:870-881.

Vance, J.E. and Vance, D.E. (1988) Does rat liver Golgi have the capacity to synthesize phospholipids for lipoprotein secretion? J. Biol. Chem. 263:5898-5909.

Watkins, J.D. and Kent. C. (1990) Phosphorylation of CTP:phosphocholine cytidylyltransferase in vivo:Lack of effect of phorbol ester treatment in HeLa cells. J. Biol. Chem. 265:2190-2197.

Yao, Z., Jamil, H. and Vance, D.E. (1990) Choline deficiency causes translocation of CTP:phosphocholine cytidylyltransferase from cytosol to endoplasmic reticulum in rat liver. J. Biol. Chem. 265:4326-4331.

Yao, Z. and Vance, D.E. (1988) The active synthesis of phosphatidylcholine is required for very low density lipoprotein secretion from rat hepatocytes. J. Biol. Chem. 263:2998-3004.

Yao, Z. and Vance, D.E. (1989) Head group specificity in the requirement of phosphatidylcholine biosynthesis for very low density lipoprotein secretion from cultured hepatocytes. J. Biol. Chem. 264:11373-11380.

Yao, Z. and Vance, D.E. (1990) Reduction in VLDL, but not HDL, in plasma of rats deficient in choline. Biochem. Cell Biol. 68:552-558.

Young, S.G. (1990) Recent progress in understanding apolipoprotein B. Circulation 82:1574-1594.

PLASTID ENVELOPE MEMBRANES AND THE DYNAMICS OF PLANT MEMBRANE BIOGENESIS

Roland Douce, Maryse A. Block, Eric Maréchal, Albert-Jean Dorne and Jacques Joyard
Laboratoire de Physiologie Cellulaire Végétale, URA CNRS n·576
Département de Biologie Moléculaire et Structurale
Centre d'Etudes Nucléaires et Université Joseph Fourier
85 X, F-38041 Grenoble-cedex
France

INTRODUCTION

In higher plants, the process of photosynthesis occurs within specific membrane-bound organelles called chloroplasts. These large (5-10 μm diameter) organelles, which are probably the descendants of prokaryotic ancestors, are found in the cytosol of leaf cells in apposition to the cytoplasmic and tonoplasmic membranes. Electron micrographs of higher plant cells show that chloroplasts present the three following major structural regions: (a) a highly organized internal membrane network, formed of flat compressed vesicles and called thylakoids; (b) an amorphous background rich in soluble proteins and ribosomes, called the stroma; (c) a pair of outer membranes known as the chloroplast envelope.

Chloroplasts are crucial indeed for plant cell metabolism: they are the site for carbon dioxide reduction and its assimilation into carbohydrates, aminoacids, fatty acids and terpenoid compounds. They are also the site for nitrite and sulfate reduction and their assimilation into aminoacids. Envelope membranes, which are at the border between plastids and their surrounding cytosol, are therefore essential for the control of uptake of raw material for all synthesis occuring in the plastids and for the regulation of the export to the cytosol of newly synthesized molecules necessary for the cell metabolism: the envelope membranes separate spatially and temporally the light-dependent energy conversion processes from the other major cellular activities.

Some of the most interesting functions of the plastid envelope membranes concern their participation to plastid biogenesis. The dynamics of the plastid envelope membrane are important for the formation of thylakoids during development. The inner envelope membrane is the site of biosynthesis for typical plastid components such as glycerolipids, pigments and prenylquinones. In addition, the outer envelope membrane plays a key role in the sorting of plastid proteins that are coded for by nuclear DNA and are synthesized on cytoribosomes. Therefore, envelope membranes provide a unique model system to study membrane biogenesis and sensory transduction, i.e. the sensing and initiation of the plastid response to changing cytosolic conditions.

NATO ASI Series, Vol. H 63
Dynamics of Membrane Assembly
Edited by J. A. F. Op den Kamp
© Springer-Verlag Berlin Heidelberg 1992

PLASTID ENVELOPE MEMBRANES AS A PERMANENT STRUCTURE IN PLANT CELLS

In addition to chloroplast, which are localized in leaves and other green tissues, a large variety of plastid types are found in plant cells (Kirk & Tilney-Bassett 1978). Meristematic cells contain small (0.5-1μm diameter) undifferentiated plastids or proplastids, which have little internal structure apart from a few flattened sacs which occasionnally have continuity with the inner envelope. They ensure the continuance of plastids within a species from generation to generation and are capable of considerable structural and metabolic transformations which give rise to more mature plastids. Leucoplasts are undifferentiated proplastids which have lost their capability to form chloroplasts. Amyloplasts, which contain large (from 1 to 50 μm diameter) starch grains, are storage plastids found in stems, roots and tubers. Carotenoid-rich plastids or chromoplasts are present in petals, fruits and even roots. When grown in the dark, leaves contain etioplasts, which can be transformed into chloroplasts upon illumination. All these plastids contribute considerably to the general metabolism of the cell.

Despite their wide size range, their considerable structural and physiological diversity, all these plastids are bounded by the two concentric envelope membranes. During all the interconversions involving non-green plastids and chloroplasts, dramatic changes occur within the organelles, with the development or the regression of the internal membrane systems (thylakoids, prolamellar bodies...), but at all stages of these transformations, the two envelope membranes remain, apparently identical.

Plastids, which contain DNA and all the biosynthetic machinery for protein synthesis, do not arise *de novo*. In dividing vegetative plant cells, plastids are formed by division of preexisting plastids and transmitted to the daughter cells, their multiplication is necessary to maintain the plastid number to a final value per cell that is characteristic of the tissue and the species. Kirk & Tilney-Bassett (1978), Sears (1980) and Whatley (1982) have traced the evolution of plastid inheritance in the plant kingdom. There is a general trend toward strict uniparental (maternal) inheritance in angiosperms versus biparental inheritance in lower plants. Unlike the nuclear envelope, which is reversibly disassembled during mitosis, the plastid envelope membranes remain apparently intact during plastid division. Normally, division occurs by constriction in the center of the plastid, the mechanism of plastid division and its control are largely unknown, but could involve a circular bundle of actin-like filaments, each about 4-5 nm in diameter, called the "plastid-dividing ring", located in the cytosol and associated with the outer envelope membrane (Kuroiwa 1989). The study of the envelope membranes illustrates the concept of autonomy and continuity of plastids, originated from microscopy studies and further supported by numerous biochemical analyses (Kirk & Tilney-Bassett 1978).

ENVELOPE MEMBRANES, LIKE ALL PLASTID MEMBRANES, CONTAIN SPECIFIC LIPID CONSTITUENTS

Glycerolipid Composition. In contrast to extraplastidial membranes, plastid membranes (envelope and thylakoids) are characterized by a low phospholipid content and by the presence of glycolipids (Douce *et al* 1973). The major plastid lipid compounds are galactolipids, which contain one or two galactose molecules attached to the *sn*-3 position of the glycerol backbone, corresponding respectively to monogalactosyldiacylglycerol (or MGDG) and digalactosyldiacylglycerol (or DGDG).

Galactolipids in thylakoid and envelope membranes contain a high amount of polyunsaturated fatty acids: up to 95 % (in some species) of the total fatty acids is linolenic acid. In non-green plastids, 18:3 is still a major component although appreciable amounts of 18:1 and 18:2 are present. Therefore, the most abundant molecular species of MGDG and DGDG have 18:3 at both *sn*-1 and *sn*-2 positions of the glycerol backbone. Some plants, such as pea, having almost only 18:3 in MGDG are called "18:3 plants". Other plants, such as spinach, contain important amounts of 16:3 in MGDG, they are called "16:3 plants" (Heinz 1977). The positional distribution of 16:3 in MGDG is highly specific: this fatty acid is only present at the *sn*-2 position of glycerol, and is almost excluded from *sn*-1 position. Therefore, two major structures are found in galactolipids, one with C18 fatty acids at both *sn* position and one with C18 and C16 fatty acids respectively at *sn*-1 and *sn*-2 position. The first one is typical of "eukaryotic" lipids (such as phosphatidylcholine or PC) and the second one corresponds to a "prokaryotic" structure. These differences are probably due to galactolipid biosynthetic pathways (see below).

Plastid membranes sometimes contain galactolipids with 3 (tri-GDG) and 4 (tetra-GDG) galactoses. They are formed by an enzymatic galactose exchange between MGDG and DGDG, owing to a galactolipid:galactolipid galactosyltransferase (Van Besouw & Wintermans 1978, Heemskerk & Wintermans 1987) which catalyzes the following reactions:

$$2 \text{ MGDG} \rightarrow \text{DGDG} + \text{diacylglycerol}$$
$$\text{MGDG} + \text{DGDG} \rightarrow \text{Tri-GDG} + \text{diacylglycerol}$$
$$2 \text{ DGDG} \rightarrow \text{tetra-GDG} + \text{diacylglycerol}$$

Dorne *et al* (1982) have demonstrated that this enzyme (a) is located on the cytosolic side of the outer envelope membrane, (b) is susceptible to proteolytic digestion by thermolysin, a non-penetrant protease and (c) is present in all plastids (chloroplasts and non-green plastids). Consequently, in order to obtain a glycerolipid composition which could represent the *in vivo* situation within plastid membranes, the galactosyltransferase should be destroyed prior to fractionation of plastids.

<u>Table 1</u>: Glycerolipid composition of membranes from plant mitochondria and plastids.

ORGANELLE	MGDG	DGDG	SL	PC	PG	PI	PE	DPG
MITOCHONDRIA								
. sycamore cells								
total membranes	0	0	0	43	3	6	35	13
inner membrane	0	0	0	41	2.5	5	37	14.5
outer membrane	0	0	0	54	4.5	11	30	0
PLASTIDS								
. spinach chloroplasts								
thylakoids*	57	27	7	0	7	1	0	0
total envelope	32	30	6	20	9	4	0	0
inner membrane*	55	29	5	0	9	1	0	0
outer membrane	17	29	6	32	10	5	0	0
. pea etioplasts								
total envelope	34	31	6	17	5	4	0	0
. cauliflower proplastids								
total envelope	31.5	27.5	6	20	9	4.5	1	0

Plastid envelope membranes were analyzed after thermolysin treatment of intact plastids. *: Values for thylakoids and inner envelope membrane were recalculated to account for contamination by outer envelope membrane (Dorne *et al* 1990). *Abbreviations* MGDG, monogalactosyldiacylglycerol; DGDG, diga-lactosyldiacylglycerol; SL, sulfolipid; PC, phosphatidylcholine; PG, phosphatidylglycerol; PI, phosphatidyli-nositol; PE, phosphatidylethanolamine; DPG, diphosphatidylglycerol.

The glycerolipid composition of envelope membranes from chloroplasts and non-green plastids is given in table 1. As expected after thermolysin treatment, no diacylgly-cerol was detected (whereas it represents about 10 % of the total envelope glycerolipids from non-treated plastids). The most striking feature is that the glycerolipid pattern is almost identical in envelope membranes from chloroplasts, etioplasts or other non-green plastids. The high amount of MGDG reflects the presence of the inner envelope membrane which has a glycerolipid composition close to that of thylakoids as shown in pea (Cline *et al* 1981) or spinach (Block *et al* 1983): in the inner envelope membrane and in thylakoids, MGDG is the major component. The presence of PC in envelope membranes reflects the presence of the outer envelope membrane: using phospholipase C digestion of intact chloroplasts, we have demonstrated that no PC is present in thylakoids (Dorne *et al* 1990). In contrast, PC represents about 30-35 % of the outer envelope glycerolipids, where it is concentrated in the outer leaflet of the membrane (Dorne *et al* 1985). The presence of this typical eukaryotic lipid in the outer membrane leaflet facing the cytosol could be due to the functioning of a cytosolic phospholipid transfer protein. The major phospholipid in the inner envelope membrane and in thylakoids is phosphatidylglycerol (PG), which is unique because of a $16:1_{trans}$ fatty acid at sn-2 position of the glycerol backbone. This phospholipid is therefore different from that found in extra-plastidial membranes. Finally, table 1 also confirms the observation that plastid membranes are devoid of phosphatidylethanolamine (PE), which is a major component (together with PC) of mitochondrial, endoplasmic reticulum or other extra-plastidial membranes.

Pigment and Prenylquinone Composition. In contrast to thylakoids, envelope membranes from chloroplasts or non-green plastids are yellow, due to the presence of carotenoids and the absence of chlorophyll. The major carotenoid in envelope membranes from chloroplasts and from non-green plastids is violaxanthin, in contrast, thylakoids are rich in β-carotene and zeaxanthin (Douce *et al*/1984).

Although devoid of the most conspicuous plastid pigment (chlorophyll) some non-green plastids contain chlorophyll precursors such as protochlorophyllide. Envelope membranes from mature spinach chloroplasts contain low amounts of protochlorophyllide and chlorophyllide (Pineau *et al*/1986). In contrast, chlorophylls are concentrated in thylakoids but their precursors could not be detected (Pineau *et al*/1986).

In addition to pigments, envelope membranes from spinach chloroplasts contain prenylquinones as genuine components. Both envelope membranes contain plastoquinone-9, γ-tocopherol and phylloquinone (Soll *et al*/1985). However, in contrast to thylakoids, the major prenylquinone in envelope membranes is γ-tocopherol whereas it is plastoquinone-9 in thylakoids (Soll *et al*/1985).

DYNAMICS OF THE OUTER ENVELOPE MEMBRANE

The Outer Envelope Membrane as a Site of Interaction with Cytoplasmic Membranes. Based on ultrastructural studies, it has been suggested that the outer envelope interacts with a number of extraplastidial membranes. Close associations of outer envelope membrane with endoplasmic reticulum or tonoplast have been reported in some tissues (for a review, see Douce & Joyard 1979). Such structural relationship have often been taken as an argument for a structural continuity of the outer envelope membrane with the cell endomembrane system as has been suggested by Morré (1975). As discussed by Douce & Joyard (1979) and Douce *et al* (1984), appropriate biochemical investigations do not support the generalization of the endomembrane concept to the plastid outer envelope membrane although direct continuities of the outer envelope membrane with cytoplasmic membranes, and especially with endoplasmic reticulum, may occur in some specialized tissues. In these specialized tissues endoplasmic reticulum behave simply as a simple passive corridor that transfers harmful compounds from their biosynthetic site (envelope membranes) to the accumulation site without contact with the cytoplasm (Douce *et al* 1984). For instance, in secretory cells such as the resin canal from pine needles, multilobate leucoplasts are sheathed by a layer of fenestrated endoplasmic reticulum associated with the envelope membrane (Carde & Bernard-Dagan 1982). Terpenes synthesized at the level of envelope membranes are discharged into the endoplasmic reticulum, where they accumulate before being released in the periplasmic space of the cell (Carde & Bernard-Dagan 1982).

The Outer Envelope Membrane and the Supply of Fatty Acids for the Biosynthesis of Extra-plastidial Membranes. Plastids provide fatty acids for all plant membranes (Stumpf 1987). Intact spinach chloroplasts incorporate [^{14}C]-acetate into 16:0-ACP and 18:1-ACP, which can be either incorporated into plastid glycerolipids by the envelope acyltransferases, or exported to the cytosol for phospholipid synthesis by extraplastidial membranes (Joyard & Douce 1987). The plastid acyltransferases and the soluble acyl-ACP thioesterase compete for the same substrate: acyl-ACP. In addition, ACP exists as distinct isoforms, ACP I and ACP II, which probably have different role: acyl-ACP I is the prefered substrate for the acyl-ACP thioesterase and acyl-ACP II for the acyltransferases (Ohlrogge 1987). Thus, the channelling of fatty acids between esterification to glycerol in the envelope (see below) or release by hydrolysis for export in the cytosol may be regulated by expression or acylation of ACP isoforms (Ohlrogge 1987, Joyard & Douce 1987). However, fatty acids are not soluble in aqueous phases and it is often suggested that acyl-CoA are better candidates for transport of fatty acids. The formation of acyl-CoA in envelope membranes is due to an acyl-CoA synthetase (Joyard & Douce 1977) localized in the outer envelope membrane (Dorne *et al* 1982), in contrast with an acyl-CoA thioesterase which is localized in the inner envelope membrane (Dorne *et al* 1982). Oleic acid (the major fatty acid exported) is one of the best substrate for the acyl-CoA synthetase (Joyard & Stumpf 1980, 1981). It is therefore possible that the outer envelope acyl-CoA synthetase could be involved in the export of fatty acids towards the cytosol.

THE INNER ENVELOPE MEMBRANE AND THE DYNAMICS OF PLASTID MEMBRANE BIOGENESIS

The Inner Envelope Membrane as a Site of Glycerolipid Synthesis. The first observation that envelope membranes could be involved in the biosynthesis of plastid components was provided by Douce (1974): envelope membranes purified from spinach chloroplasts were able to catalyze MGDG synthesis from diacylglycerol at high rates. Following the obser-vations of Douce & Guillot-Salomon (1970) who established the presence of the enzymes from the Kornberg-Pricer pathway in chloroplasts and non-green plastids, a considerable body of data has been accumulated which clearly demonstrate that plastids are able to incorporate *sn* –glycerol 3–phosphate into lysophosphatidic acid, phosphatidic acid and diacylglycerol, and then into MGDG (after addition of UDP–galactose). Glycerolipid biosynthesis requires the assembly of three parts (Joyard & Douce, 1987): fatty acids, glycerol and a polar head group (galactose, for galactolipids; sulfoquinovose, for sulfolipid; and phosphorylglycerol, for PG). The whole pathway for this biosynthesis, which takes place in the inner envelope membrane, is presented in Figure 1. Only the major features will be presented in this article. For a more detailed study, the reader is refered to several reviews (Heinz 1977, Roughan & Slack 1982, Joyard & Douce 1987, Douce & Joyard 1990, Joyard *et al* 1991).

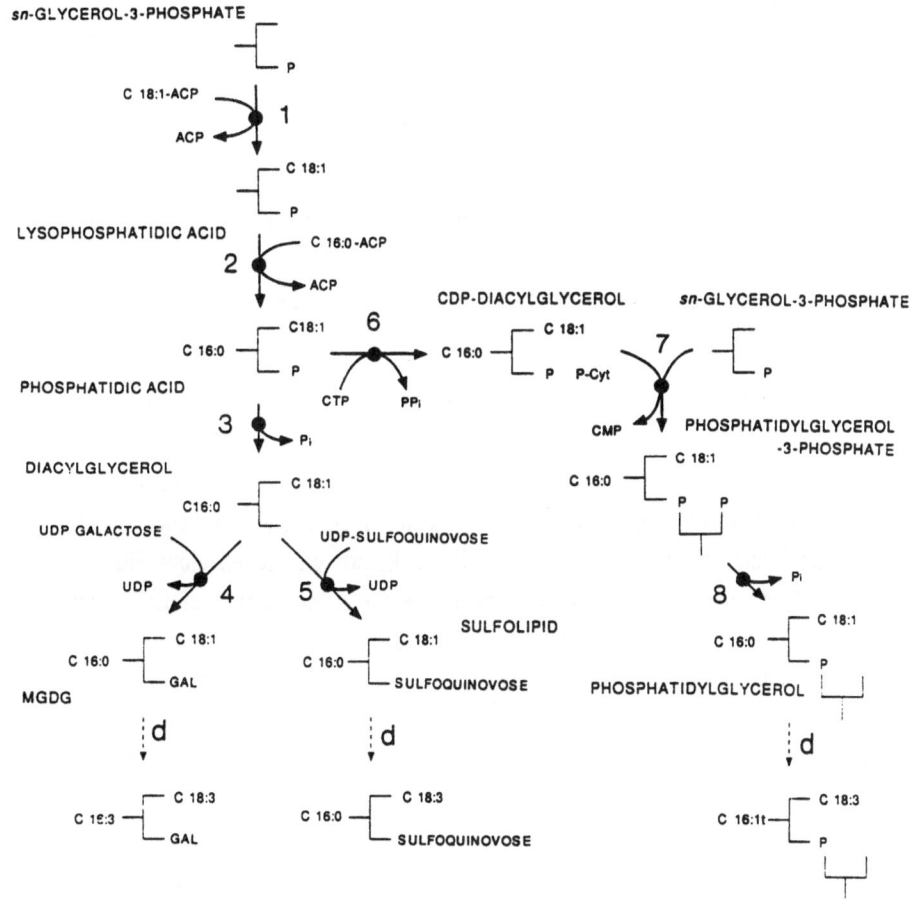

Figure 1: Biosynthesis of plastid glycerolipids by the inner envelope membrane. The enzymes involved are: (1) glycerol–3–phosphate acyltransferase, (2) 1–acylglycerol–3–phosphate acyltransferase, (3) phosphatidate phosphohydrolase, (4) monogalactosyldiacylglycerol synthase, (5) sulfolipid synthase, (6) phosphatidate cytidyltransferase, (7) CDP–diacylglycerol–glycerol–3–phosphate 3–phosphatidyltransferase, (8) phosphatidylglycerophosphatase, (d) desaturases. This biosynthetic pathway lead to the synthesis of monogalactosyldiacylglycerol, sulfolipid and phosphatidylglycerol with unsaturated C 18 and C 16 fatty acids at the *sn*–1 and *sn*–2 positions of glycerol respectively.

The first important point is that the two acyltransferases have distinct specificities and selectivities for acylation of *sn* -glycerol 3-phosphate (Frentzen *et al* 1983, Joyard 1979): together, they led to the formation of phosphatidic acid having 18:1 and 16:0 fatty acids respectively at *sn* -1 and *sn* -2 position of the glycerol backbone (figure 1). This structure is typical for the so-called prokaryotic glycerolipids which are found in 16:3 plants (see above). In contrast, in 18:3 plants, plastid glycerolipids (MGDG) do not contain C16 fatty acids at *sn* -2 position, and therefore probably not derive directly from the envelope Kornberg-Pricer pathway, their origin is still poorly understood.

The phosphatidic acid formed in the inner envelope membrane can be used either for the synthesis of phosphatidylglycerol (Mudd *et al* 1987) or for the formation of diacylglycerol (Joyard & Douce 1977) which is the substrate for MGDG and sulfolipid biosynthesis (Figure 1).

The level of phosphatidate phosphatase activity in envelope membranes could be responsible for the difference observed between 18:3 and 16:3 plants at the level of MGDG synthesis: in contrast to 16:3 plants, 18:3 plants have a rather low phosphatidate phosphatase activity and therefore are almost unable to form diacylglycerol for further galactosylation (Heinz & Roughan 1983). Therefore, a limiting step in the synthesis of MGDG could be the formation of diacylglycerol.

Chloroplast envelope membranes isolated from 16:3 and 18:3 plants, as well as non-green plastids have the same capacity to form MGDG when supplied with diacylglycerol and UDP-galactose (UDP-gal). The reaction is catalyzed by a UDP-galactose:diacylglycerol galactosyltransferase (or MGDG synthase). Despite its importance in synthesizing MGDG, the most abundant polar lipid on earth, very little is known on the envelope MGDG synthase. Covès *et al* (1986) have (a) described a procedure for solubilization of this enzyme from spinach chloroplasts envelope membrane, (b) developped an assay for measuring enzyme activity after solubilization, and (c) achieved a partial purification of the enzyme. Covès *et al* (1988) have demonstrated a strong lipid requirement for the MGDG synthase. Acidic glycerolipids, and especially PG, were shown to be the best activators of the enzyme. We have achieved further purification of the enzyme. After a series of chromatographic steps on Hydroxyapatite, Biogel-P6, Blue Dextran agarose and Zn^{++}-Chelating Sepharose, we have prepared between 0.5 to 1 μg protein containing a huge activity (up to 50 μmol galactose incorporated/h/mg protein) from about 100 mg envelope proteins. A major 19,000 daltons-polypeptide together with a series of minor compounds were present in the fraction (Figure 2). The results obtained strongly suggest that the 19,000 daltons-polypeptide could be associated with MGDG synthase activity, but the possibility that one of the minor polypeptides is indeed the enzyme cannot be ruled out. These observation demonstrate that

MGDG synthase is a very minor protein component (between 0.1 to 0.5 %) in the inner envelope membrane, and has an enormous specific activity which is able to sustain the rates necessary for MGDG synthesis during the formation of plastid membranes.

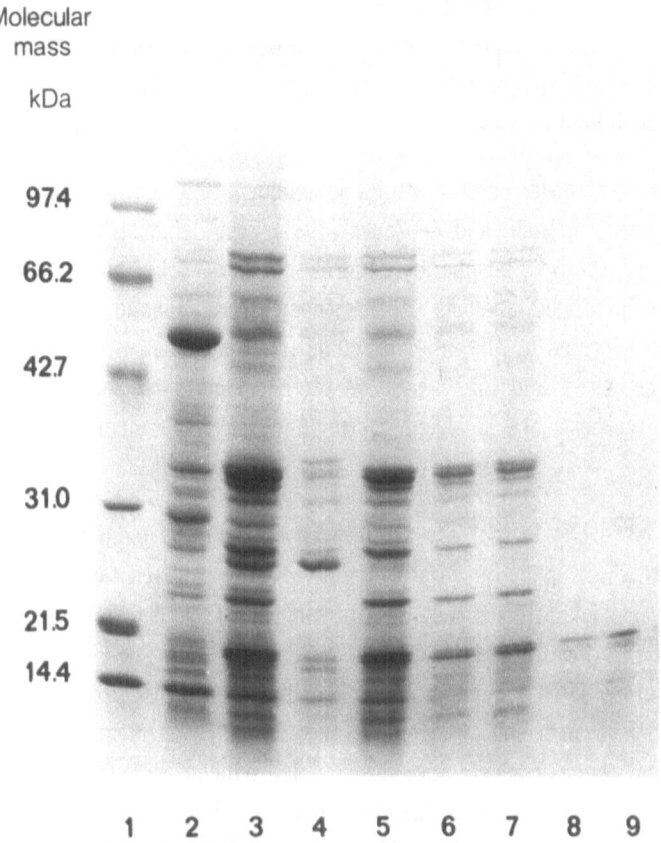

Figure 2: Analyses of envelope polypeptides during MGDG synthase purification. The polypeptides were separated by SDS-polyacrylamide gel electrophoresis (7.5–15 %) according to Block et al (1983). Lane 1: Standards; lane 2: total envelope; lane 3: fraction. containing the highest MGDG synthase activity. obtained by hydroxyapatite chromatography (see Covès et al/1986); lane 4: void volume of Blue–Dextran agarose chromatography; lane 5: fraction. containing the highest MGDG synthase activity. obtained by Blue–Dextran agarose chromatography; lanes 6 & 7: void volume of Zn^{++}-Chelating sepharose chromatography; lanes 8 & 9: fraction. containing the highest MGDG synthase activity. obtained by Zn^{++}-Chelating sepharose chromatography. Proteins loaded in each well were: lanes 2 to 5: 20 µg; lanes 6 & 7: 6 µg; the proteins present in wells 8 & 9 were estimated to less than 0.5 µg by using a scanner. Total MGDG synthase activity loaded in each well (from lane 2 to 9) was respectively: 1.8. 74. 5. 101. 8.5. 8.5. 5.2. 8.2 nmol/h. A major 19.000 daltons–polypeptide is enriched in fractions 8 and 9. but several other minor polypeptides are also present in the fraction.

Figure 1 also demonstrates that all the enzymes for sulfolipid and PG synthesis are associated with the envelope membrane. MGDG and sulfolipid synthesis share a common pool of diacylglycerol (Joyard et al/1986). Under these conditions. the steric placement of these enzymes within the membrane as well as their kinetic parameters may be relevant to the outcome of the competition. Heinz *et al* (1989) synthesized different nucleoside 5'-diphospho-sulfoquinovoses and demonstrated that both UDP- and GDP-sulfoquinovose significantly increased sulfolipid synthesis by spinach chloroplasts and by isolated envelope membranes, UDP-sulfoquinovose being twice as active as the GDP derivative. Therefore, sulfolipid synthesis in envelope membranes is probably due to a UDP-sulfoquinovose:1,2-diacylglycerol 3-β-sulfoquinovosyltransferase (i.e. 1,2-diacylglycerol 3-β-sulfoquinovosyl-transferase or sulfolipid synthase).

As presented above, chloroplast PG is unique because it contains a $16:1_{trans}$ fatty acid at *sn*-2 position. Mudd and de Zacks (1981) first demonstrated the ability of intact chloroplasts to synthesize phosphatidylglycerol. Further work from Mudd's laboratory (see Mudd *et al* 1987) refined analysis of this biosynthesis. Plastid PG is synthesized from phosphatidic acid formed by the inner envelope (Figure 1), i.e. with C18 fatty acid and C16 fatty acid at the *sn*-1 and *sn*-2 position of the glycerol respectively, and the enzymes involved in its synthesis are localized in the inner envelope membrane (Andrews & Mudd 1985).

Finally, glycerolipids that are synthesized on envelope membranes contain 16:0 and 18:1 fatty acids which must be desaturated to polyunsaturated fatty acids. There is evidence from several laboratories that desaturation of C16 and C18 fatty acids occurs when esterified to MGDG (for reviews, see refs. Roughan & Slack 1982, Joyard & Douce 1987). However, the nature of the enzymes involved in the desaturation of plastid fatty acids are not well established. Andrews *et al* (1989) and Schmidt and Heinz (1991) have proposed that desaturation in chloroplasts involves soluble reduced ferredoxin (a stromal enzyme) and ferredoxin-NADP reductase (a thylakoid enzyme), but since envelope membranes are involved in oleoyl desaturation (Schmidt & Heinz 1991), it is likely that the enzymes involved are operating at the inner surface of the inner envelope membrane.

The inner envelope membrane as a site of pigment and prenylquinone synthesis. The localization of carotenoid synthesis within chloroplasts is not clearly established. It is generally assumed that thylakoids, and not envelope membranes, play the central role in carotenogenesis (Britton 1988). Since (a) carotenoid synthesis occurs in all plastids, (b) all the last steps in carotenoid biosynthesis involve membrane-bound enzymes and (c) carotenoids are present in given proportions in all plastid envelope membranes (which are almost the only membrane in some plastids such as proplastids and amyloplasts), the participation of envelope membranes from all plastid types in carotenogenesis cannot be excluded. This hypothesis is supported by several lines of evidence. First, chromoplast membranes, assumed to derive from the inner envelope membrane, are very active in carotenoid biosynthesis. In addition, phytoene synthase and desaturase (Lütke-Brinkhaus *et al* 1982) and zeaxanthine epoxidase (Costes *et al* 1979) activities were demonstrated in envelope membranes from mature spinach chloroplasts.

A possible role of envelope membranes in chlorophyll biosynthesis was recently proposed (Pineau *et al* 1986, Joyard *et al* 1990). Chloroplast envelope membranes contain a light- and NADPH-dependent protochlorophyllide reductase (Pineau *et al* 1986, Joyard *et al* 1990). In etioplasts, this enzyme is concentrated in much larger amounts in the prolamellar body, and is barely detectable in envelope membranes. Envelope protochlorophyllide reductase is a very minor 37,000 daltons-protein which can be separated from the major E37 only by two-dimensional polyacrylamide gel electrophoresis (Joyard *et al* 1990). Recently, Matringe et al (1991) have demonstrated that envelope membranes from mature spinach chloroplasts contain another enzyme of the chlorophyll biosynthetic pathway: protoporphyrinogen oxidase, the target enzyme for diphenyl ether herbicides. From these data, one can propose the following hypothesis: if the first steps in chlorophyll biosynthesis, namely those from (2)δ−aminolevulinic acid to protoporphyrinogen IX, occur in the soluble phase of chloroplasts (Castefranco & Beale 1983), it is possible that all the subsequent steps in protoporphyrinogen IX transformation to chlorophyllide are catalyzed by envelope-bound enzymes. The final step of chlorophyll formation involves addition of the phytoyl moiety and is specifically associated with thyla-koids (Block *et al* 1980). Thus, the pathway for chlorophyll biosynthesis in mature chloroplasts seem to involve a very complex cooperation between all the chloroplast compartments. This hypothesis is supported by indirect evidence from Fuesler *et al* (1984) who demonstrated that in intact chloroplasts, Mg-chelatase (the enzyme following protoporphyrinogen oxidase) was accessible to molecules unable to pass through the inner envelope membrane, thus suggesting that Mg-chelatase could be present in the envelope membranes.

Prenylquinones, and especially plastoquinone-9, are essential compounds for electron and proton transfer during photosynthesis. All the enzymes involved in prenylquinone biosynthesis have been localized in the inner envelope membrane of spinach chloroplasts (Soll *et al* 1985). As shown in Figure 3, tocopherols are synthesized from homogentisic acid and a C20-prenyl unit to form 2-methyl-6-prenyl quinol. By a series of methylations and cyclization, the 2-methyl-6-prenylquinol gives rise successively to 2,3-dimethyl-6-prenylquinol, γ-tocopherol (or γ-tocotrienol) and finally α-tocopherol (α-tocotrienol). Plastoquinone-9 is synthesized from homogentisic acid and solanesyl-pyrophosphate to form 2-methyl-6-solanesylquinol, which is methylated and oxidized to form successively plastoquinol-9 and plastoquinone-9.

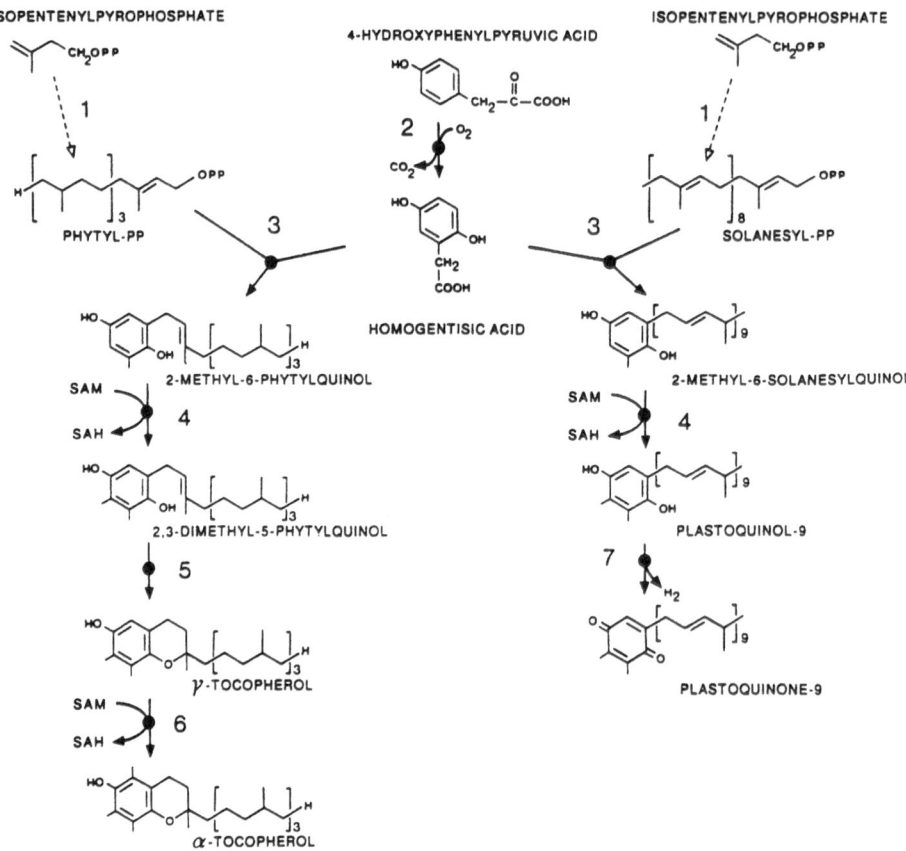

Figure 3: Biosynthesis of α–tocopherol and plastoquinone–9 by the inner envelope membrane. Phytyl pyrophosphate (Phytyl–PP) and solanesyl pyrophosphate (solanesyl–PP) are synthesized from isopentenylpyrophosphate by a series of enzymes (1) localized in the plastid stroma and in the plastid envelope (Block *et al* 1980). Homogentisic acid is probably synthesized in the envelope via a 4-hydroxyphenylpyruvate dioxygenase. α–Tocopherol and plastoquinone–9 are then synthesized by a series of prenyltransferases (3), C–methyltransferases (4,6), cyclase (5) and dehydrogenase (7) that are all associated with the inner envelope membrane (Soll *et al* 1985).

Intraplastidial Lipid Transfer. All the observations summarized above demonstrate that (a) the inner envelope membrane is the site of synthesis for MGDG, sulfolipid, PG, α–tocopherol, plastoquinone–9 and, probably, of some pigments such as carotenoids and chlorophyll, and (b) that thylakoids contain the largest amount of these plastid constituents. Therefore, transport of these molecules from their site of synthesis (envelope membranes) to their site of accumulation (thylakoids) should take place during development. The mechanisms involved (vesicular transport, transfer of lipid monomers through the stroma either by protein–facilitated transport or by spontaneous diffusion of free monomers, lateral diffusion of lipids between membranes at regions of direct intermembrane contact, etc...) is still unknown and only speculations can be presented. It is possible that import of thylakoid membrane lipids may occur through the stroma en route to the thylakoids *via* vesicles derived from the inner membrane (Wellburn 1982), but this remains to be demonstrated (see below). In contrast, recent evidences (Cline 1986, 1988, Smeekens *et al* 1986, Hageman *et al* 1986...) suggest translocation of thylakoid- (chlorophyll a/b binding protein, etc...) or lumen- (plastocyanin, 33 kDa oxygen-evolving protein, etc...) directed proteins may occur through the stroma as soluble proteins. These results strongly suggest that the proteins do not depart from the inner membrane by a bulk flow process.

Electron microscopy studies have suggested however that numerous vesicles, probably deriving from the inner envelope membrane, are formed during development (see for instance Carde *et al* 1982). Since the inner envelope membrane and thylakoids do not differ in polar lipid composition (Cline *et al* 1981, Block *et al* 1983), natural fusion of vesicles produced by the inner envelope membrane with growing thylakoids is possible. In addition, MGDG forms hexagonal type II structures (rather than Lα bilayers) (Gounaris & Barber 1983) which could behave as intermediates in membrane fusion. Therefore, MGDG (which represents almost half of the lipid content of inner envelope vesicles and thylakoids) could favor fusions between plastid membranes. Nothing is known of how vesicle buds or how its selective fusion with thylakoids is programmed and catalyzed, nor is it known how vesicles deriving from the inner membrane select their content while rejecting the bulk constituents of the inner membrane from which they bud. If the vesicles deriving from the inner envelope represent rudimentary thylakoids, considerable modifications must take place following the initial step of invagination: vesicles deriving from the inner envelope membrane should be modified by both degradation of inner envelope proteins (unless they consist only of a lipid matrix) and addition of new polypeptides to acquire unique thylakoid characteristics. For instance, during development of young maize plastids (with a single prothylakoid) into mature chloroplasts, the activity of the two photosystems was detected only in thylakoids (even at early stages of development) but never in envelope membranes (Wrischer 1989). However, it is well established that membrane synthesis never occurs *de nova* a membrane always derives from a preexisting membrane and thylakoids are probably not an exception to this rule. Therefore, vesicles deriving from the inner envelope should be added to a preexisting membrane (pre- thylakoids), whose nature is totally unknown. The discrete templates for thylakoids can be either modified inner envelope membrane vesicles or genuine distinct prethylakoids. Little information is available to solve this problem, probably because the rules governing membrane identity and function are largely unknown. It is clear, therefore, that the inner membrane of the plastid envelope is both a stable and dynamic structure, the identity of envelope membranes is likely to be governed by a sophisticated set of structurally interactive resident nuclear-coded protein and polar lipids. While this assembly confers specific functional properties, it clearly does not imped the extensive flow of lipids mediated by carrier vesicles that continuously bud at the inner membrane surface. Both the biochemical and molecular details of many answered questions related to envelope structure and the dynamic of vesicle formation and reassociation (budding, targeting and fusion) will undoubtely emerge from future studies.

REFERENCES

Andrews J, Schmidt H, Heinz E (1989) Interference of electron transport inhibitors with desaturation of monogalactosyl diacylglycerol in intact chloroplasts. *Arch Biochem Biophys* 270:611–622

Andrews J, Mudd JB (1985) Phosphatidylglycerol synthesis in pea chloroplasts. Pathway and localization. *Plant Physiol* 79:259–265

Block MA, Dorne AJ, Joyard J, Douce R (1983) Preparation and characterization of membrane fractions enriched in outer and inner envelope membranes from spinach chloroplasts. II – Biochemical characterization. *J Biol Chem* 258:13281–13286

Block MA, Joyard J, Douce R (1980) Site of synthesis of geranylgeraniol derivatives in intact spinach chloroplasts. *Biochim Biophys Acta* 631:210–219

Britton G (1988) Biosynthesis of carotenoids. In *Plant Pigments*, Goodwin TW (ed) pp 133–182, Academic Press London

Carde JP, Bernard-Dagan C (1982) Compartimentation de la synthèse terpénique chez le Pin maritime. *Bull Soc Bot Fr* Actual Bot 129:53–70

Carde JP, Joyard J, Douce R (1982) Electron microscopic studies of envelope membranes from spinach plastids. *Biol Cell* 44:315–324

Castelfranco PA, Beale SI (1983) Chlorophyll biosynthesis: Recent advances and areas of current interest. *Ann Rev Plant Physiol* 34:241–278

Cline K (1986) Import of proteins into chloroplasts. Membrane integration of a thylakoid precursor protein reconstituted in chloroplast lysates. *J Biol Chem* 261:14804–14810

Cline K, Andrews J, Mersey B, Newcomb EH, Keegstra K (1981) Separation and characterization of inner and outer envelope membranes of pea chloroplasts. *Proc Natl Acad Sci USA* 78:3595–3599

Costes C, Burghoffer C, Joyard J, Block MA, Douce R (1979) Occurence and biosynthesis of violaxanthin in isolated spinach chloroplast envelope. *FEBS Lett* 103:17–21

Covès J, Block MA, Joyard J, Douce R (1986) Solubilization and partial purification of UDP-galactose:diacylglycerol galactosyltransferase activity from spinach chloroplast envelope. *FEBS Lett* 208:401–406

Covès J, Joyard J, Douce R (1988) Lipid requirement and kinetic studies of solubilized UDP-galactose:diacylglycerol galactosyltransferase activity from spinach chloroplast envelope membranes. *Proc Natl Acad Sci USA* 85:4966–4970

Dorne AJ, Block MA, Joyard J, Douce R (1982) The galactolipid:galactolipid galactosyltransferase is located on the outer membrane of the chloroplast envelope. *FEBS Lett* 145:30–34

Dorne AJ, Joyard J, Block MA, Douce R (1985) Localization of phosphatidylcholine in outer envelope membrane of spinach chloroplasts. *J Cell Biol* 100:1690–1697

Dorne AJ, Joyard J, Douce R (1990) Do thylakoids really contain phosphatidylcholine? *Proc Natl Acad Sci USA* 87, 71–74

Douce R (1974) Site of biosynthesis of galactolipids in spinach chloroplasts. *Science* 183:852–853

Douce R, Guillot-Salomon T (1970) Sur l'incorporation de la radioactivité du *sn*-glycérol 3-

phosphate-^{14}C dans le monogalactosyldiglycéride des plastes isolés. *FEBS Lett* 11:121-126

Douce R, Joyard J (1979) Structure and function of the plastid envelope. *Adv Bot Res* 7:1-116

Douce R, Joyard J (1990) Biochemistry and function of the plastid envelope. *Ann Rev Cell Biol* 6:173-216

Douce R, Holtz RB, Benson AA (1973) Isolation and properties of the envelope of spinach chloroplasts. *J Biol Chem* 248:7215-7222

Douce R, Block MA, Dorne AJ, Joyard J (1984) The plastid envelope membranes: Their structure, composition, and role in chloroplast biogenesis. *Subcell Biochem* 10:1-86

Frentzen M, Heinz E, McKeon TA, Stumpf PK (1983) Specificities and selectivities of glycerol-3-phosphate acyltransferase and monoacylglycerol-3-phosphate acyltransferase from pea and spinach chloroplasts. *Eur J Biochem* 129:629-636

Fuesler TP, Wong YS, Castelfranco PA (1984) Localization of Mg-chelatase and Mg-protoporphyrin IX mono methyl ester (oxidative) cyclase activities within isolated developing cucumber chloroplasts. *Plant Physiol* 75:662-664

Gounaris K, Barber J (1983) Monogalactosyldiacylglycerol: The most abundant polar lipid in Nature. *Trends Biochem Sci* 8:378-381

Hageman J, Robinson C, Smeekens S, Weisbeek P (1986) A thylakoid processing protease is required for complete of the lumen protein plastocyanine. *Nature* 324:567-569

Heemskerk JWM, Wintermans JFGM (1987) Role of the chloroplast in the leaf acyl-lipid synthesis. *Physiol Plantarum* 70:558-568

Heinz E (1977) Enzymatic reactions in galactolipid biosynthesis. In *Lipids and Lipid Polymers*, Tevini M, Lichtenthaler HK (eds), pp 102-120, Springer Berlin

Heinz E, Roughan PG (1983) Similarities and differences in lipid metabolism of chloroplasts isolated from 18:3 and 16:3 plants. *Plant Physiol* 72:273-279

Heinz E, Schmidt H, Hoch M, Jung KH, Binder H, Schmidt RR (1989) Synthesis of different nucleoside 5'-diphospho-sulfoquinovoses and their use for studies on sulfolipid biosynthesis in chloroplasts. *Eur J Biochem* 184:445-453

Joyard J (1979) L'enveloppe des chloroplastes. Thèse de Doctorat d'Etat. Université de Grenoble.

Joyard J, Douce R (1977) Site of synthesis of phosphatidic acid and diacylglycerol in spinach chloroplasts. *Biochim Biophys Acta* 486:273-285

Joyard J, Douce R (1987) Galactolipid biosynthesis. In *The Biochemistry of Plants. Lipids: Structure and function*, Vol 9, Stumpf PK (ed) pp 215-274, Academic Press New York

Joyard J, Stumpf PK (1980) Characterization of an acyl-CoA thioesterase associated with the envelope of spinach chloroplasts. *Plant Physiol* 65:1039-1043

Joyard J, Stumpf PK (1981) Synthesis of long-chain acyl-CoA in chloroplasts envelope membranes. *Plant Physiol* 67:250-256

Joyard J, Blée E, Douce R (1986) Sulfolipid synthesis from $^{35}SO_4^{2-}$ and [1-^{14}C]-acetate in isolated intact spinach chloroplasts. *Biochim Biophys Acta* 879:78-87

Joyard J, Block MA, Pineau B, Albrieux C, Douce R (1990) Envelope membranes from mature spinach chloroplasts contain a NADPH:protochlorophyllide reductase on the cytosolic side of the outer membrane. *J Biol Chem* 265:21820-21827

Joyard J, Block MA, Douce R (1991) Molecular aspects of plastid envelope biochemistry. *Eur J Biochem*, in press

Kirk JTO, Tilney-Bassett RAE (1978) *The Plastids. Their Chemistry, structure, growth and inheritance*. 2nd ed, Elsevier Amsterdam

Kuroiwa T (1989) The nuclei of cellular organelles and the formation of daughter organelles by the "plastid-dividing ring". *Bot Mag Tokyo* 102:291-329

Lütke-Brinkhaus F, Liedvogel B, Kreuz K, Kleinig H (1982) Phytoene synthase and phytoene dehydrogenase associated with envelope membranes from spinach chloroplasts. *Planta* 156:176-180

Matringe M, Camadro JM, Block MA, Joyard J, Scalla R, Labbe, P, Douce R (1991) Localization within spinach chloroplasts of protoporphyrinogen oxidase, the target enzyme for diphenylether herbicides. *submitted for publication*

Morré DJ (1975) Membrane biogenesis. *Ann Rev Plant Physiol* 26:441-481

Mudd JB, Andrews JE, Sparace SA (1987) Phosphatidylglycerol synthesis in chloroplasts. *Methods Enzymol* 148:338–345

Mudd JB, de Zacks R (1981) Synthesis of phosphatidylglycerol by chloroplasts from leaves of *Spinacia oleracea* L. (spinach). *Arch Biochem Biophys* 209:584–591

Ohlrogge JB (1987) Biochemistry of plant acyl carrier proteins. In *The Biochemistry of Plants. Lipids: Structure and function*, Vol 9, Stumpf PK (ed), pp 137–157, Academic Press New York

Pineau B, Dubertret G, Joyard J, Douce R (1986) Fluorescence properties of the envelope membranes from spinach chloroplasts. Detection of protochlorophyllide. *J Biol Chem* 261:9210–9215

Roughan PG, Slack CR (1982) Cellular organization of glycerolipid metabolism. *Ann Rev Plant Physiol* 33:97–132

Sears BB (1980) Elimination of plastids: spermatogenesis and fertilization in the plant kingdom. *Plasmid* 4:233–255

Smeekens S, Bauerle C, Hageman J, Keegstra K, Weisbeek P (1986) The role of the transit peptide in the routing of precursors toward different chloroplast compartments. *Cell* 46:365–375

Soll J, Schultz G, Joyard J, Douce R, Block MA (1985) Localization and synthesis of prenyl-quinones in isolated outer and inner envelope membrane from spinach chloroplasts. *Arch Biochem Biophys* 238:290–299

Stumpf PK (1987) The Biosynthesis of saturated fatty acids. In *The Biochemistry of Plants. Lipids: Structure and function*, Vol 9, Stumpf PK (ed), pp 121–136, Academic Press New York

Van Besouw A, Wintermans JFGM (1978) Galactolipid formation in chloroplast envelopes. 1. Evidence for two mechanisms in galactosylation. *Biochim Biophys Acta* 529:44–53

Wellburn AR (1982) Bioenergetic and ultrastructural changes associated with chloroplast development. *Int Rev Cytol* 80:133–191

Whatley JM (1982) Ultrastructure of plastid inheritance: green algae to angiosperms. *Biol Rev* 57, 527–569

Wrischer M (1989) Ultrastuctural localization of photosynthetic activity in thylakoids during chloroplast development in maize. *Planta* 177:18–23

TRANSBILAYER AND LATERAL MOTIONS OF FLUORESCENT ANALOGS OF PHOSPHATIDYLCHOLINE AND PHOSPHATIDYLETHANOLAMINE IN THE PLASMA MEMBRANE OF BOVINE AORTIC ENDOTHELIAL CELLS

J.F. TOURNIER, M. JULIEN, J.F. TOCANNE
Centre de Recherche de Biochimie et de Génétique Cellulaires
118, route de Narbonne
31062 Toulouse Cédex. France

ABSTRACT

The apical plasma membranes of vascular endothelial cells present a non thrombogenic surface to the blood vessel, *in vivo*, or to the culture medium, *in vitro*. In the view of the catalytic role attributed to the anionic phospholipids in the blood coagulation cascade, we have checked if exogenous added fluorescent analog of phosphatidylethanolamine in the outer leaflet of the apical plasma membrane of these cells have different transmembrane and/or lateral movements than the same fluorescent analog of phosphatidylcholine. We have found that the fluorescent phosphatidylcholine derivative remains located at least one hour on the outer monolayer of the plasma membrane at 0°C as well as at 20°C, while the phosphatidylethanolamine analog is rapidly "internalized" at 20°C (half time approximatively 30 min) but not at 0°C. Evidences are presented suggesting that this "internalization" corresponds mainly to the transfert of the phosphatidylethanolamine from the outer to the inner leaflet of the membrane lipid bilayer, without labeling of intracellular particles. Furthermore, the lateral diffusion rate of the phosphatidylethanolamine probe was found to raise fourfold after these fluorescent probes have "flip" from the outer to the inner leaflet of the plasma membrane, demonstrating that the lateral motion of the lipids are considerably different on the two leaflets of the apical plasma membrane of aortic endothelial cells.

NATO ASI Series, Vol. H 63
Dynamics of Membrane Assembly
Edited by J. A. F. Op den Kamp
© Springer-Verlag Berlin Heidelberg 1992

INTRODUCTION

Recent studies have demonstrated the existence of an aminophospholipid translocase activity in the plasma membrane of blood circulating cells like erythrocytes (Zachowski et al. 1985), lymphocytes (Zachowski et al. 1987) and platelets (Sune et al. 1987, Sune and Bienvenüe 1988). This activity is presently considered to be implicated in the asymmetric distribution of aminophospholipids in the plasma membrane of these cells. More precisely, phosphatidylethanolamine and particularly phosphatidylserine are almost exclusively located in the inner face of the membrane lipid bilayer as is well demonstrated in the case of erythrocytes (Rawlyer et al. 1985) and blood platelets (Perret et al. 1979). Beside the physical properties that these phospholipids can confer to the plasma membrane, their asymmetric distribution is implicated in peculiar phenomenons such as morphological changes of erythrocytes and blood platelets (Daleke and Huestis 1985, Sune and Bienvenüe 1988, Daleke and Huestis 1989), cell-cell interactions (van Deenen et al. 1988, Lubin et al. 1981, McEvoy et al. 1986, Schlegel 1985, Schlegel and Williamson 1987, Tanaka and Schroit 1983, Wali et al. 1988....) and platelet activation (Bevers et al. 1983).

Two sequential reactions of the coagulation cascade are dramatically accelerated in the presence of an anionic phospholipid surface (Rosing et al. 1985, Gerads et al. 1990); therefore it was of interest to check whether differential transbilayer movements exist between anionic and neutral phospholipids in the apical plasma membrane of vascular endothelial cells, which form the inner nonthrombogenic lining of the large blood vessel.

For this purpose we have compared the localization and also the lateral motion of two fluorescent phospholipids probes, namely the 1-acyl-2-(N-4-nitrobenzo-2-oxa-1,3-diazole) phosphatidylcholine and phosphatidylethanolamine (NBD-PC and NBD-PE), inserted into the apical plasma membrane of bovine aortic endothelial cells, *in vitro*.

The results reported in the next paragraphs are included in a submited publication, where more detailed experimental data will be available. We summarize here the most striking features obtained on vascular endothelial cells by comparison with other cell types. In peculiar, this type of study is presently considered to be difficult because of the occurrence of an important membrane traffic in this type of cells (pinocytose, endocytose, transcytose...), when compared for instance to red blood cells.

RESULTS

Localization of the fluorescent lipid probes in the endothelial cell membranes:

When liposomes of dioleoyl phosphatidylcholine containing the fluorescent phospholipid of interest were incubated with monolayers of confluent endothelial cells 10 minutes at 0°C, spontaneous transfert of the probes from the liposomes to the cells occured, resulting in a prominent labeling of the apical plasma membrane. We have first measured the initial fluorescence intensities originating from the cell surface by the aid of an apparatus using a fluorescence microscope, as described in more details elsewhere (Lopez et al. 1988). These measurements were made on small cell plasma membrane areas (3.9 μm in diameter) and Fig.1 shows that a great heterogeneity of membrane labeling could be observed by this method. This is not due to the apparatus used because the same measurement repeated 10 times on the same cell area gave a value with a standard deviation of less than 5% (not shown). Back-exchange experiments were then performed to determine more precisely the localization of the probes in the cell membrane i.e. in the outer or inner leaflet of the lipid bilayer. This method consists in determining the quantity of the probes which can be removed from the plasma membrane, in the presence of an "acceptor" in the culture medium, after cell labeling in the conditions of interest. The probe molecules which can be removed by this mean are considered to be located in the outer leaflet of the plasma membrane. In our hands, liposomes of dioleoyl phosphatidylcholine were found to be no good acceptors for removing the fluorescent lipids from the labeled cells. Instead, culture medium containing 10% calf serum was used and permited the removal of 98% and 74% of NBD-PC and NBD-PE respectively, just after the cells were labeled 10 min at 0°C (Fig.1). By this method, NBD-PC was found to remain exchangeable, and then located in the outer leaflet of the plasma membrane, at least 1 hour at 0°C, as well as at 20°C (Fig.1). On the contrary, the same experiments performed with NBD-PE revealed that this phospholipid remained in the outer leaflet of the apical plasma membrane only when the cells were maintained at 0°C. It became rapidly "internalized" when the temperature was raised to 20°C after the labeling step (Fig.1; half time approx. 30 minutes). This "internalization" of NBD-PE corresponds to its transfer into the inner leaflet of the membrane lipid bilayer, as demonstrated by the following experiments. First, fluorescence microscopy did not show any labeling of

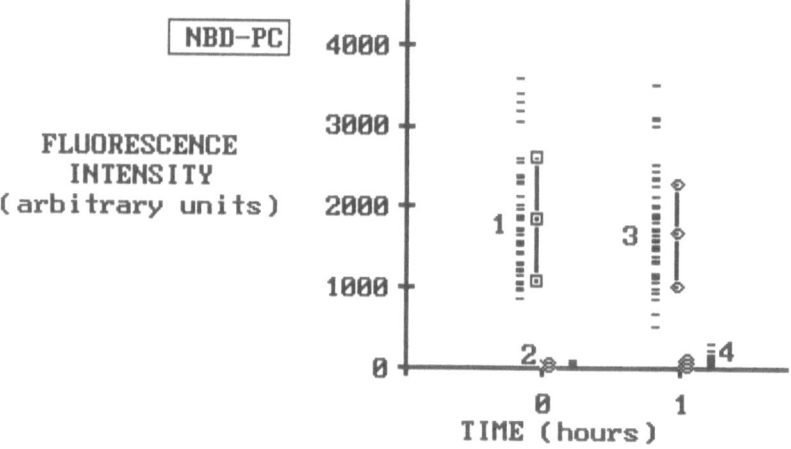

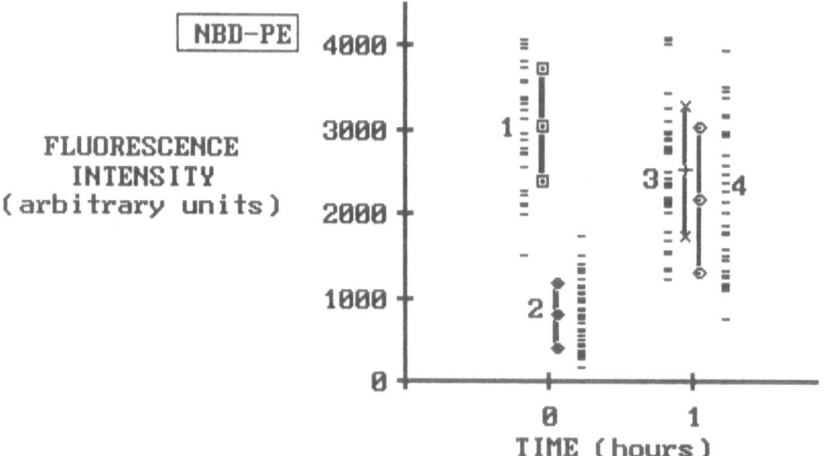

Figure 1: Back-exchange experiments on vascular endothelial cells labeled with NBD-phospholipids.

Dashes show individual fluorescence intensity values measured on small spots of 3.9 μm in diameter on confluent cultures, for each experimental condition . Just aside are ploted the mean value and the standard deviation (solid lines).

Upper and lower schemes show the data obtained with cells labeled with NBD-PC and NBD-PE respectively, under the following conditions:

1-just after the labeling step (10 min at 0°C)

2-after back-exchange procedure just after the labeling step

3-after the cells were maintained 1 hour at 20°C after the labeling step

4-after back-exchange experiments performed on cultures incubated 1 hour at 20°C after labeling.

intracellular particles when the cells were incubated 1 hour at 20°C, after labeling either with NBD-PC or NBD-PE (not shown). Furthermore, the two probes remained mainly located in the plasma membrane, as demonstrated below by the fact that the fractions of these probes which were free to lateraly diffuse were not greatly reduced when the cells were incubated 1 hour at 20°C, after the labeling step.

In a first attempt to characterize this aminophospholid translocase activity (to determine the enzymatic nature of this process), endothelial cells monolayers were incubated in a phosphate buffer saline, 5 hours at 20°C, in conditions which presumably reduce their intracellular content in ATP. After labeling of these cells with NBD-PE, this probe was found to remain exchangable at least one hour at 20°C (not shown), indicating that in these conditions of incubation, the inward movement of the probe to the inner leaflet of the plasma membrane did not occur. Precise measurements of the intracellular contents in ATP in these conditions remain to be made, with special care to the technical difficulties of the currently employed method (Martin and Pagano 1987)

Lateral diffusion parameters of NBD-PC and NBD-PE in the endothelial cell plasma membranes:

We have used Fluorescence Recovery After Photobleaching experiments (FRAP) to determine the rate of lateral diffusion (D) and the percentage of these probes (mobile fraction, M) which are free to diffuse in the endothelial cell plasma membrane. For experimental details and statistical analysis of the data see Tournier et al. 1989a. As shown in Fig.2, the measured mobile fractions were reduced by less than 15% when the cells were incubated 1 hour at 20°C in the culture medium (without serum) after labeling with either lipid probes. Surprisingly this indicates a rather limited internalization of the probes in these conditions. Furthermore, the lateral diffusion coefficient of NBD-PC remained nearly constant before and after exposure of the cells 1 hour at 20°C (D = 1.2 to $2.0x10^{-9}cm^2/s$ respectively), while that of NBD-PE raised from $2.8x10^{-9}cm^2/s$ at 4°C (outer leaflet) to $9.1x10^{-9}cm^2/s$ at 20°C (inner leaflet). These results extend to aortic endothelial cells previous data obtained on erythrocytes showing that the rapid "flip" of aminophospholipids on the inner leaflet of the plasma membrane is accompagnied by a rise in their lateral diffusion coefficient (Rimon et al. 1984, Morrot et al. 1986, Cribier et al. 1990).

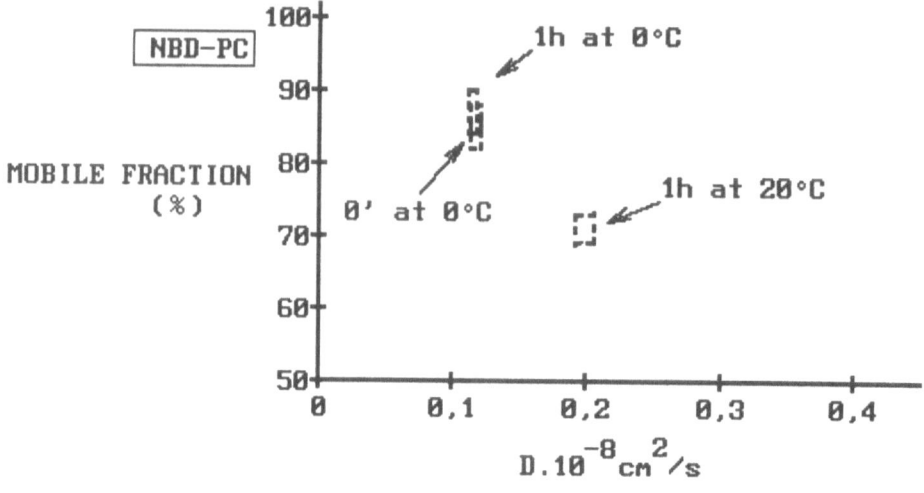

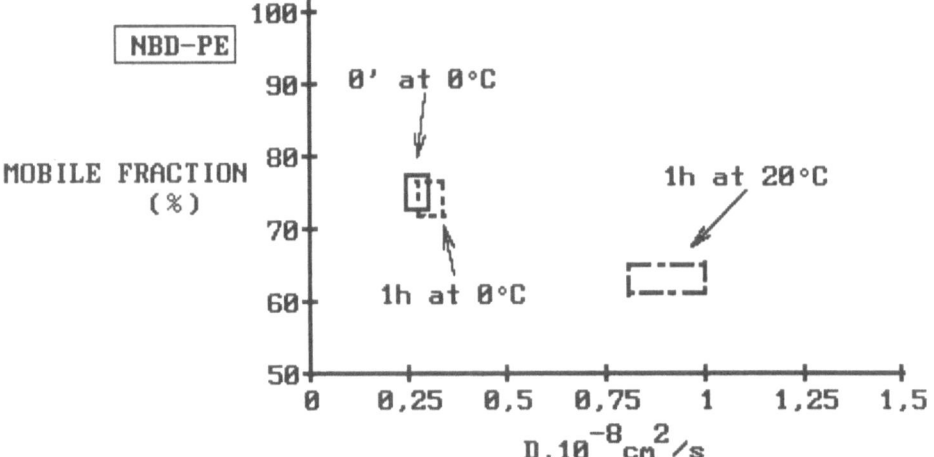

Figure 2: Lateral diffusion coefficients (D) and mobile fractions (M) for NBD-PC (upper) and NBD-PE (lower) in the membrane of labeled vascular endothelial cells.

Using Snedecor-Fisher's law, we have determined the minimum and maximum values of these two parameters at a 95% level of confidence (for details see Tournier et al. 1989b). This allowed us to determine rectangular surfaces of confidence, delimitated on the vertical side by the M_{min} and M_{max} values, and on the horizontal side by the D_{min} and D_{max} values. From a strict statistical point of view, the solutions can be considered as being significantly different from each others, only in the absence of overlap between the rectangular areas defined by this manner. After cell labeling (see text), D and M values were measured at time 0 (0' at 0°C), and after the cells were incubated 1 hour at 0°C or 1h at 20°C in the culture medium (without serum).

DISCUSSION

This study was a first attempt to identify an aminophospholipid translocase activity in the vascular endothelium which assures many important physiological functions like regulation of blood pressure, nutriments and hormones traffic between the blood and the tissues, and present a nonthrombogenic membrane surface to the blood stream.   The main results obtained in the present study are:

(*i*) controled incorporation of the NBD-labeled phospholipid probes in the apical membrane of endothelial cell results in an heterogeneous membrane labeling. This suggests the occurrence of lateral heterogeneity of the cell surface, as recently described by Rodgers and Glaser (1991) and reviewed by Tocanne et al. (1989).

(*ii*) Surprisingly, the obtained data indicate a great stability in the labeling of the plasma membrane by the fluorescent phospholipid probes in the outer or the inner leaflet. This result has to be compared to previous works indicating less stability for certain cell lines (Martin and Pagano 1987, Kobayashi and Arakawa 1991, Kok et al. 1990, Sleight and Abanto 1989). In this respect and using the same lipid probes, exogenous added fluorescent phospholipids probes in the outer leaflet of the plasma membrane behave differently in regard of their polar head groups: NBD-PE "flip" rapidly to the inner leaflet of the plasma membrane while NBD-PC remains located in the outer leaflet in the same period of time (at least one hour at 20°C).   This inward movement of NBD-PE is temperature and presumably ATP dependent. This has to be related to other works demonstrating an enzymatic mechanism of transbilayer movement of aminophospholipids (for reviews see Devaux 1990, Zachowski and Devaux 1989, Zachowski and Devaux 1990)

(*iii*) the two fluorescent probes used:

-do not lateraly diffuse at the same rate in the outer leaflet of the apical plasma membrane of these endothelial cells

-when NBD-PE moved in the inner leaflet of the plasma membrane, the measured lateral diffusion coefficient of this probe increased approximatively fourfold (at 20°C).

These results raise many questions of great interest which have to be related to the fact that lipids in cell plasma membranes are in a high dynamic state. This includes membrane biogenesis, maintenance of membrane structure (with peculiar lipid macro and microdomains and lipid transverse asymmetry), membrane traffic (endocytose, pinocytose, transcytose ...) which implies membrane recycling.

The precise elucidation of each of these phenomenons and their physiological functionnal integrations more than likely depends of the cell type studied and constitutes a major task in cell biology.

REFERENCES

Bevers E M, Comfurius P, Zwall R F A (1983) Changes in membrane phospholipid distribution during platelet activation. Biochim Biophys Acta 736:57-66
Cribier S, Morrot G, Neumann J M, Deveaux P F (1990) Lateral diffusion of erythrocyte phospholipids in model membranes comparison between inner and outer leaflet components. Eur Biophys J 18:33-41
Daleke D L, Huestis W H (1985) Incorporation and translocation of aminophospholipids in human erythrocytes. Biochemistry 24:5406-5416
Daleke D L, Huestis W H (1989) Erythrocyte morphology reflects the transbilayer distribution of incorporated phospholipids. J Cell Biol 108:1375-1385
van Deenen L L M, Roelofsen B, Op den Kamp J (1988) Transbilayer organization and mobility of phospholipids in normal and pathologic states. In Membrane Biogenesis (Op den Kamp J A F Ed.). NATO ASI Series Vol. H16 pp 15-28, Springer-Verlag, Berlin/Heidelberg.
Deveaux P H (1990) The aminophospholipid translocase: transmembranelipid pump-physiological significance. NIPS 5:53-58
Gerads I, Govers-Riemslag J W P, Tans G, Zwaal R F A, Rosing J (1990) Prothrombin activation on membranes with anionic lipids containing phosphate, sulfate or carboxyl groups. Biochemistry 29:7967-7974
Kobayashi T, Arakawa Y (1991) Transport of exogenous fluorescent phosphatidylserine analogue to the golgi apparatus in cultured fibroblasts. J Cell Biol 113(2):235-244
Kok J W, ter Beest M, Scherphof G, Hoekstra D (1990) A non-exchangeable phospholipid analog as a membrane traffic marker of the endocytic pathway. Eur J Cell Biol 53:173-184
Lopez A, Dupou L, Altibelli A, Trotard J, Tocanne J F (1988) Fluorescence recovery after photobleaching (FRAP) experiments under conditions of uniform disk illumination. Critical comparison of analytical solutions, and a new mathematical method for calculation of diffusion coefficient D. Biophys J 53:963-970
Lubin B, Chiu D, Bastacky J, Roelofsen B, van Deenen L L M (1981) Abnormalities in membrane phospholipid organization in sickled erythrocytes. J Clin Invest 67:1643-1649
Martin O, Pagano R E (1987) Transbilayer movement of fluorescent analogs of phosphatidylserine and phosphatidylethanolamine at the plasma membrane of cultured cells. Evidence for a protein-mediated and ATP-dependant process. J Biol Chem 262:5890-5898
McEvoy L, Williamson P, Schlegel R A (1986) Membrane phospholipid asymmetry as a determinant of erythrocyte recognition by macrophages. Proc Natl Acad Sci USA 83:3311-3315
Morrot G, Cribier S, Deveaux P F, Geldwerth D, Davoust J, Bureau J F, Fellmann P, Hervé P, Frilley B (1986) Asymmetric lateral mobility of phospholipids in the human erythrocyte membrane. Proc Natl Acad Sci USA 83:6863-6867

Perret B, Chap H, Douste-Blazy L (1979) Asymetric distribution of arachidonic acid in the plasma membrane of human platelets. Biochim Biophys Acta 556:434-446

Rimon G, Meyerstein N, Henis Y I (1984) Lateral mobility of phospholipids in the external and internal leaflets of normal and hereditary spherocytic human erythrocytes. Biochim Biophys Acta 775:283-290

Rodgers W, Glaser M (1991) Characterization of lipid domains in erythrocyte membranes. Proc Natl Acad Sci USA 88:1364-1368

Rosing J, van Rijn J L M, Bevers E M, van Dieijen G, Comfurius P, Zwaal R F A (1985) The role of activated platelets in prothrombin and factor X activation. Blood 65:319-332

Schlegel R A, Prendergast T W, Williamson P (1985) Membrane phospholipid asymmetry as a factor in erythrocyte-endothelial cell interactions. J Clin Invest 123:215-218

Schlegel R A, Williamson P (1987) Membrane phospholipid organization as a determinant of blood cell-reticuloendothelial cell interactions. J Cell Physiol 132:381-384

Sleight R G, Abanto M N (1989) Differences in intracellular transport of a fluorescent phosphatidylcholine analog in established cell lines. J Cell Science 93:363-374

Sune A, Bette-Bobillo P, Bienvenüe A, Fellman P, Devaux P F (1987) Selective outside-inside translocation of aminophospholipids in human platelets. Biochemistry 26:2972-2978

Sune A, Bienvenüe A (1988) Relationship between the transverse distribution of phospholipids in plasma membrane and shape change of human platelets. Biochemistry 27:6794-6800

Tanaka Y, Schroit A J (1983) Insertion of fluorescent phosphatidylserine into the plasma membrane of red blood cells: recognition by autologous macrophages. J Biol Chem 258(18):11335-11343

Tocanne J F, Dupou-Cézanne L, Lopez A, Tournier J F (1989) Lipid lateral diffusion and membrane organization. FEBS Lett 257(1):10-16

Tournier J F, Lopez A, Tocanne J F (1989a) Effect of cell substratum on the lateral mobility of lipids in the plasma membrane of vascular endothelial cells. Exp Cell Res 181:105-115

Tournier J F, Lopez A, Gas N, Tocanne J F (1989b) The lateral motion of lipid molecules in the apical plasma membrane of endothelial cells is reversibly affected by the presence of cell junctions. Exp Cell Res 181:375-384

Wali R K, Jaffe S, Kumar D, Kalra V K (1988) Alterations in organization of phospholipids in erythrocytes as factor in adherence to endothelial cells in diabetes mellitus. Diabetes 37:104-111

Zachowski A, Craescu C T, Galacteros F, Devaux P F (1985) Abnormality of phospholipid transverse diffusion in sickle erythrocytes. J Clin Invest 75:1713-1719

Zachowski A, Herrmann A, Paraf A, Devaux P F (1987) Phospholipid outside-inside translocation in lymphocyte plasma membrane is a protein mediated phenomenon. Biochim Biophys Acta 897:197-200

Zachowski A and Devaux P F (1989) Bilayer asymmetry and lipid transport across biomembranes. Comments Mol Cell Biophys 6:63-90

Zachowski A and Devaux P F (1990) Transmembrane movements of lipids. Experientia 46:644-656

INTERORGANELLE TRAFFICKING OF PHOSPHATIDYLSERINE AND PHOSPHATIDYLETHANOLAMINE

Jean E. Vance
Lipid and Lipoprotein Research Group
University of Alberta
Edmonton
Alberta T6G 2S2
CANADA

Our knowledge of the mechanisms by which membrane lipids are transported from their sites of synthesis, principally the endoplasmic reticulum (ER), to other membranes, such as the plasma membrane, mitochondria, and nucleus, lags far behind that of how proteins are transported and targeted to membranes. Several mechanisms of lipid trafficking have been proposed and each is supported by some experimental evidence (Voelker, 1985a).

One potential mechanism involves the free diffusion of lipids through the aqueous medium between the membranes. Apart from a few lipids such as free fatty acids, phosphatidic acid and CDP-diacylglycerol, this mechanism is not feasible for most lipids because of their limited solubility in water. More likely is a mechanism in which a protein carrier mediates the movement of a lipid molecule from the donor to the acceptor membrane. Several proteins that could fulfil this role have been isolated and studied in detail. These are the lipid transfer proteins which have been identified from a variety of cell types, mainly in the cytosol. Individual lipid transfer proteins have defined lipid specificities. For example, one protein is specific for the binding/transfer of phosphatidylcholine and phosphatidylinositol, another is specific for phosphatidylcholine, whereas another, the non-specific lipid transfer protein (also called sterol carrier protein 2), has a broad specificity and is able to transport most phospholipids as well as

NATO ASI Series, Vol. H 63
Dynamics of Membrane Assembly
Edited by J. A. F. Op den Kamp
© Springer-Verlag Berlin Heidelberg 1992

cholesterol. These proteins can accelerate the movement of lipids between membranes *in vitro*, but whether or not they catalyse this process *in vivo* has still not been established.

A third potential mechanism of lipid translocation between membranes is via vesicles. Vesicles bud from the donor membrane and fuse with the acceptor membrane, thereby delivering lipid. Vesicle movement among the ER, Golgi and plasma membrane has been well documented because this is the mechanism by which proteins are transferred from their sites of synthesis on the rough ER to the Golgi, through the Golgi stacks, and from Golgi to the plasma membrane. Since these protein-containing vesicles contain lipids, transport via these vesicles would be an efficient way of transferring lipids at the same time as proteins. A massive flux of vesicles occurs in this manner. Rothman and coworkers (Wieland *et al.*, 1987) have estimated that 50% of the lipids of the ER leave this membrane every 10 min *en route* to the Golgi.

A final mechanism, frequently disregarded, is one in which lipids are transferred between membranes via contact, perhaps transient, perhaps involving a protein bridge, between the donor and acceptor membranes. Most of the evidence supporting this mechanism is morphological.

Of course, not all lipids, or even all phospholipids, are necessarily transported via the same mechanism, and mechanisms of transport between different membranes are probably different. For example, transfer of a phospholipid between the ER and mitochondria might occur by a mechanism distinct from that between the ER and plasma membrane.

I shall present some recent experiments from my laboratory concerning the transport of (1) phosphatidylserine (PS) from the ER to mitochondria, and (2) phosphatidylethanolamine (PE) from mitochondria and the ER to the cell surface.

1. Transport of PS from the ER to the mitochondria

The enzyme PS synthase (exchange enzyme), that synthesizes PS, is located primarily on ER membranes (Vance and Vance, 1988) and catalyses the exchange of the choline or ethanolamine moiety of phosphatidylcholine (PC) or PE, respectively, for serine. Mitochondria contain negligible PS synthase activity (Vance and Vance, 1988). In contrast, the enzyme that converts PS to PE resides solely on the outer leaflet of the inner mitochondrial membrane (Dennis and Kennedy, 1972). Consequently, for the formation of PS-derived PE, the PS must undergo an obligatory translocation from the ER (the site of PS synthesis) to the mitochondria (mito, the site of PS decarboxylation), as shown below.

$$\text{serine} + \text{PC or PE} \longrightarrow \text{PS}_{ER} \longrightarrow \text{PS}_{Mito} \longrightarrow \text{PE} + \text{CO}_2$$

ER   Mito

For many years it has been predicted that PS translocation between the ER and mitochondria involved the participation of a cytosolic phospholipid transfer protein. However, Voelker and associates have studied the mechanism of synthesis, translocation and decarboxylation of PS in intact BHK cells (Voelker,1985b), isolated organelles from rat liver (Voelker,1989a) and permeabilized CHO cells (Voelker,1989b; Voelker 1990). The conclusions from their studies are that PS translocation from ER to mitochondria in intact cells and permeabilized cells is dependent upon ATP but is independent of cytosolic phospholipid transfer proteins or ongoing synthesis of PS. The experiments are consistent with a transport mechanism that involves vesicles or a close juxtapostion between the ER and mitochondrial membranes.

In my laboratory two *in vitro* systems of rat liver mitochondria and microsomes have been developed in which synthesis, translocation and decarboxylation of PS have been reconstituted (Vance, 1991). Cytosolic proteins do not stimulate this process and newly-made PS, rather than pre-existing PS of the microsomal membranes, is preferred for conversion to PE. In the first type of reconstituted system, rat liver microsomes (100 μg protein) and mitochondria (500 µg protein) were mixed in a test tube in a buffer containing 25 mM Hepes (pH 7.4), 5 mM dithiothreitol, 0.3 M sucrose and 10 mM calcium chloride. [3-^{3}H]Serine was added and the biosynthesis of radiolabeled PS and PE occurred (Vance, 1991). As shown in Figure 1A, radioactivity was rapidly incorporated into PS. After 20 min of incubation at 37°C the radioactivity in PS gradually declined with a concomitant increase in radioactivity in PE. After 60 min the distribution of radioactivity was 64% in PE and 36% in PS. The conversion of PS to PE was not affected by inclusion of cytosolic proteins in the incubation mixture. The data demonstrate that PS had been synthesized in microsomal membranes and efficiently translocated from the ER to mitochondria, wherein the PS was decarboxylated to PE.

A second type of reconstitution system for PS transfer to mitochondria was developed (Vance, 1991). Rat liver microsomes were isolated and incubated with [3-^{3}H]serine for 1 h. The membranes, which had become labeled with [^{3}H]PS, were re-isolated by centrifugation and washed free of [^{3}H]serine. The small amount of microsomal [^{3}H]PE could be accounted for by contamination of the ER by mitochondrial membranes, as measured by the activity of the marker enzyme, cytochrome c oxidase. The PS-labeled microsomes were subsequently incubated with mitochondria under exactly the same conditions as used for the first reconstitution system, except that [^{3}H]serine was not included. As shown in Figure 1B, after 1 h incubation only 15% of the radioactivity was in PE whereas 85% was in PS. The extent of conversion of PS to PE was therefore markedly less than for the experiment shown in Figure 1A. As a control that pre-incubation and re-isolation of the microsomes had not affected their competence for PS translocation, unlabeled microsmes were subjected to the identical procedure (i.e. 1 h pre-incubation and subsequent ultracentrifugation), as used for the experiment shown in Figure 1B). The re-isolated microsomes were then incubated with [^{3}H]serine, as for the experiment depicted in Figure 1A. The result was as shown in Figure 1A, therefore the re-isolated microsomes remained competent for efficient PS transfer.

TABLE 1

Effect of nucleoside phosphates on formation of PS and PE
In the reconstituted mixture used for the experiment depicted in Figure 1A, rat liver microsomes and mitochondria were incubated for 45 min with [3-^{3}H]serine. Unless otherwise stated the concentration of nucleotide was 2 mM. The data are means ± S.D. of 3 measurements. [Adapted from J.E. Vance, *J. Biol. Chem.* (1991) **266**:89-97.]

Nucleotide added	PS	PE
	x 10^3 dpm	
None	18.7 ± 2.6	11.1 ± 1.2
ATP	16.3 ± 0.9	16.6 ± 0.6
GTP	24.0 ± 4.4	19.9 ± 1.9
GTP (0.2 mM)	17.9 ± 2.4	12.5 ± 1.7
UTP	18.4 ± 1.7	22.8 ± 1.9
CTP	22.6 ± 0.1	12.8 ± 1.6
ADP	21.6 ± 1.6	22.8 ± 0.9
GDP	14.9 ± 1.2	25.4 ± 1.2
UDP	13.5 ± 1.9	41.5 ± 2.7
CDP	15.9 ± 1.5	15.6 ± 2.7
CMP	12.0 ± 0.9	14.0 ± 1.4
AMP	18.7 ± 2.8	11.3 ± 0.3
App(NH)p	22.3 ± 0.7	14.1 ± 2.0

TABLE 2

Newly-made PS is preferentially converted into PE

Rat liver microsomes were prepared and incubated for 1 h with [^{14}C]serine. The [^{14}C]microsomes were re-isolated and incubated with mitochondria in the presence of [3-^{3}H]serine. Radioactivity was measured in PS and PE.

x 10^3 dpm

Time, min	PS			PE		
	^{3}H	^{14}C	^{3}H/^{14}C	^{3}H	^{14}C	^{3}H/^{14}C
0	2.75	21.54	0.13	0.45	0.36	1.25
1 0	12.14	15.01	0.81	5.48	0.69	7.99
3 0	10.07	12.44	0.81	10.09	1.07	9.45

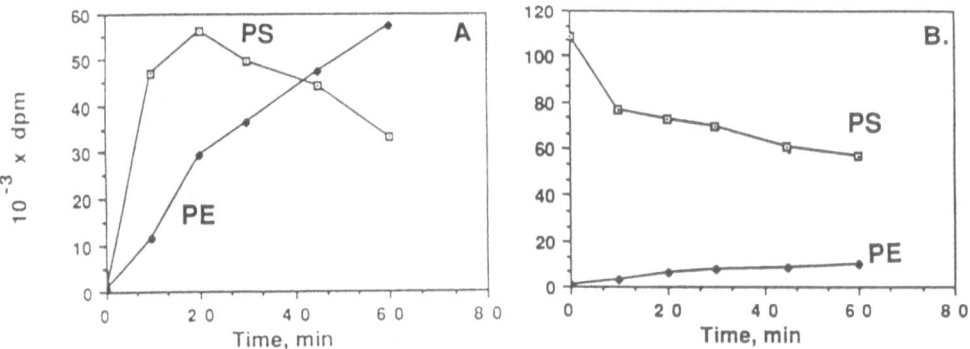

Figure 1. **A.** Coordinated synthesis of PS and PE from [3-^{3}H]serine in microsomes reconstituted with mitochondria. **B.** Conversion of [^{3}H]PS to [^{3}H]PE by mitochondria incubated with prelabeled, microsomal PS. [Adapted from J.E. Vance, *J. Biol. Chem.* (1990) **256**:7248-7256.]

The data shown in Figures 1A and 1B suggest that newly-synthesized PS was more readily translocated to mitochondria and converted into PE than was "old" PS, pre-existing in the microsomal membranes. Moreover, since the addition of cytosolic protein to the incubation did not increase the conversion of PS to PE, the translocation did not require the participation of cytosolic transfer proteins. The data do not exclude the involvement of membrane-bound protein factors in the translocation process.

Theoretically in these incubations it would also be possible for PE synthesized in mitochondria to be transported to the microsomes, the site of PE methyltransferase activity, so that serine-derived PE would be methylated to PC. Initially when PC was isolated from the thin-layer chromatography plate after the incubation, significant radioactivity was detected in the PC band. However, further investigation showed that the majority of the radiolabeled product was not PC, but was lyso PE, which co-migrated with PC. Thus, the requirements for the movement of PE from the mitochondria to the microsomes have not been established.

When ATP (2 mM) was added to the first type of reconstitution system (as for Fig. 1A) the conversion of PS to PE was stimulated. The effect was not specific for ATP, however, since several other nucleoside mono-, di- and tri-phosphates, as well as the non-hydrolysable ATP analogue, adenyl-5'-yl imidodiphosphate [App(NH)p], stimulated the reaction as efficiently as did ATP (Table 1). In particular, UDP was the most active in the process. The nucleoside phosphates were also tested for their effect on the conversion of PS to PE in the second type of reconstituted system (i.e. from pre-labeled PS); none of the compounds appreciably stimulated the conversion of PS to PE. These data suggest that there is not a specific requirement for ATP hydrolysis for PS translocation in these reconstituted systems.

As confirmation that newly-made PS was preferred for translocation to the mitochondria and/or decarboxylation to PE a double-labeling experiment was performed. Rat liver microsomes were pre-labeled with [^{14}C]serine, then incubated in the presence of [3-^{3}H]serine and mitochondria. In this system, "old" PS was marked with ^{14}C, whereas "new" PS was derived from [^{3}H]serine. PS and PE were isolated from the incubation and the ratio of ^{3}H/^{14}C was measured in PS and PE. As shown in Table 2, at all times the ratio of ^{3}H/^{14}C was higher in PE than in PS. For example, after incubation for 30 min, the ratio was approximately 11 times higher in PE than in PS. These data confirm that new PS, rather than pre-exiting PS, was preferentially converted into PE.

The results of these reconstitution studies demonstrate that newly-made PS is translocated from microsomes to mitochondria in the absence of cytosolic proteins. Interesting in this respect is the recent finding (van Heusden *et al.* 1990) that a CHO cell mutant lacking the non-specific lipid transfer protein, which had previously been suggested as a candidate for assisting the translocation of PS from ER to mitochondria, has a normal rate of PE biosynthesis from PS. All the data combined suggest that soluble cytosolic phospholipid transfer proteins are not involved in PS translocation to mitochondria. Instead, a mechanism compatible with these data would be a vesicle-mediated mechanism or a collision-based mechanism in which there was a close juxtaposition or contact between the ER and mitochondrial membranes. However, there is no evidence that proteins travel to the mitochondria via vesicles. Proteins destined for the mitochondria are synthesized on free ribosomes in the cytosol and the movement of proteins to mitochondria apparently does not require the concomitant movement of lipids

From our study, in which newly-made PS was translocated from ER to mitochondria more efficiently than was old PS, we can conclude that the collision process alone is not sufficient for maximal PS transfer. Apparently there was not an immediate mixing of new and old pools of PS. Consequently, new PS is, for some reason, more accessible for translocation than is old PS. Compartmentalization of new and old pools of phospholipids (Vance, 1989), and also of pools of phospholipids derived from different biosynthetic origins, has previously been suggested in the utilization of phospholipids for assembly into hepatic lipoproteins (Vance and Vance, 1986). Perhaps newly-synthesized PS, which is made on the outer (cytosolic) surface of the ER bilayer, is transferred from the ER surface to mitochondria before becoming completely integrated into the bilayer. Also possible is that older PS molecules may have undergone transbilayer movement to the inner (luminal) surface of the bilayer and therefore may not be so readily accessible for transport to other membranes. An alternative explanation is that translocation of PS to mitochondria may be linked to PS synthesis, although the recent experiments of Voelker in permeabilized CHO cells tend to discount this possibility (Voelker, 1990). In contrast, however, PS translocation from the outer to the inner mitochondrial membrane of yeast is reported to be driven by synthesis (Simbeni et al., 1990).

In support of a contact mechanism for transfer of PS from the ER to the mitochondria we have recently isolated a unique membrane fraction from rat liver (Vance, 1990). This membrane fraction, called "fraction X", was originally isolated as a contaminant of mitochondria. A crude mitochondrial pellet was obtained from rat liver and applied to a Percoll gradient from which purified mitochondria were isolated. A second membrane-containing fraction, floating above the mitochondria on the Percoll gradient, was isolated, and fraction X was obtained from this by centrifugation (Figure 2). Fraction X was examined for standard membrane marker enzyme activities. The activity of the mitochondrial marker enzyme, cytochrome c oxidase, in fraction X was 19% of that in mitochondria. The specific activity of the traditional ER marker enzyme, NADPH:cyt c reductase, in fraction X was approximately one third that in microsomes, whereas the specific activity of another ER marker enzyme, glucose-6-phosphate phosphatase, in fraction X was approximately double that in microsomes. When microsomes were centrifuged on the Percoll gradient no fraction with the properties of fraction X was obtained. Marker enzymes for plasma membrane, Golgi, lysosomes and peroxisomes had very low activities in fraction X compared with membrane preparations enriched in these fractions.

The phospholipid composition of fraction X was compared with that of microsomes and mitochondria (Vance, 1990). As expected,

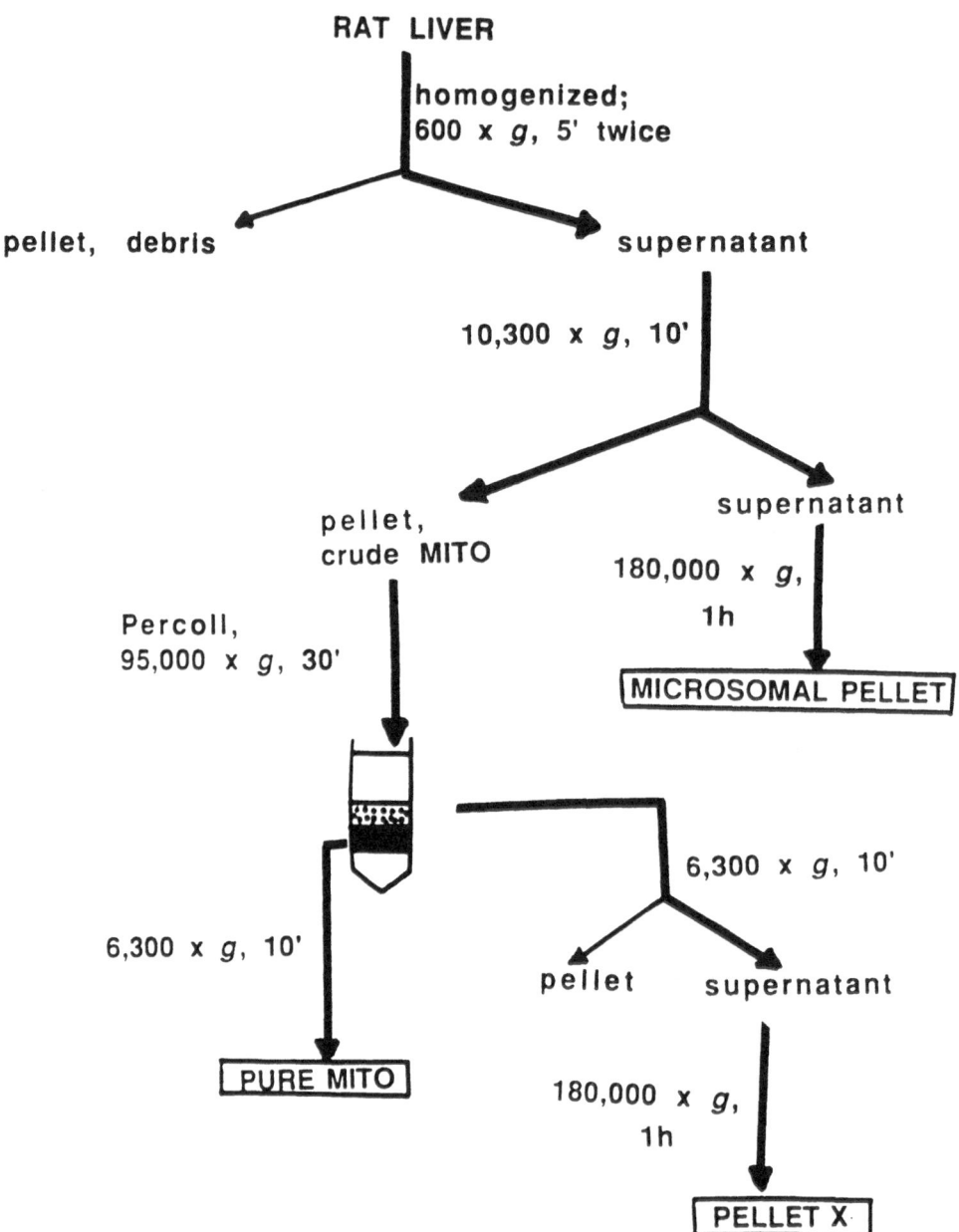

Figure 2.   Preparation of fraction X from rat liver.

microsomes and mitochondria had appreciably different phospholipid compositions, with the content of PE being higher in mitochondria. The composition of fraction X resembled that of microsomes.

Polyacrylamide gel electrophoresis (on 3 to 15% gradient gels containing 0.1% sodium dodecyl sulfate) was performed on rough and smooth ER fractions (Vance, 1990; Croze and Morre, 1984), also on fraction X, microsomes, intact mitochondria, and membranes enriched in inner and outer mitochondrial membranes. The protein profile of fraction X was markedly different from that of purified mitochondrial membranes, or inner and outer mitochondrial membranes. Although the proteins of fraction X were qualitatively similar to those of microsomes there were significant quantitative differences in the two types of membranes. There were also differences among the protein profiles of fraction X and rough and smooth ER, although the proteins of fraction X were more similar to those of the smooth, than the rough, ER.

The activities of several phospholipid biosynthetic enzymes were detected in fraction X (Vance, 1990). Notably, the specific activity of PS synthase was 1.5- to 2-fold higher in fraction X than in microsomes (Table 3). Fraction X also contained PE N-methyltransferase, cholinephosphotransferase and ethanolaminephosphotransferase activities that were of similar specific activities to those in microsomes. In contrast, the activity of PS decarboxylase in fraction X was only 13% of that in purified mitochondria.

From these properties it is apparent that fraction X is distinct from mitochondria, lysosomes, peroxisomes, plasma membranes, Golgi (trans) and microsomes. In many respects, fraction X resembles microsomes. We propose that fraction X is a specialized domain of the ER that co-purifies with mitochondria (Vance, 1990). Fraction X may be a region of the ER that comes into contact with, or has a high affinity for, mitochondria. Several previous reports have suggested that there are regions of membrane contact between the ER and mitochondria (Katz et al., 1983; Meier et al., 1981). There is also evidence that the ER is not homogeneous with respect to its distribution of proteins (Gierow and Jergil, 1989; Lewis and

TABLE 3

Phospholipid biosynthetic enzymes in fraction X, microsomes and mitochondria

Subcellular membrane fractions were prepared from rat liver and assayed for phospholipid biosynthetic enzyme activities. The units of enzyme activity for PS synthase are nmol PS/h/mg protein. The activities of the other enzymes are expressed as nmol/min/mg protein. All values are means ± S.D. The numbers in parentheses (n) give the number of individual subcellular fractionations. [Adapted from J.E. Vance, *J. Biol. Chem.* (1990) **256**:7248-7256.]

Enzyme	Mitochondria	Microsomes	Fraction X
PS synthase	0.14±0.07	1.76±0.69	2.70±0.09
	(n=9)	(n=13)	(n=15)
PS decarboxylase	0.30±0.09	0.04±0.01	0.04±0.02
	(n=6)	(n=3)	(n=5)
PE methyltransferase	0.25±0.12	2.64±0.56	2.74±0.62
	(n=8)	(n=12)	(n=10)
Cholinephosphotransferase	0.31±0.13	2.95±0.65	2.83±0.67
	(n=3)	(n=4)	(n=5)
Ethanolaminephospho-transferase	0.06±0.01	1.31±0.41	1.50±0.20
	(n=3)	(n=4)	(n=5)

Tata,1973; Glaumann and Dallner, 1970). The close physical juxtaposition of a region of the ER rich in PS synthase activity with the mitochondria containing PS decarboxylase activity, would provide an efficient mechanism for the synthesis, transfer and decarboxylation of PS.

2. Transport of PE from its sites of synthesis in the ER and mitochondria to the cell surface

Proteins destined for secretion or for residence at the plasma membrane are transported via vesicles from their sites of synthesis on the rough ER, through the Golgi apparatus to the plasma membrane. In this scheme, protein-containing vesicles derived from Golgi membranes fuse with the plasma membrane thereby delivering their cargo of proteins. It has been predicted that during this delivery of proteins, lipids are concomitantly transferred to the plasma membrane. (The plasma membrane itself has little capacity for lipid synthesis (Vance and Vance, 1988; Jelsema and Morre, 1978)). Some recent findings, however, have suggested that lipid transport to the plasma membrane may not be a simple bulk transfer of lipids from the protein-carrying vesicles. In several instances the arrival of phospholipids at the plasma membrane has been shown to occur more rapidly than the delivery of proteins (Sleight and Pagano, 1983; Kaplan and Simoni, 1985). Moreover, phospholipid transport to the plasma membrane is independent of ATP (Sleight and Pagano, 1983) and continues normally at 15° C (Sleight and Pagano, 1983), all of which factors indicate that vesicle movement of phospholipids to the plasma membrane is unlikely.

Studies on the movement of proteins and lipids to the cell surface have recently benefitted from the discovery of brefeldin A (BFA), a fungal metabolite. When hepatocytes and other cell types are treated with BFA the movement of proteins through the Golgi is disrupted (Misumi *et al.*, 1986; Doms *et al.*, 1989; Lippincott-

Schwartz *et al.*, 1989). BFA causes disassembly of the Golgi apparatus and redistribution of Golgi proteins to the ER (Lippincott-Schwartz *et al.* 1989). Interestingly, when CHO cells were treated with BFA the transport of vesicular stomatitis virus G protein to the plasma membrane was blocked, whereas the movement of cholesterol was unaffected (Urbani and Simoni, 1990). These data are consistent with the the idea that cholesterol and proteins travel to the cell surface by different routes.

We have used BFA to investigate the mechanism of movement of PE to the cell surface of rat hepatocytes (Vance *et al.*, 1991). In these cells, PE is biosynthesized by three different routes, the majority being made via the CDP-ethanolamine pathway (Kennedy and Weiss, 1956) and from the decarboxylation of PS (Dennis and Kennedy, 1972). A minor fraction of the PE is made by the base exchange reaction (Bjerve, 1973). We postulated that if ethanolamine-derived PE moved to the plasma membrane from its site of synthesis on the ER via the Golgi, this movement would be inhibited by BFA. On the other hand, PE synthesized in the mitochondria via decarboxylation of PS might move to the cell surface independently of the Golgi apparatus, thus this movement would not be affected by BFA.

Monolayer cultures of rat hepatocytes were incubated with [^{3}H]leucine in the presence or absence of different concentrations of BFA (Vance *et al.*, 1991). Protein secretion into the medium was monitored as a function of BFA concentration. As shown in Figure 3A BFA caused a dose-dependent reduction in protein secretion. At a BFA concentration of 10 μM the secretion of proteins was inhibited by 65%. In another experiment, albumin was immunoprecipitated from the culture medium of cells treated with different concentrations of BFA (0 to 10 μM) (Figure 3B). BFA inhibited albumin secretion at all concentrations used; the inhibitory effect was greatest and most persistent with 10 μM BFA. At the same time, BFA did not affect the viability of the cells as judged by Trypan blue exclusion and leakage of lactate dehydrogenase into the medium. Nor did BFA affect the rate of cellular protein synthesis (Vance *et al.*, 1991).

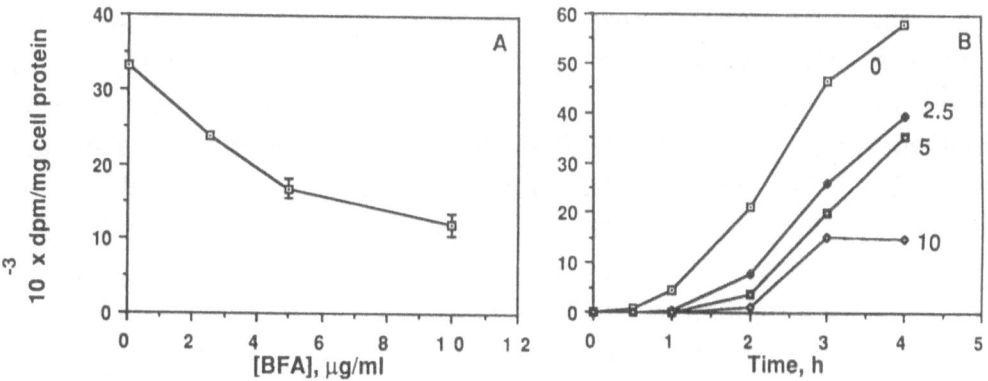

Figure 3. **A.** Effect of 4h treatment with BFA (10 μM) on secretion of proteins by rat hepatocytes. **B.** Effect of 4h treatment with BFA on secretion of albumin (immunoprecipitated). The numbers 0, 2.5, 5, 10 refer to the concentration of BFA (μM) used. [Adapted from J.E. Vance *et al. J. Biol. Chem.* (1991) **266**:8242-8247.]

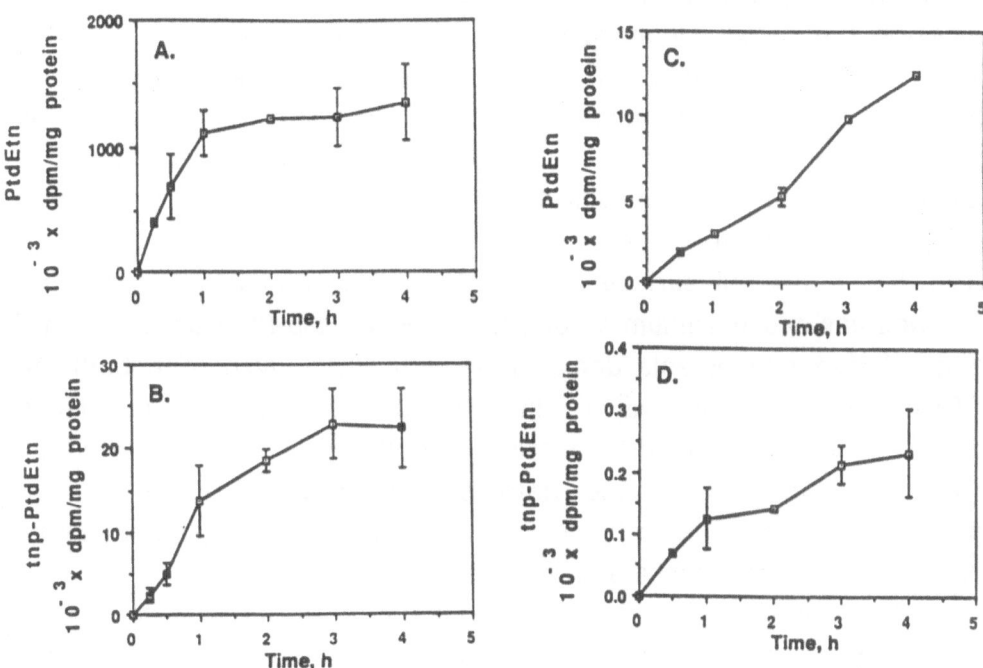

Figure 4. Incorporation of [1-^{3}H]ethanolamine and [3-^{3}H]serine into intracellular PE and cell surface tnp-PE. Hepatocytes were incubated with either [^{3}H]ethanolamine (panels A and B) or [^{3}H]serine (panels C and D). At the indicated times cells were reacted with TNBS, and the [^{3}H] content of PE and tnp-PE was measured. The data are averages ± S.D. of 4 independent experiments. [Adapted from J.E. Vance *et al. J. Biol. Chem.* (1991) **266**:8241-8247.]

The movement of newly-synthesized PE from its two sites of synthesis (ER and mitochondria) to the outer leaflet of the plasma membrane was studied by incubation of hepatocytes with radiolabeled precursors of PE (i.e. serine or ethanolamine) (Vance *et al.*, 1991). The arrival of PE at the cell surface was monitored by reaction of intact cells with the reagent trinitrobenzene sulfonate (TNBS) (Sleight and Pagano, 1983). TNBS reacts with primary amino groups with formation of their trinitrophenyl (tnp) derivatives. PE molecules exposed on the surface of the cells, therefore, would react with TNBS, but intracellular PE molecules would not be accessible for reaction. Rat hepatocytes were incubated with either [1-^{3}H]ethanolamine or [3-^{3}H]serine so that newly-made PE was labeled. After various time intervals up to 4 h the cells were carefully treated with TNBS and lipids were extracted. PE and tnp-PE were separated by thin-layer chromatography and radioactivity incorporated into both PE and tnp-PE was measured. Thus, the movement of newly-synthesized PE to the cell surface (i.e. tnp-PE) was monitored (Figure 4). As shown in Figures 4A and 4B [^{3}H]ethanolamine was incorporated into PE and tnp-PE, as was [^{3}H]serine (Figures 4C and 4D). However the kinetics and extent of labeling of tnp-PE was different for the two labeled precursors (Figure 5). For ethanolamine-labeled PE the maximum % of tnp-PE (i.e. PE at the cell surface) was 1.8% of the total cellular PE after 3h, whereas the maximum % of serine-derived tnp-PE was 4.1% after 1 h. These experiments demonstrate that PE derived from both the CDP-ethanolamine pathway in the ER, and from the decarboxylation of PS in the mitochondria, were efficiently transported to the cell surface. Moreover, these data confirm previous findings that phospholipids in hepatocytes do not rapidly equilibrate into a single homogeneous pool (Vance and Vance, 1986).

The effect of BFA on the movement of PE derived from both biosynthetic origins was investigated (Vance *et al.*, 1991). Hepatocytes were incubated with either [^{3}H]serine or [^{3}H]ethanolamine in the presence or absence of 10 µM BFA. The appearance of PE at the cell surface was monitored by reaction with TNBS. The results are shown in Figure 6. Clearly, the movement of

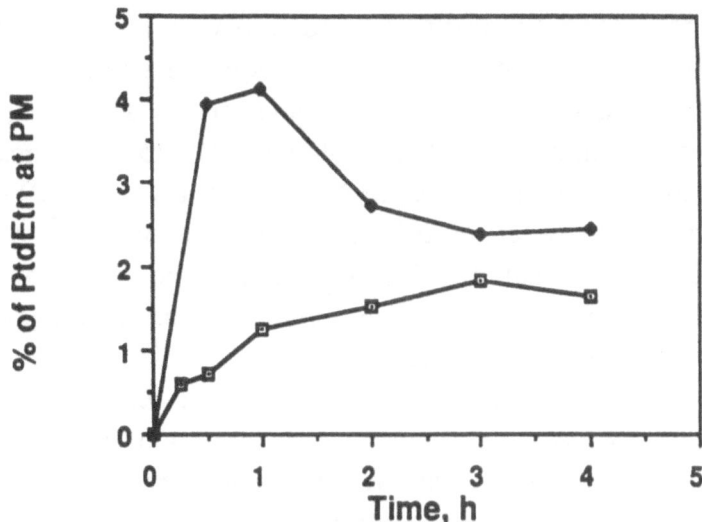

Figure 5. Comparison of movement of [³H]ethanolamine- (open symbols) and [³H]serine-derived (closed symbols) PE to the cell surface. Data are taken from the experiments depicted in Figure 4. [Adapted from J.E. Vance *et al. J. Biol. Chem.* (1991) **266**:8241-8247.]

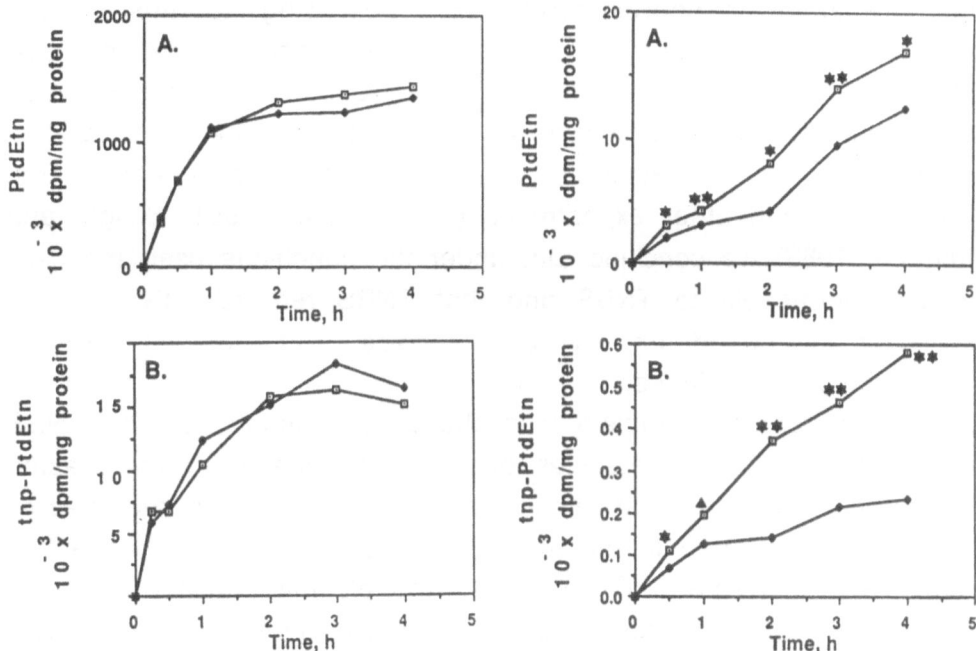

Figure 6. Effect of BFA on PE synthesis and movement to the cell surface. Incorporation of [³H]ethanolamine into PE (A) and tnp-PE (B). Incorporation of [³H]serine into PE (C) and tnp-PE (D). Open symbols = 10 µM BFA; closed symbols = no BFA. [Adapted from J.E. Vance *et al. J. Biol. Chem.* (1991) **266**:8241-8247.]

ethanolamine-derived PE to the cell surface was not affected by BFA treatment (Figures 6A and 6B), even though the secretion of albumin and the other proteins was greatly diminished as shown in Figure 3. Nor was the movement of [³H]serine-labeled PE to the outside of the plasma membrane inhibited by BFA (Figures 6C and 6D). On the contrary, in the presence of BFA the formation of serine-derived PE and tnp-PE was stimulated. The mechanism and significance of this stimulation are not understood; the incorporation of [³H]serine into PS was unaffected by BFA treatment.

That TNBS did not penetrate the cells was crucial for the experiments. As one control that the formation of tnp-PE in the intact hepatocytes was not the result of leakage of small amounts of TNBS into the cells, mitochondria were isolated from [³H]ethanolamine-labeled hepatocytes that had been exposed to TNBS. In mitochondria the percentage of the dpm incorporated into tnp-PE compared with PE was 0.07% in cells incubated with or without BFA. Since the contamination of mitochondria by plasma membranes was approximately 5% (according to the plasma membrane marker enzyme 5'-nucleotidase) and the % of total ethanolamine-labeled cellular PE that reacted with TNBS was approximately 2% (see Figure 4), all tnp-PE in the mitochondrial preparation was the result of contamination by plasma membranes. From these and other experiments (Vance *et al.*, 1991; Sleight and Pagano, 1983) we conclude that under the conditions used the cells were impermeable to TNBS and that TNBS reacted only with PE molecules exposed on the cell surface, but not with PE in intracellular membranes.

One potential problem with these experiments is that TNBS reacts only with PE molecules on the outer leaflet of the plasma membrane whereas most likely the majority of PE in the plasma membrane is located in the inner leaflet of the bilayer. An active transbilayer movement of PE across the plasma membrane bilayer via a specific aminophospholipid translocase does, however, occur (Devaux, 1988). Indeed, such a mechanism of flip-flop from inside to outside must exist since newly-synthesized PE was detected at the cell surface in the form of tnp-PE within minutes of synthesis (Vance *et al.*, 1991; Sleight and Pagano, 1983).

Our experiments with BFA disagree with some aspects of the hypothesis of bulk membrane flow, since BFA inhibited the movement of proteins but not PE to the cell surface. The experiments with BFA (Vance *et al.*, 1991) suggest that the majority of PE is transported to the plasma membrane independently of proteins and by a route not involving transit through the Golgi. Vesicle trafficking of the majority of PE seems unlikely for the following reasons. (i) Neither BFA nor monensin (Sleight and Pagano, 1983) inhibit the movement of PE to the cell surface. (ii) The kinetics of movement of phospholipids (Kaplan and Simoni, 1985; Sleight and Pagano, 1983), cholesterol (DeGrella and Simoni, 1990) and proteins are different. (iii) Depletion of ATP from cells does not inhibit transport of PE to the plasma membrane (Sleight and Pagano, 1983). (iv) Low temperatures, which inhibit vesicle-mediated processes, do not affect PE transport to the cell surface (Sleight and Pagano, 1983). (v) In mitotic CHO cells, in which vesicular transport was inhibited, the transport of PE to the plasma membrane was not affected (Kobayashi and Pagano, 1989).

Although a unique type of vesicular transport of PE to the plasma membrane cannot be eliminated more likely mechanisms are either a process catalysed by cytosolic phospholipid transfer proteins or a direct transfer by membrane contact, perhaps transient, between the ER and plasma membrane, as has been suggested by electron microscopy studies (Scow and Blanchette-Mackie, 1986). An enigma that arises from these findings is that vesicles apparently deliver proteins to the plasma membrane yet apparently do so without delivering their complement of PE.

REFERENCES

Bjerve, K.S. (1973) The Ca^{2+}-dependent biosynthesis of lecithin, phosphatidylethanolamine and phosphatidylserine in rat liver subcellular particles. *Biochim. Biophys. Acta* **296:**549-562.

Croze, E.M. and Morré, J. (1984) Isolation of plasma membrane, Golgi apparatus, and endoplasmic reticulum fractions from single homogenates of mouse liver. *J. Cell. Physiol.* **119:**46-57.

DeGrella, R.F. and Simoni, R.D. (1982) Intracellular transport of cholesterol to the plasma membrane. *J. Biol. Chem.* **257:**14256-14262.

Dennis, E.A. and Kennedy, E.P. (1972) Intracellular sites of lipid synthesis and the biogenesis of mitochondria. *J. Lipid Res.* **13:**263-267.

Devaux, P.F. (1988) Phospholipid flippases. *FEBS Letts.* **234:**8-12.

Doms, R.V., Russ, G. and Yewdell, J.W. (1989) Brefeldin A redistributes resident and itinerant golgi proteins to the endoplasmic reticulum. *J. Cell Biol.* **109:**61-72.

Gierow, P. and Jergil, B. (1989). Heterogeneity of smooth endoplasmic reticulum from rat liver studied by two-phase monitoring. *Biochem. J.* **262:**55-61.

Glaumann, H. and Dallner, G. (1970). Subfractionation of smooth microsomes from rat liver. *J. Cell Biol.* **47:**34-48.

Jelsema, C.L. and Morré, D.J. (1978) Distribution of phospholipid biosynthetic enzymes among cell components of rat liver. *J. Biol. Chem.* **253:**7960-7971.

Kaplan, M.R. and Simoni, R.D. (1985). Intracellular transport of phosphatidylcholine to the plasma membrane. *J. Cell Biol.* **101:**441-445.

Katz, J., Wals, P.A., Golden, S. and Raijman, L. (1983) Mitochondrial-reticular cytostructure in liver cells. *Biochem. J.* **214:**795-813.

Kennedy, E.P. and Weiss, S.B. (1956) The function of cytidine coenzymes in the biosynthesis of phospholipides. *J. Biol. Chem.* **222:**193-214.

Kobayashi, T. and Pagano, R.E. (1989) Lipid transport during mitosis: Alternative pathways for delivery of newly synthesized lipids to the cell surface. *J. Biol. Chem.* **264:**5966-5973.

Lewis, J.A. and Tata, J.R. (1973) A rapidly sedimenting fraction of rat liver endoplasmic reticulum. *J. Cell Sci.* **13:**447-459.

Lippincott-Schwartz, J., Yuan, L.C., Bonifacino, J.S. and Klausner, R.D. (1989) Rapid redistribution of Golgi proteins into the ER in cells treated with brefeldin A: evidence for membrane cycling from Golgi to ER. *Cell* **56:**801-813.

Meier, P.J., Spycher, M.A. and Meyer, U.A. (1981) Isolation and characterization of rough endoplasmic reticulum associated with mitochondria from normal rat liver. *Biochim. Biophys. Acta* **646:**283-297.

Misumi, Y., Misumi, Y., Miki, K., Takatsuki, A., Tamura, G. and Ikehara, Y. (1986) Novel blockade by brefeldin A of intracellular transport of secretory proteins in cultured rat hepatocytes. *J. Biol. Chem.* **261:**11398-11403.

Scow, R.O. and Blanchette-Mackie, E.J. (1985) Why fatty acids flow in cell membranes. *Prog. Lipid Res.* **24:**197-241.

Urbani, L. and Simoni, R.D. (1990) Cholesterol and vesicular stomatitis virus G protein take separate routes from the endoplasmic reticulum to the plasma membrane. *J. Biol. Chem.* **265:**1919-1923.

Simbeni, R., Paltauf, F. and Daum, G. (1990) Intramitochindrial transfer of phospholipids in the yeast, *Saccharomyces cerevisiae. J. Biol. Chem.* **265:**281-285.

Sleight, R.G. and Pagano, R.E. (1985) Transbilayer movement of a fluorescent phosphatidylethanolamine analogue across the plasma membranes of cultured mammalian cells. *J. Biol. Chem.* **260:**1146-1154.

Vance, J.E. (1989) The use of newly synthesized phospholipids for assembly into secreted hepatic lipoproteins. *Biochim. Biophys. Acta*1006:59-69.

Vance, J.E. (1990) Phospholipid synthesis in a membrane fraction associated with mitochondria. *J. Biol. Chem.* **265**:7248-7256.

Vance, J.E. (1991) Newly made phosphatidylserine and phosphatidylethanolamine are preferentially translocated between rat liver mitochondria and endoplasmic reticulum. *J. Biol. Chem.* **266**:89-97.

Vance, J.E. and Vance, D.E. (1986) Specific pools of phospholipids are used for lipoprotein secretion by cultured rat hepatocytes. *J. Biol. Chem.* **261**:4486-4491.

Vance, J.E. and Vance, D.E. (1988) Does rat liver Golgi have the capacity to synthesize phospholipids for lipoprotein secretion? *J. Biol. Chem.* **263**:5898-5909.

Vance, J.E., Aasman, E.J. and Szarka, R. (1991) Brefeldin A does not inhibit the movement of phosphatidylethanolamine from its sites of synthesis to the cell surface. *J. Biol. Chem.* **266**:8241-8247.

van Heusden, G.P.H., Bos, K., Raetz, C.R.H. and Wirtz, K.W.A. (1990) Chinese hamster ovary cells deficient in peroxisomes lack the nonspecific lipid transfer protein (sterol carrier protein 2). *J. Biol. Chem.* **265**:4105-4110.

Voelker, D.R. (1985a) in Biochemistry of Lipids and Membranes (Vance, D.E. and Vance, J.E., eds) pp. 475-502, Benjamin-Cummings Publishing Co. Inc., Menlo Park, CA.

Voelker, D.R. (1985b) Disruption of phosphatidylserine translocation to the mitochondria in baby hamster kidney cells. *J. Biol. Chem.* **260**:14671-14676.

Voelker, D.R. (1989a) Phosphatidylserine translocation to the mitochondria is an ATP-dependent process in permeabilized animal cells. *Proc. Natl. Acad. Sci. USA* **86**:9921-9925.

Voelker, D.R. (1989b)   Reconstitution of phosphatidylserine import into rat liver mitochondria.   *J. Biol. Chem.* **264**:8019-8025.

Voelker, D.R. (1990)   Characterization of phosphatidylserine synthesis and translocation in permeabilized animal cells. *J. Biol. Chem.* **265**:14340-14346.

Wieland, F.T., Gleason, M.L., Serafini, T.A. and Rothman, J.E. (1987) The rate of bulk flow from the endoplasmic reticulum to the cell surface.   *Cell* **50**:289-300.

METABOLIC INCORPORATION OF A NOVEL FLUORESCENT FATTY ACID 7-METHYL-
BODIPY-1-DODECANOIC ACID TO LIPIDS OF BHK CELLS

J.Kasurinen and P. Somerharju

Department of Medical Chemistry
University of Helsinki
Siltavuorenpenger 10 A
SF-00170 Helsinki
Finland

Introduction

Transport of newly synthesized lipids from the sites of their
synthesis to the target membranes is an intriguing problem in cell
biology. Fluorescent lipid molecules offer a versatile tool to
approach this question, because the metabolism of lipids can be
quantitated with very high sensitivity (pmol-fmol range) from cell
extracts and, most importantly, the localization and distribution
of lipids can be observed in intact cells in fluorescence
microscope (Pownall and Smith, 1989; Pagano and Sleight 1985)

From the standpoint of metabolic and microscopic studies a good
fluorescent lipid analogue should fulfill certain requirements: it
should have a high quantum yield and preferably emit fluorescence
light in the visible wavelength range, it should not be prone to
photo-chemical decomposition (bleaching), it should not be toxic to
cells, and finally, it should mimick the behaviour of its natural
counterparts as closely as possible.

NBD-labelled short-chain phospholipids and ceramides were used by
Pagano and coworkers in an elegant series of studies on
phospholipid transbilayer movement, intracellular translocation and

NATO ASI Series, Vol. H 63
Dynamics of Membrane Assembly
Edited by J. A. F. Op den Kamp
© Springer-Verlag Berlin Heidelberg 1992

metabolism (Martin and Pagano 1987; Pagano and Sleight 1985; Koval and Pagano 1989). NBD-lipids have also been valuable tools in studies on epithelial cell lipid polarization (Simons and van Meer, 1988). However, NBD-labelled lipids may not be the optimal probes, because NBD-moiety tends to localize to aqueous interphase (Chattopadhyay and London, 1990). This, along with short acyl chain length, results in unnaturally rapid transfer from one membrane to another within the cell. Also, it has been shown recently that in the red blood cells NBD-aminophospholipids have an anomalous spontaneous as well as aminophospholipid translocase mediated transmembrane movement (Colleau et al, 1991).

Another widely used fluorescent moiety is pyrene, which has a high quantum yield and unusually long fluorescence lifetime. Pyrene-moiety, coupled to fatty acyl chains of almost all major lipid classes, has been used extensively as a reporter molecule in in vitro studies aimed at elucidating membrane physico-chemical phenomena like phase transitions, phase separation, lateral diffusion, fusion, lipid-protein interactions in membranes, lipid movement between model membranes and protein-mediated lipid transfer (Pownall and Smith, 1989; Galla and Hartman, 1985; Somerharju et al, 1987). Pyrene fatty acids seem to mimick natural fatty acids since they are taken up and incorporated to lipids by different cell lines (Morand et al, 1984; Radom et al, 1990; Kasurinen and Somerharju, 1991, manuscript in preparation). A drawback of pyrenyl chromophore is that the monomer emission of pyrene is in the ultraviolet region of spectrum and thus invisible to the naked eye.

Recently, novel fluorescent fatty acid analogues have become commercially available as BODIPY fatty acids (a trademark of Molecular Probes). BODIPY has spectral properties very similar to NBD, but it is less polar than NBD which should make it a more close analogue of the natural fatty acids. However, to use BODIPY fatty acids as analogues of natural fatty acids in metabolic and lipid transport studies, their metabolic fate in vivo should be known. Thus in this study we have compared the metabolism of 7-

methyl-BODIPY-dodecanoic acid with that of pyrene decanoic acid in
BHK cells.

Materials

Lipids.   Pyrene decanoic acid (P-10) was synthesized in our
laboratory and 7-methyl-BODIPY-1-dodecanoic acid (BODIPY-12) was
from Molecular Probes (Eugene, USA). The concentrations of P-10
and BODIPY-12 were determined in ethanol using 42000 as the molar
extinction coefficient for pyrene and 72000 for BODIPY.   1-
palmitoyl -sn-glycero-3 -phosphocholine (PyrPC) and pyrene-labelled
fatty acid anhydride were synthesized and purified as described
earlier (Somerharju et al, 1985) Pyrene-diglycerides were obtained
by phospholipase C catalyzed hydrolysis of pyrene labelled PC
(PyrPC).   Cholesterol esters were prepared of cholesterol and
pyrene fatty acid anhydride as in PyrPC synthesis.   All these
reference lipids have been analyzed by HPLC and found to be more
than 98% pure.   The molecular structure and the spectral
characteristics of BODIPY-12 is shown in the figure 1.

       BODIPY-12
      Excitation wavelength 488 nm
      Emission wavelength 519 nm (monomer); 620 nm (excimer)
      Extinction coefficient 72000 M^{-1}

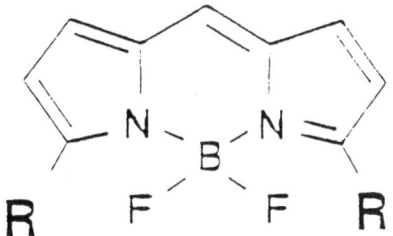

Figure 1. BODIPY-12: structure and spectral characteristics

Reagents. Lipid solvents as well as HPTLC plates were from Merck (Germany).

Methods

Cell labelling procedure. BHK cells were grown on plastic tissue culture dishes (Nunc, Denmark) in Dulbecco's MEM supplemented with 10 % fetal calf serum (Gibco,England), 10 mM glutamine and antibiotic-antimycotic solution (penicillin 100 units/ml, streptomycin 100 ug/ml and Amphotericin B 0.25 ug/ml) in 5 % CO_2 at 37 °C. Cells were passaged with trypsin-EDTA treatment. To label the cells they were incubated in a medium into which P-10 fatty acid or BODIPY-12 fatty acid had been added in dimethyl-sulfoxide (DMSO) or ethanol, respectively. The final concentrations of the probes in the medium was 10 uM. DMSO as well as ethanol concentrations were less than 0.1%. After indicated time periods the labelling medium was removed and cells washed thrice with 2 ml of PBS at room temperature. The cells were scraped from the dishes with a rubber policeman into 0.8 ml PBS and the lipids extracted subsequently according to Bligh and Dyer (1959). The lipid extracts were stored in chloroform below -20 °C.

Lipid analyses. Neutral lipids were analyzed by one-dimensional thin-layer chromatography on HPTLC plates using hexane:diethyl ether:acetic acid (80:30:5, by vol.) as a solvent. The lipids were visualized under UV light, scraped into test tubes and extracted from the silica with chloroform:methanol:water:acetic acid

(100:100:5:0.1, by vol.). The fluorescence intensity of the
extracts was measured at appropriate wavelengths to quantitate the
probe in each lipid class.

For qualitative analysis phospholipid classes were separated by
two-dimensional thin-layer chromatography (Esko and Raetz, 1980).
Lipids were identified by running unlabelled natural and synthetic
pyrene-labelled phospholipids under similar conditions as a
reference. For quantitative analysis plates were run only in one
dimension using chloroform:methanol:acetic acid (65:35:10, by
volume) as a solvent. Phospholipids were extracted from the plate
and their fluorescence measured as described above for neutral
lipids.

Results

Microscopical observations. The uptake of BODIPY-12 was fast,
already after a 5 minute incubation at $^{+}$37 °C the cells displayed
a strong uniform green fluorescence in a fluorescence microscope
(figure 2A). After two hours small yellowish-green spherical
structures with a slight reddish hue appeared (figure 2B). These
resembled the structures seen with P-10 (not shown) and which are
known to compose of tri- or diglycerides (Radom et al, 1987). With
BODIPY-12 these spherical structures were most probably not
triglycerides, because there was little incorporation to
triglycerides at 2 hours (see below), but merely either BODIPY-12
fatty acid or PC, the amount of which were rather high in the end
of the labelling period.

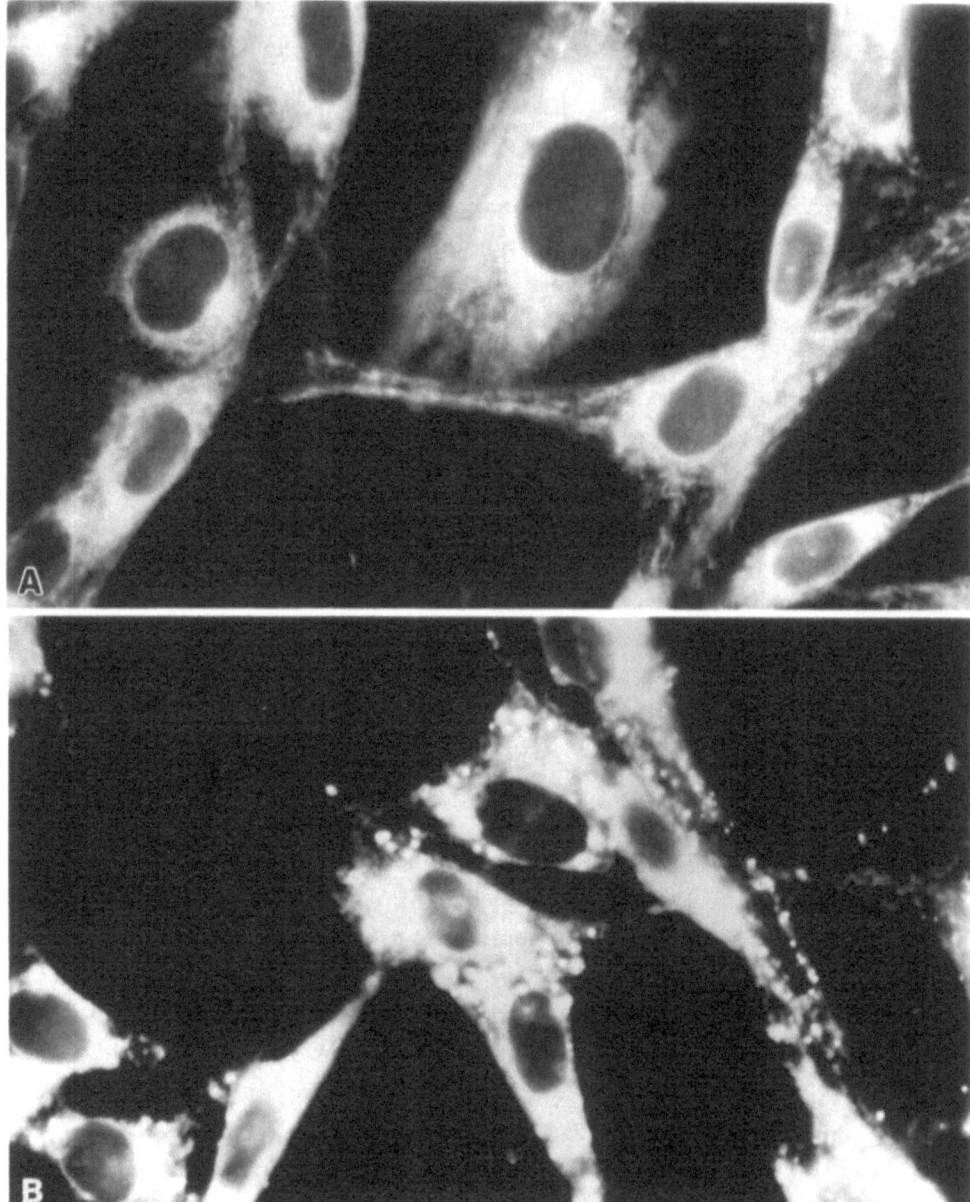

Figure 2. Fluorescence microscopic images of BHK cells labelled 5 minutes (A) or two hours (B) with BODIPY-12. Cells were labelled as was described in Methods. A Zeiss fluorescence microscope was used with the following filter combination: exciter filter BP 450-490, dichroic beam splitter FT 510, barrier filter LP 520.

<u>Incorporation of BODIPY-12 and P-10 fatty acids to neutral lipids.</u>
The distribution of BODIPY-12 and P-10 fatty acids between
different neutral lipid classes was analyzed by TLC (figure 3).
The identification of different bands was based on migration
pattern of P-10 labelled neutral lipids (figure 3).

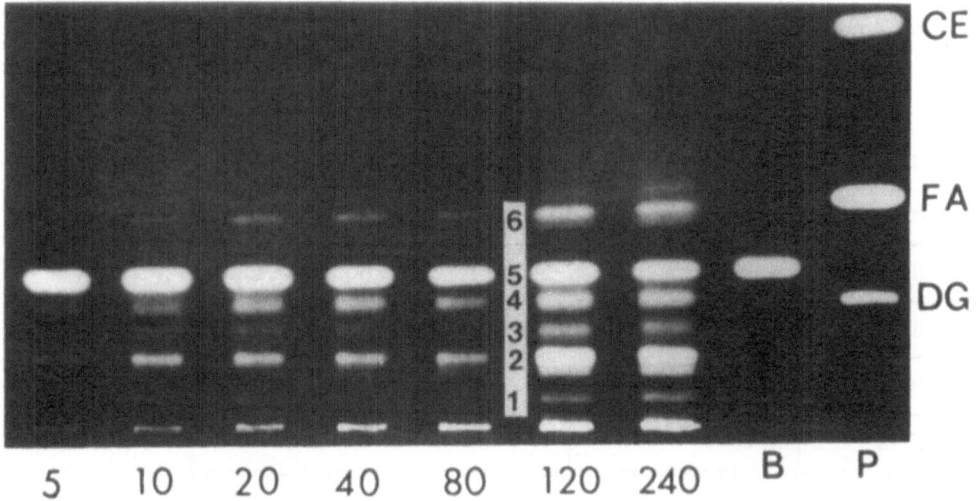

Figure 3. TLC separation of BODIPY-12 labelled neutral lipids of
BHK cells. X-axis is incubation time in minutes and B is BODIPY-12
fatty acid standard and P is a mixture of synthetic P-10
diglyceride (DG), P-10 fatty acid (FA) and cholesterol ester of P-
10 (CE). The neutral lipid bands are numbered from the first band
near origin.

The BODIPY-12 labelled lipids migrated on TLC plate more slowly
than the P-10 labelled lipids (or natural lipids) which made the
identification of BODIPY-12 bands only tentative and would suggest
that BODIPY-moiety is more polar than pyrene-moiety. Only the band
5 could be firmly identified as BODIPY-12 fatty acid. Band 6 is
most likely a triglyceride containing BODIPY-12. Band 2 is most
probably a diglyceride and band 1 a monoglyceride of BODIPY-12.
Bands 3 and 4 could be different molecular species of diglycerides
of BODIPY-12.

The distribution of BODIPY-12 and P-10 in neutral lipids of BHK cells after two hours incubation is shown in the table I.   If fatty acid fraction is excluded, BODIPY-12 is mostly in diglyceride fraction (band 2), whereas with P-10 most of the label is in triglyceride fraction (table I).

BODIPY-12	BODIPY-12 percent of fluorescence	P-10 percent of fluorescence	P-10
BAND 6	6.4	11.4	CE
BAND 5	70.0	66.6	MPTG
BAND 4	6.5	14.5	DPTG+FA
BAND 3	1.7	7.5	MPDG
BAND 2	14.3		
BAND 1	1.0		

Table 1: Distribution of BODIPY-12 and P-10 in the neutral lipids of BHK cells (incubation time 2 hours). Numbers of the bands refer to the figure 3: band 6 is most probably BODIPY-12 triglyceride, band 5 BODIPY-12 fatty acid and band 2 BODIPY-12 diglyceride CE: cholesterol ester of P-10, MPTG: mono-P-10-triglyceride, DPTG: di-P-10-triglyceride, MPDG: mono-P-10-diglyceride.

The time-course of BODIPY-12 incorporation to neutral lipids is shown in figure 4.   BODIPY-12 fraction (band 5) decreased significantly from 40 minutes to 4 hours, whereas the other neutral lipid fractions increased steadily during the whole incubation period. This is contrast to phospholipids (figure 5), where the incorporation increased linearly after a one hour lag period.

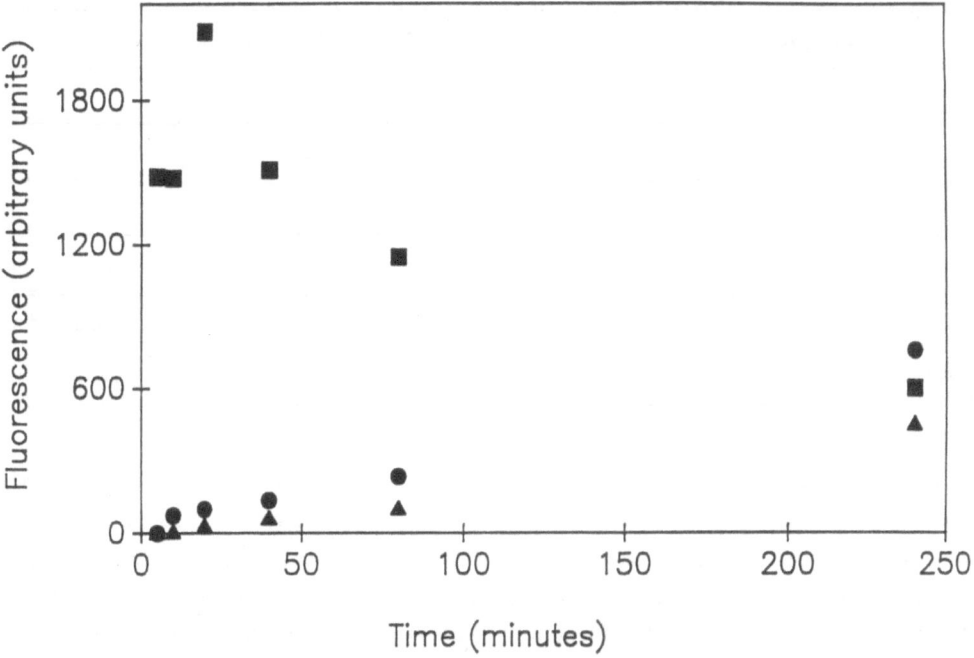

Figure 4. Time-course of BODIPY-12 incorporation to neutral lipids.Symbols:■ band 5 (FA),● band 2 (DG),▲ band 6 (TG)

The incorporation of BODIPY-12 to phospholipids. From two-dimensional TLC analyses it was evident that besides PC, all major phospholipids were labelled, including PE, PS, PI, PA and possibly PG (not shown). To analyze the distribution of BODIPY-12 in phospholipid classes, the lipid extract of 80 minutes incubation time sample was run in one-dimensional TLC, the phospholipid bands were extracted from silica and their fluorescence measured as described in methods (table 2).

phospholipid	percent of phospholipids
PC	94.8
PE	1.2
PI+PS	0.7
PA+CL	3.3

Table 2. The distribution of BODIPY-12 fluorescence in phospholipids of BHK cells (80 minutes incubation at 37 °C). PC is phosphatidylcholine, PS phosphatidylserine, PE phosphatidyl-ethanolamine, PA phosphatidic acid, PI phosphatidylinositol, CL cardiolipin.

However, although different phospholipid classes were labelled, PC was exclusively the most prominent fraction (see table 2), 94.8 percent of BODIPY-12 label being in PC. The time course of BODIPY-12 incorporation to PC is shown in the figure 5. Similar time-course was obtained with PE (not shown). A substantially long lag period in the incorporation can be observed. No lag was observed in tentatively as diglyceride identified neutral lipid fractions, which would indicate that BODIPY-12 labelled PC is synthesized from BODIPY-12 labelled diglycerides.

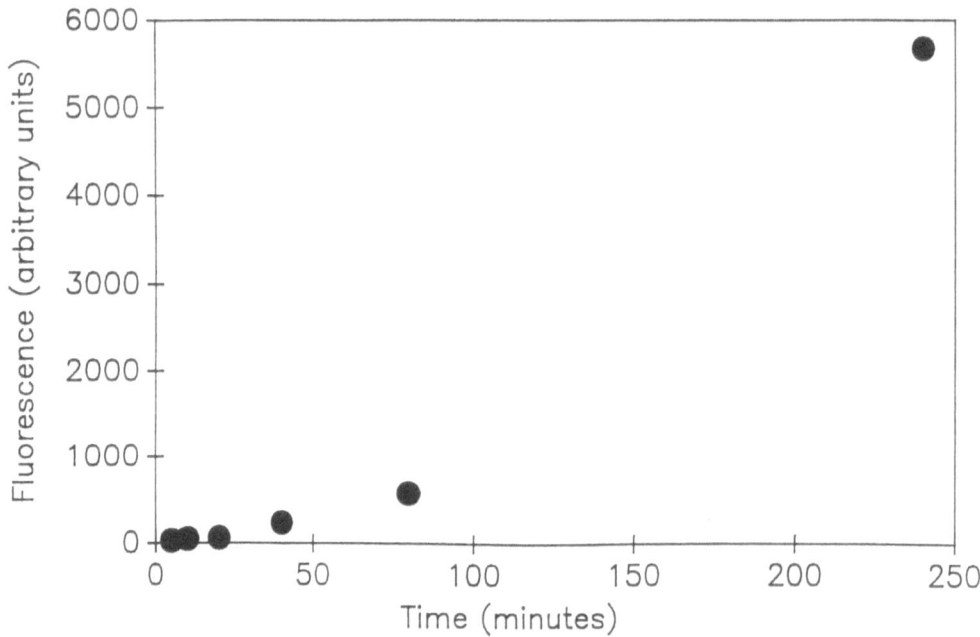

Figure 5. Time-course of BODIPY-12 incorporation to PC of BHK cells

Discussion

Several different fluorescent analogues of lipids have been
designed in order to provide researchers versatile tools to tackle
the problems of lipid cell biology. Many of them have found to
behave in model membranes in vitro as their natural counterparts.
However, the physico-chemical properties alone are not enough to
characterize a lipid analogue. Obviously the major criterion is
how intact cells transport and metabolize these analogues.

In this study we have used BHK cells to investigate the
incorporation to lipids of a novel fluorescent fatty acid analogue
7-methyl-BODIPY-1-dodecanoic (BODIPY-12) acid. P-10 was used as a
reference, because its metabolism in BHK cells has been
characterized in a considerable detail (Radom et al, 1990;
Kasurinen and Somerharju, 1991, manuscript in preparation).

BODIPY-moiety has favourable fluorescent properties: it was found
to be rather slowly bleached in microscopy examination, it has a
high quantum yield (extinction coefficient is 72000 M⁻1) and the
emission spectrum is in the most sensitive wavelength range of
vision. BODIPY is less bulky, but on the other hand, more polar
than pyrene as demonstrated by the slower mobility of BODIPY-12
neutral lipids on TLC. Nor pyrene neither BODIPY was found to be
toxic to BHK cells at ten micromolar concentration used.

When cells were incubated only five minutes in the presence of
BODIPY-12, cells were uniformly labelled with yellowish-green
fluorescence. Some reticular structures (most probably ER) could
also be observed. With P-10 this was far too short a labelling
period to detect any fluoresence by naked eye due to UV range
monomeric pyrene fluorescence. At longer incubation times P-10 was
also visible, now due to appearance of excimer (emission at 475 nm)
fluorescence. It has been shown with human leukemic myeloid cells
that pyrene fatty acids accumulate as triglyceride droplets during
prolonged incubation (Morand et al, 1984). With BODIPY-12 numerous

spherical structures were also observed at longer incubation times. These spherical structures were visualized as reddish spots on a green background indicating a high local concentration of BODIPY-12 in these structures since the excimer fluorescence maximum of BODIPY is at 620 nm. However, the nature of these spherical structures may be different in BODIPY-12 and P-10 labelled cells, since the former was mainly in PCs and the latter in triglycerides at longer incubation times.

BODIPY-12 was found in all major phospholipid and neutral lipid fractions, although PC, diglyceride and fatty acid fraction were by far the most prominent. In this respect BODIPY-12 resembles pyrene butyric acid, which was shown to incorporate mainly to phospholipids in multisystemic storage myopathy affected fibroblasts (Radom _et al_, 1990) and BHK cells (Kasurinen and Somerharju, 1991, manuscript in preparation).

The analysis of the time-course incorporation of BODIPY-12 to BHK cell lipids revealed that while neutral lipids were continuously labelled, the labelling of PC was preceded by over a one hour long lag time, which would indicate that BODIPY-12 is not incorporated to phospholipids by acylation-reacylation cycle, but more likely as an activated diglyceride.

BODIPY is a new fluorescent moiety which is now commercially available as 7-methyl-BODIPY-1-dodecanoic acid. It has very favourable spectral properties, and here we have shown that it is not toxic to cells and at least BHK cells can incorporate it to all major lipid classes. As compared to NBD-moiety BODIPY seems to be almost a probe of choice in metabolic and microscopic studies, because it is metabolized by cells and it is less prone to bleaching. However, more studies are needed to estimate the true value of BODIPY-12 as an analogue of natural lipids in lipid transport.

References

Bligh and Dyer (1959) A rapid method of total lipid extraction and purification. Can J Biochem Physiol 37:911-917

Chattopadhyay A and London E (1990) Spectroscopic and ionization properties of N(-7-nitrobenz-2-oxa-1,3-diazol-4-yl)-labeled lipids in model membranes. Biochim Biophys Acta 938: 24-34

Christie WW (1986) Rapid separation and quantification of lipid classes by high-performance liquid chromatography and mass (light-scattering) detection. J Lipid Res 26: 507-512

Colleau M, Hervè P, Fellmann P, Devaux PF (1991) Transmembrane diffusion of fluorescent phospholipids in human erythrocytes. Chem Phys Lipids 57:29-37

Esko JD and Raetz CRH (1980) Mutants of chinese hamster ovary cells with altered membrane phospholipid composition. J Biol Chem 255: 4474-4480

Galla H-J and Hartman W (1985) Excimer-forming lipids in membrane research. Chem Phys Lipids 27:199-219

Koval M and Pagano RE (1989) Lipid recycling between the plasma membrane and intracellular compartments: transport and metabolism of fluorescent sphingomyelin analogues in cultured fibroblasts. J Cell Biol 108:2169-2181

Martin OC and Pagano RE (1987) Transbilayer movement of fluorescent analogs of phosphatidylserine and phosphatidylethanolamine at the plasma membrane of cultured cells. J Biol Chem 262:5890-5898

Morand O, Fibach E, Livni N, Gatt S (1984) Induction of lipid storage in cultured leukemic myeloid cells by pyrene-dodecanoic acid. Biochim Biophys Acta 793:95-104

Pagano RE and Sleight RG (1985) Defining lipid transport pathways in animal cells. Science 229: 1051-1057

Pownall HJ and Smith LC (1989) Pyrene-labeled lipids: versatile probes of membrane dynamics in vitro and in living cells. Chem Phys Lipids 50: 191-211

Radom J, Salvayre R, Levade T, Douste-Blazy L (1990) Influence of chain length of pyrene fatty acids on their uptake and metabolism by Epstein-Barr-virus-transformed lymphoid cell lines from a patient with multisystemic lipid storage myopathy and from control subjects. Biochem J 269: 107-113, 1990

Simons K and van Meer G (1988) Lipid sorting in epithelial cells. Biochemistry 27:6197-6202

Somerharju PJ, Virtanen JA, Eklund KK, Vainio P, Kinnunen PKJ (1985) 1-palmitoyl-2-pyrenedecanoyl glycerophospholipids as membrane probes:evidence for regular distribution in liquid-crystalline phosphatidylcholine bilayers. Biochemistry 24: 2773-2781

Somerharju PJ, van Loon D, Wirtz KWA (1987) Determination of the acyl chain specificity of the bovine liver phosphatidylcholine transfer protein. Application of pyrene-labeled phosphatidylcholine species. Biochemistry 26:7193-7199

OXIDATIVE DAMAGE AND REPAIR OF HUMAN ERYTHROCYTE LIPIDS

Jeroen J.M. van den Berg and Frans A. Kuypers

Children's Hospital Oakland Research Institute
747 Fifty Second St.
Oakland, CA 94609
U.S.A.

INTRODUCTION

Biological membranes are composed of a complex and highly organized mixture of proteins and lipids. The complicated architecture of a membrane appears necessary for its optimal function, with each individual component an important and integral part in preserving structure and/or function of the membrane and of the cell as a whole. Most cells possess a very elaborate and ingenious apparatus to synthesize membrane components and to transport them to their site of destination. Routes of synthesis and transport of proteins and lipids are discussed by other contributors. We will focus on mechanisms that maintain the phospholipid molecular species composition and organization under conditions of oxidative stress in a relatively simple cell membrane, that of the red blood cell (RBC) or erythrocyte.

The RBC is a cell with no internal organelles. Basically, it consists of a plasma membrane and a cytosol that contains large amounts of hemoglobin, which binds and releases oxygen in this cell's function as oxygen transporter. The RBC contains no machinery for *de novo* synthesis of proteins or lipids, and has only a limited ability to repair proteins that are damaged. For membrane phospholipids, on the other hand, an elaborate remodeling system appears to be present to regulate the membrane composition and to repair lipid oxidative damage. Although there are only 5 major classes of phospholipids, more than 200 different glycerophospholipid molecular species and 50 sphingomyelin species have been identified (Myher et al., 1989). Each of these phospholipid molecules has specific characteristics and their relative amount as well as organization within the bilayer appear to be carefully controlled (Kuypers et al., 1987). Obviously, the phospholipid and fatty acyl composition of the membrane is important in determining lipid-lipid and lipid-protein interactions, and therefore it is important to define the underlying mechanisms for phospholipid homeostasis in normal and oxidized RBCs.

In this chapter, we will discuss some aspects of oxidative damage and repair of human RBC phospholipids. First, studies are described that were undertaken to characterize systems to

NATO ASI Series, Vol. H 63
Dynamics of Membrane Assembly
Edited by J. A. F. Op den Kamp
© Springer-Verlag Berlin Heidelberg 1992

oxidatively damage RBCs *in vitro* with respect to the kinetics of oxidative damage and the primary site of attack (membrane/cytosol). Subsequently, we will present some results that were obtained in studies of fatty acid turnover and repair in RBCs in the absence and presence of oxidative stress.

HYDROPEROXIDE-INDUCED OXIDATIVE STRESS IN RED BLOOD CELLS

The suggested involvement of oxidative stress and lipid oxidative damage in various pathological conditions has led to an ever increasing scientific interest in the nature of oxidative processes. In oxidative damage research, the RBC is a popular and physiologically relevant subject. Structure and function of the RBC destine it to be under a constant threat of oxidative damage to its lipid and protein membrane constituents (Clemens & Waller, 1987). As in mitochondria, there is a continuous production of oxygen radicals and other reactive oxygen species, starting with the formation of superoxide radical (O_2°) from a hemoglobin-oxygen complex (Sadrzadeh et al., 1984). Membrane phospholipids with polyunsaturated fatty acyl chains, present at high concentrations, are very susceptible to radical attack, while hemoglobin (Hb) may act as a powerful promoter of oxidative damage.

In the experiments described here, RBCs were incubated with commonly used hydroperoxides that generate radicals upon their Hb-dependent decomposition. Hydrogen peroxide (H_2O_2) is water-soluble and also formed *in vivo*, and cumene hydroperoxide (cumOOH) is a lipid-soluble lipid hydroperoxide analogue (lipid hydroperoxides are an intermediate product of the reaction sequence started by the attack of oxygen radicals on unsaturated fatty acyl chains. In the presence of transition metal ions they can decompose and form radicals again). We previously developed a method in which we use the fluorescent polyunsaturated fatty acid (PUFA), parinaric acid (PnA) to report on the fate of fatty acids in a membrane under oxidative stress (Fig.1) (Van den Berg et al., 1988, 1990, 1991a). As for other PUFAs, oxidative attack on PnA involves reaction of radicals with double bonds. The fluorescent properties of PnA depend on the presence of an intact conjugated double bond system. Oxidative modification of these double bonds can be monitored directly as a loss of fluorescence. In addition to PnA measurements, a number of other oxidation parameters was determined, of which Hb oxidation and vitamin E degradation are shown in this section.

Fig.2 shows the different kinetics of PnA oxidation in intact RBCs challenged with either cumOOH or H_2O_2. The PnA degradation curves with cumOOH as oxidant indicate an oxidative stress on the RBC membrane that is initially low, but increases in time and eventually leads to oxidation of all PnA present. In contrast, oxidative stress on the RBC membrane when

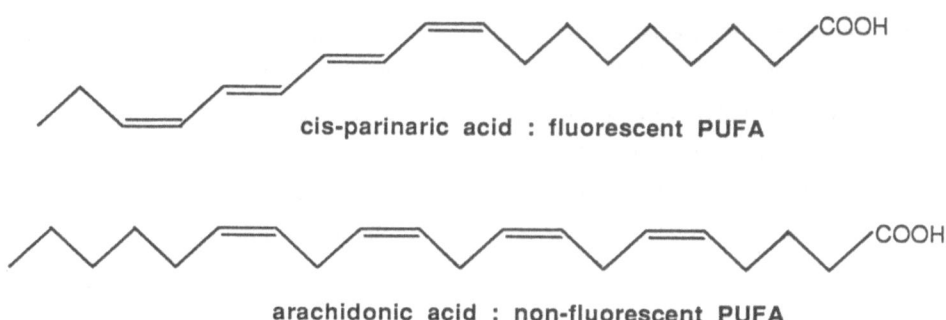

cis-parinaric acid : fluorescent PUFA

arachidonic acid : non-fluorescent PUFA

Fig.1. Comparison of the molecular structures of *cis*-parinaric acid (9,11,13,15-*cis,trans,trans,cis*-octadecatetraenoic acid) and arachidonic acid. Oxidative destruction of the conjugated double bond system of PnA immediately results in a loss of fluorescence, which is the rationale behind its use as a probe molecule for oxidative processes.

induced by H_2O_2 appears to be at its highest directly after addition of this oxidant. Furthermore, PnA degradation is not complete, but levels off at a concentration-dependent plateau value, indicating that already after a few minutes the oxidative stress on the membrane has stopped.

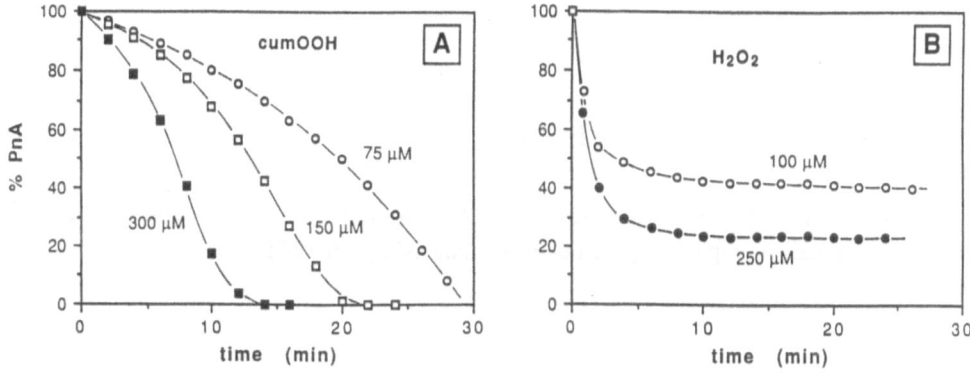

Fig.2. Oxidation kinetics of PnA in intact RBCs upon challenge with hydroperoxide. RBCs were incubated at 25 °C as a 1 % suspension in Hanks' balanced salt solution. The experiments with H_2O_2 were performed in the presence of 1 mM sodium azide to inhibit catalase. Before addition of oxidant, 1.12 µM *cis*-PnA was added from an ethanolic solution, after which PnA spontaneously incorporates into the RBC membranes. A. PnA degradation in RBCs challenged with various concentrations of cumOOH. B. PnA degradation in RBCs challenged with 2 concentrations of H_2O_2.

Hb is not only a promoter of peroxidation processes (in our experiments through the Hb-dependent decomposition of hydroperoxides yielding radical species), but at the same time a prime target for radicals. Transfer of an electron from Hb-Fe^{2+} to oxygen generates superoxide ($O_2^{\underline{o}}$) and metHb (Hb-Fe^{3+}). Upon further reaction, various Hb degradation products may be formed. The oxidation of Hb was determined qualitatively and quantitatively by spectrophotometric analysis of RBC lysates (Winterbourn, 1990; Van den Berg et al, 1991b). Fig.3 shows the formation of Hb degradation products upon incubation of RBCs with hydroperoxide. The kinetics of Hb oxidation are clearly different for cumOOH and H_2O_2, as was also noted in the PnA experiments. In addition, these two hydroperoxides induce the formation of different Hb degradation products (cumOOH: hemichrome; H_2O_2: ferrylHb). The extent of Hb oxidation is much larger for H_2O_2 than for cumOOH.

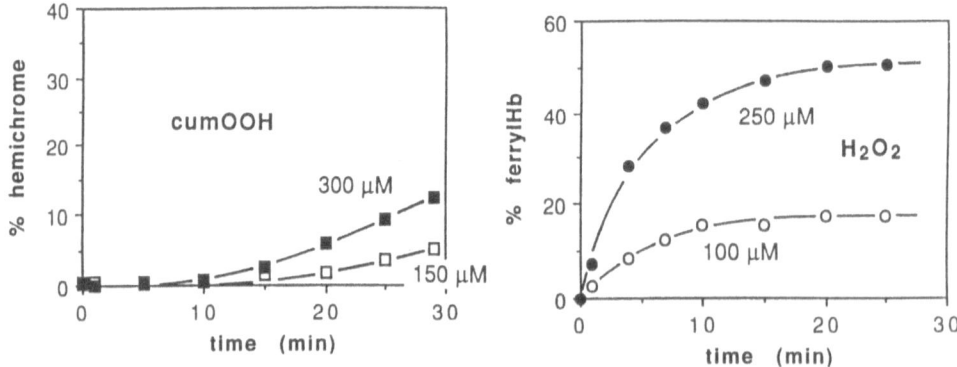

Fig.3. Hemoglobin oxidation in RBCs under oxidative stress. Experimental conditions as in Fig.2.

Vitamin E is a membrane antioxidant that converts radicals into non-radical species, e.g. lipid peroxyl radicals into lipid hydroperoxides. The kinetics of cumOOH- and H_2O_2-induced degradation of endogenous RBC vitamin E (Fig.4) closely resemble the pattern obtained for PnA: initially rapid, but incomplete degradation when induced by H_2O_2 as opposed to initially slow, but eventually complete degradation when induced by cumOOH.

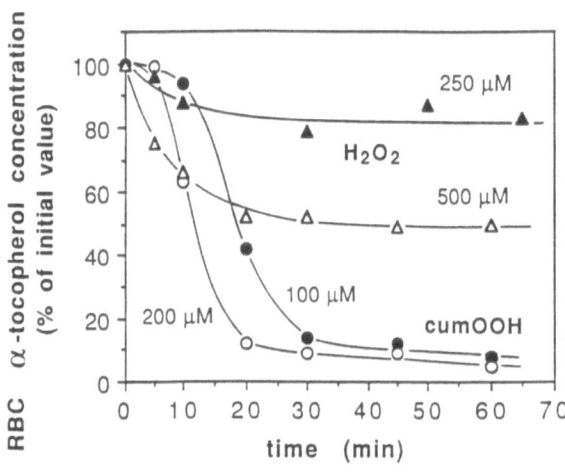

Fig.4. The kinetics of vitamin E degradation in intact RBCs challenged with hydroperoxide. General incubation conditions as in Fig.2. Vitamin E in a lipid extract was determined as α-tocopherol by HPLC (Van den Berg et al., 1991b).

These data clearly illustrate that differences in solubility characteristics and reactivity of hydroperoxides and the radicals derived from them determine the primary target (cumOOH primarily causes damage to the membrane, whereas H_2O_2-derived radicals attack cytosolic components) of oxidative stress generated by these compounds. These differences also account for the different oxidation kinetics observed in our experiments. When studying effects of oxidative stress, one should be aware of the vastly different mechanisms by which different oxidants may operate, especially when focusing on one aspect of oxidative damage.

THE REPAIR MECHANISM FOR OXIDATIVELY DAMAGED PHOSPHOLIPIDS

Throughout its lifespan, the RBC is exposed to reactive oxygen species that can cause damage to both cytosolic and membrane components. Oxidative damage to phospholipids will alter the molecular species composition of the membrane affecting its structure and function (Kuypers et al., 1987, 1990). As a first line of defense against oxidative damage, several antioxidant enzymes and other compounds in the RBC cytosol (superoxide dismutase, catalase, glutathione peroxidase) function to remove reactive oxygen species before they can inflict damage upon RBC components (Clemens & Waller, 1987). A second part of the protective system is involved in the repair of oxidative damage to phospholipids (Fig.5). A first step in a possible phospholipid repair route could be the conversion of a lipid peroxyl radical (lipid-OO•) into the more stable lipid hydroperoxide (lipid-OOH) by the membrane-soluble antioxidant vitamin E. After recognition of the site of damage and cleavage of the altered acyl chain by a phospholipase, the fatty acid hydroperoxide can be reduced further by a glutathione peroxidase to a fatty acid hydroxide and expelled to plasma (Sevanian & Kim, 1985).

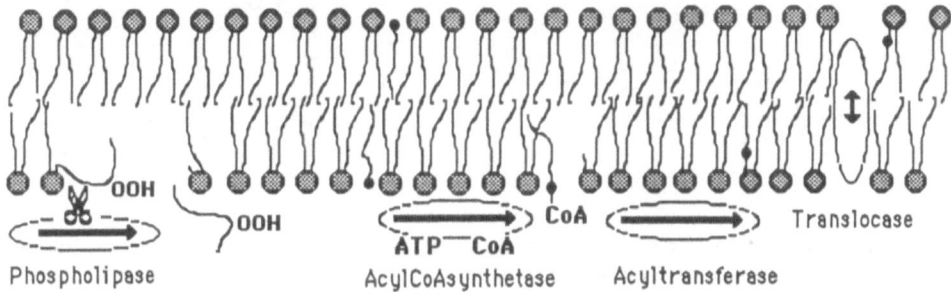

Fig.5. Enzymes proposed to be involved in the repair of lipid oxidative damage and the restoration of molecular species composition and organization.

An acylCoAsynthetase activates an intact fatty acid obtained from the plasma in order for an acyltransferase to incorporate this fatty acid into the remaining lysophospholipid, thereby repairing the lipid oxidative damage. Lipid transfer systems such as a translocase are furthermore involved in relocating the repaired phospholipid to its appropriate location in the inner or outer membrane leaflet (Devaux, 1991).

The repair enzymes (phospholipase(s), acyltransferase(s), and translocase(s)) involved in maintaining RBC phospholipid homeostasis are poorly characterized. Furthermore, little is known about their susceptibility to oxidation. In the following section, we will present some experiments designed to investigate the effects of oxidative stress on these repair enzymes and on the regulation of the phospholipid molecular species composition.

Fatty acid incorporation and turnover

In experiments performed with intact RBCs in the absence of an exogenous oxidative stress, the incorporation of trace labeled palmitic and arachidonic acid (Robinson et al., 1986) into phospholipid classes and molecular species was determined (Fig.6). It can be seen that labeled fatty acid was rapidly lost from the free fatty acid pool (FA) and incorporated into phosphatidylcholine (PC), phosphatidylethanolamine (PE) and phosphatidylserine (PS), but not sphingomyelin (SM). The rate of incorporation and the distribution of the fatty acid over the various phospholipid classes appeared to be dependent on the fatty acyl group.

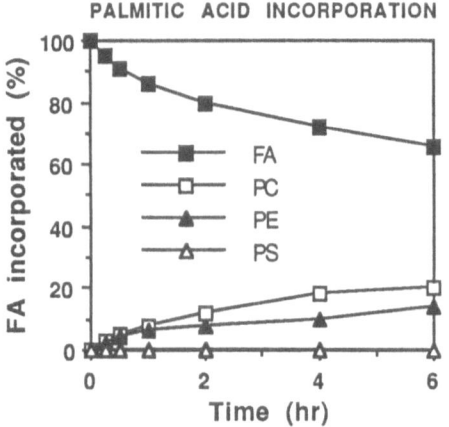 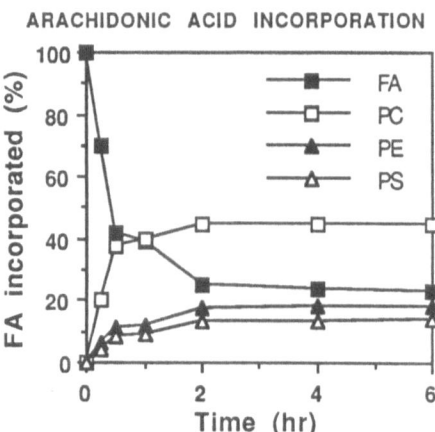

Fig.6. Rate and specificity of incorporation of palmitic acid and arachidonic acid into RBC phospholipid classes. A 10 % RBC suspension in Hanks' balanced salt solution was incubated with 3.4 nM fatty acid. Lipid extracts were made of samples taken at the indicated time points and the distribution of radioactivity over the various fractions was determined after TLC separation.

In order to study the distribution of radioactive fatty acid within molecular species, phospholipid fractions were separated by TLC and HPLC (Kuypers et al., 1991). The relative amount of radioactivity in each molecular subclass was determined. As an example, Table 1

shows analysis of the diacyl PC subclass.

Table 1. Fatty acid incorporation into diacyl PC molecular species.

SPECIES		mol (%)	palmitic cpm (%)	arachidonic cpm(%)
20:4	20:4	0.5		⎫
18:2	20:4	0.2		⎬ 9.1
16:0	20:5	0.1		⎭
16:0	22:6	2.4	6.7	
16:0	18:3	0.2	2.5	
18:1	20:4	0.6		51.0
16:0	20:4	5.7	17.2	28.7
18:1	22:4	0.8		
18:1	18:2	3.1		
16:0	18:2	34.7	39.8	
18:0	20:4	3.6		10.6
16:0	16:1	1.2	0.7	
16:0	18:1	29.3	16.4	
16:0	16:0	6.2	9.7	
18:0	18:1	4.9		

RBCs were incubated with trace labeled fatty acid as indicated in Fig.6. The relative distribution of palmitic acid (16:0) and arachidonic acid (20:4) in diacyl PC is shown. Molecular species are indicated by the number of carbon atoms and double bonds of fatty acyl groups at the sn-1 and sn-2 position of the glycerol backbone.

The patterns of incorporation for palmitic and arachidonic acid were distinct. Palmitic acid was preferentially incorporated into the sn-1 position and arachidonic acid into the sn-2 position. The incorporation of arachidonic acid into the 18:1,20:4 molecular species was very high, although the molar fraction of this phospholipid was very low. When the free fatty acid pool of labeled arachidonate was removed from the red cells by washing with defatted albumin, a rearrangement of radioactivity over the various molecular species was observed. Radioactivity in highly unsaturated species decreased while radioactivity in the major molecular species increased. These minor unsaturated species may have a high turnover rate and may be involved in the transfer of arachidonic acid from one molecular species to another. These results suggest that the incorporation of fatty acids into phospholipids is controlled at the molecular species level.

Phospholipid repair in oxidatively damaged RBCs

To study the effects of oxidative damage on phospholipid repair, cells were oxidatively damaged after which the incorporation of fatty acids into phospholipids was determined. Fig.7

shows the incorporation of trace labeled arachidonic acid into RBC phospholipid after previous exposure of the cells to oxidative stress generated by different concentrations of tert-butyl hydroperoxide (t-BOOH).

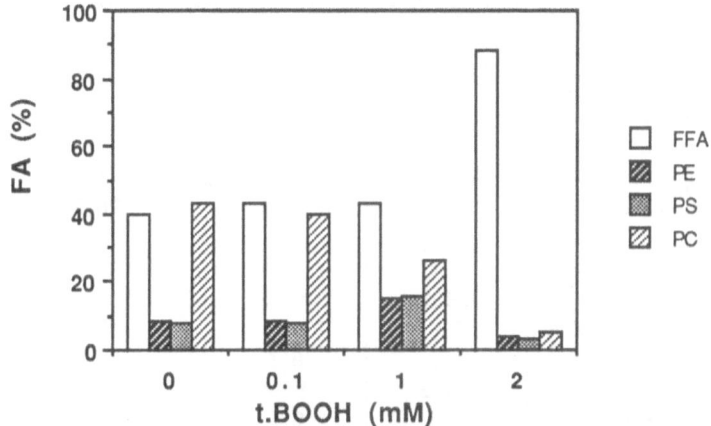

Fig.7. Arachidonic acid incorporation into oxidatively damaged RBCs. RBCs were incubated with various concentrations of t-BOOH for 1 hr, washed, and subsequently incubated with trace labeled arachidonic acid as described in Fig.6. The incorporation of fatty acid into the different fractions is expressed as the percentage of the total radioactivity in the membrane.

Both the rate and level of incorporation of arachidonic acid into phospholipids were altered in a way that was dependent on the oxidant concentration. The incorporation of arachidonic acid into PC was decreased at all concentrations of t-BOOH. In contrast, the incorporation into PE and PS was increased at 1 mM t-BOOH but was decreased at 2 mM. These findings could suggest different responses for the enzymes involved in lipid repair of each phospholipid class following oxidation. Alternatively, if PE and PS were preferentially oxidized, activation of the repair system could lead to incorporation of the available free fatty acid in the damaged PE and PS species rather than in PC.

The effects of oxidative stress on some elements of the protective system were also evaluated more directly. It was determined if n-ethyl maleimide (NEM), a sulhydryl cross-linking reagent, would affect acyltransferase activity. From human RBC ghosts, a partially purified preparation was made which retained acyltransferase activity. Increasing concentrations of NEM were seen to progressively inhibit acyltransferase activity, assayed by incubation with arachidonoylCoA and lyso-PC (Fig.8). This result indicates that the acyltransferase contains an oxidizable sulhydryl group that is essential for its activity.

149

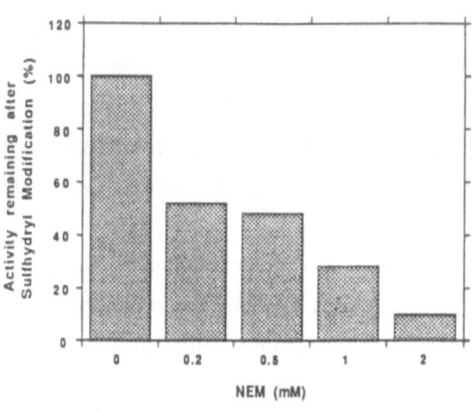

Fig.8. Effect of sulfhydryl modification on acyltransferase activity. RBC ghost membranes were preincubated with various concentrations of NEM, a sulfhydryl oxidizing reagent, before incubation with lyso-PC and arachidonoylCoA.

Oxidative damage to proteins can occur as the result of oxidative modification of susceptible sulfhydryl or tyrosine groups. Under some conditions, the oxidation of sulfhydryl groups can be reversed, a process in which glutathione (GSH), a cytosolic antioxidant compound, plays an important role. It acts either directly as a reducing agent or indirectly as a substrate for the antioxidant enzyme glutathione peroxidase. The enzyme GSH reductase is able to convert the oxidized form of GSH, glutathione disulfide (GSSG), back to the reduced form. Upon incubation of RBCs with hydroperoxide, the cellular GSH level can be observed to drop (Fig.9). At low levels of oxidative stress, activation of the GSH reductase pathway leads to a subsequent rise in the GSH level, as it is regenerated from its oxidized form. The regeneration of GSH can be seen to be dependent on the oxidative stress level. High concentrations of oxidant also result in oxidative damage to GSH reductase, thereby blocking the regeneration of GSH.

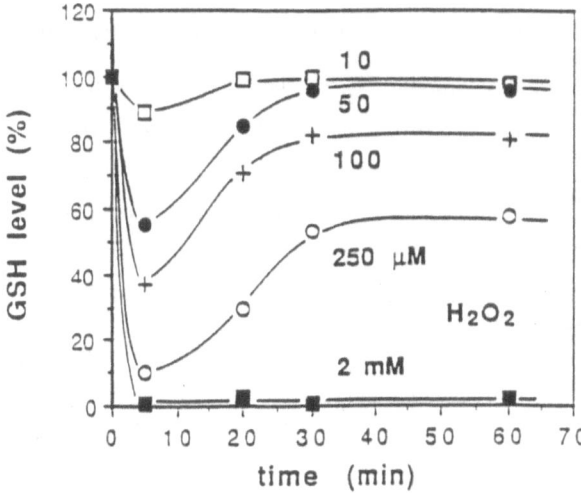

Fig.9. Activation of glutathione reductase by low oxidative stress. RBCs were incubated as a 1 % suspension in Hanks' balanced salt solution. Upon challenge of the cells by H_2O_2 samples were taken at the indicated time points and the cellular glutathione (GSH) content determined. Similar data as shown here for H_2O_2 were also obtained for other hydroperoxides.

150

The results shown in Figs.7-9 demonstrate that repair mechanisms for oxidative damage may be activated at low levels of oxidative stress. Under these conditions, lipid repair is adequate and molecular species composition is restored. However, as elements of the repair system are themselves also susceptible to oxidative damage, high levels of oxidative stress may inactivate the repair mechanism, affecting phospholipid molecular species composition and organization.

Oxidative stress can affect the cellular phospholipid molecular species composition either directly by damaging unsaturated lipid species or indirectly by damaging components of the repair mechanism, inhibiting restoration of the molecular species composition. As the phospholipid molecular species composition of the RBC can only be varied to a limited extent without affecting structure and function of the cell, it is important to define the mechanism by which the cell fine-tunes and maintains its molecular species composition. We have shown here some aspects of this mechanism regarding fatty acid specificity and response to oxidative stress. However, many details remain to be elucidated in this important area of membrane maintenance and repair.

ACKNOWLEDGMENTS

The authors are grateful to Dr. Julie Earnest, Mary Ann Schott and Charis Wagner for their respective contributions to the work shown. This work was supported by National Institute of Health grants DK 32094, HL 21061, HL 27059 and HL 20985, and NATO Travel Grant CRG 900612.

REFERENCES

Clemens MR, Waller HD (1987) Lipid peroxidation in erythrocytes. Chem Phys Lipids 45:251-268
Devaux PF (1991) Static and dynamic lipid asymmetry in cell membranes. Biochemistry 30:1163-1173
Kuypers FA, Chiu D, Mohandas N, Roelofsen B, Op den Kamp JAF, Lubin B (1987) The molecular species composition of phosphatidylcholine affects cellular properties in normal and sickle erythrocytes. Blood 70:1111-1118
Kuypers FA, Scott MD, Schott MA, Lubin B, Chiu DT-Y (1990) The use of ektacytometry to assess red cell susceptibility to oxidative stress. J Lab Clin Med 116:535-545
Kuypers FA, Bütikofer P, Shackleton C (1991) The application of liquid chromatography/thermospray mass spectrometry in the analysis of glycerophospholipid molecular species. J Chromat 562:191-206
Myher JJ, Kuksis A, Pind S (1989) Molecular species of glycerophospholipids and sphingomyelins of human erythrocytes. Lipids 24:396-407
Robinson M, Blank ML, Snyder F (1986) Highly unsaturated phospholipid molecular species of rat erythrocyte membranes: selective incorporation of arachidonic acid into phosphoglycerides containing polyunsaturation in both acyl chains. Arch Biochem Biophys 250:271-279
Sadrzadeh SMH, Graf E, Panter SS, Hallaway PE, Eaton JW (1984) Hemoglobin. A biologic Fenton reagent. J Biol Chem 259:14354-14356

Sevanian A, Kim E (1985) Phospholipase A2 dependent release of fatty acids from peroxidized membranes. J Free Radical Biol Med 1:263-271

Van den Berg JJM, Kuypers FA, Qju J, Chiu D, Lubin B, Roelofsen B, Op den Kamp JAF (1988) The use of cis-parinaric acid to determine lipid peroxidation in human erythrocyte membranes. Comparison of normal and sickle erythrocyte membranes. Biochim Biophys Acta 944:29-39

Van den Berg JJM, Kuypers FA, Roelofsen B, Op den Kamp JAF (1990) The cooperative action of vitamins E and C in the protection of parinaric acid against peroxidation in human erythrocyte membranes. Chem Phys Lipids 53:309-320

Van den Berg JJM, Kuypers FA, Lubin BH, Roelofsen B, Op den Kamp JAF (1991a) Direct and continuous measurement of hydroperoxide-induced oxidative stress on the membrane of intact erythrocytes. Free Rad Biol Med, in press

Van den Berg JJM, Op den Kamp JAF, Lubin BH, Roelofsen B, Kuypers FA (1991b) Kinetics and site specificity of hydroperoxide-induced oxidative damage in red blood cells. Free Rad Biol Med, in press

Winterbourn CC (1990) Oxidative reactions of hemoglobin. Methods Enzymol 186:265-272

RECOGNITION AND ELIMINATION OF MALARIA-INFECTED ERYTHROCYTES BY A MODIFIED PHOSPHOLIPASE A

J.A.F. Op den Kamp, G. Moll, A.P. Simões and B. Roelofsen
Centre for Biomembranes and Lipid Enzymology
State University of Utrecht
Padualaan 8
3584 CH Utrecht
The Netherlands

The erythrocyte plays an essential role in a particular part of the complex life cycle of Plasmodium, the malaria causing parasite, when it serves as a host cell in which the parasite feeds, grows and multiplies asexually. This part of the cycle is generally seen to offer potentially interesting possibilities of destroying the parasite, thereby breaking the fatal cycle. Many studies therefore aim to unravel the parasite-induced changes in the structural and functional characteristics of the membrane of the infected erythrocyte. Understandably, membrane proteins, many of which newly inserted as they have been produced by the parasite, have drawn a great deal of attention (Howard 1988, Hommel and Semoff, 1988), membrane permeability is increased (Tanabe et al, 1982; Sherman, 1979, 1988) pores are formed (Ginsburg et al, 1986), and the ultrastructural appearance of the cell can be modified by the development of knobs on the membrane surface (Howard, 1988). Recently, a series of studies was undertaken to investigate the fate of the phospholipid complement of the host cell membrane and to establish possible, parasite-induced, modifications in the composition as well as the organization of these basic membrane constituents.

Phospholipids from the membranes of parasitised erythrocytes

The polar headgroup composition of the phospholipids in the monkey erythrocyte membrane does not change upon infection with *Plasmodium knowlesi*. Table I shows that sphingomyelin (SPH), phosphatidylcholine (PC), phosphatidylethanolamine (PE) and phosphatidylserine (PS) are present in similar relative amounts before and after infection.

The transbilayer distribution of glycerophospholipids in the plasma membrane of *Plasmodium knowlesi*-infected erythrocytes was studied by several techniques including chemical probes, phospholipid exchange procedures and phospholipases such as lysine-116-ε-

NATO ASI Series, Vol. H 63
Dynamics of Membrane Assembly
Edited by J. A. F. Op den Kamp
© Springer-Verlag Berlin Heidelberg 1992

Table I: Phospholipid compositions (mole %) of erythrocyte membranes of schizont stage infected erythrocytes and normal cells, both isolated on Affigel beads, and of normal erythrocytes subjected directly to lipid extraction are presented. The number of determinations is given between parentheses.

	erythrocyte membrane of schizont infected cells (2)	normal erythrocyte membrane(2)	normal erythrocytes (5)
Sphingomyelin	16.7 ± 1.5	16.2 ± 1.2	15.5 ± 0.3
Phosphatidylcholine	40.2 ± 3.2	39.3 ± 3.0	41.1 ± 0.8
Phosphatidylethanolamine	25.0 ± 2.6	26.3 ± 2.3	27.6 ± 0.5
Phosphatidylserine/-inositol	18.1 ± 1.7	18.2 ± 2.0	15.3 ± 0.3

N-palmitoyl amidinated pancreatic phospholipase A_2. Initial experiments indicated that the sphingomyelin and phosphatidylserine topologies remained unaltered during parasite growth. Similar observations were made for the phosphatidylcholine. In order to confirm this data and to establish the localization of the phosphatidylethanolamine, experiments were carried out with the palmitate containing pancreatic phospholipase A_2 derivative. As a consequence of its superior membrane penetrating capacities, this latter enzyme rapidly degrades its substrates in the outer membrane leaflet of intact erythrocytes, a property that makes the enzyme an excellent tool to study the malaria-parasitized red cell. The modified phospholipase A_2 caused a non-lytic hydrolysis of up to 12-15% of the phosphatidylethanolamine and none of the phosphatidylserine in the red cell membrane, irrespective of whether the cells harboured trophozoite and schizont stages of parasites or no parasites at all. Consequently, the results from these (Moll et al., 1990a) and previous studies (van der Schaft et al., 1987) indicate that the plasma membrane of Plasmodium-infected erythrocytes exhibit a normal transbilayer phospholipid asymmetry. This conclusion however differs from the results obtained previously by Gupta and coworkers (Joshi et al., 1987, 1988) and also more recently by Maguire et al. (1991). The possible explanations for the discrepancies are discussed in detail elsewhere (Simões et al., 1991)

In contrast to the observed similarities in phospholipid polar headgroup composition and localization, distinct changes in fatty acid profiles could be observed between lipids extracted from the membranes of infected and non-infected cells (Simões et al., 1990). From phosphatidylcholine, one of the major phospholipids in both unparasitized and P. knowlesi-parasitized monkey erythrocytes, we have studied the molecular species composition. The characterization of the molecular species composition was accomplished with Reversed Phase High Performance Liquid Chromatography, complemented with Gas Liquid Chromatography, essentially as described by Blank *et al.*, (1984). It was shown that in trophozoite-parasitized cells the PC molecular species composition is identical to that in the isolated host erythrocyte membrane. However the composition differs from the one of unparasitized erythrocytes that had been recovered from the same batch of infected blood: there is a remarkable increase in 16:0/18:2-PC which seems to be compensated for by a decrease in 18:0/20:4-, 16:0/20:3-, 18:0/18:2- and 16:0/16:0- phosphatidylcholines. These differences are observed as well in the schizont stage of parasite development. The results demonstrate that intra-erythrocytic growth of *P. knowlesi* can have an effect on the molecular species composition of phospholipids in the parasitized cells including those present in the host cell membrane.

Since different molecular species exhibit different fitting properties in the lipid bilayer, and are therefore associated with a specific lipid packing, it is conceivable that by changing the molecular species composition of the host cell membrane, the parasite is manipulating its physiological properties. This could possibly explain the observations that membrane viscosity is decreased and that the mobility of phospholipids is increased (Taraschi *et al.*, 1986; Beaumelle *et al.*, 1988; Deguercy *et al.*, 1986; van der Schaft *et al.*, 1987; Moll *et al.*, 1988; Haldar *et al.*, 1989) Whatever the explanation might be, we have shown that the difference in physico-chemical characteristics between parasitized and non-parasitized cells can be exploited to discriminate between these two cell types and can be used to eliminate the parasite.

Action of phospholipases on the erythrocyte membrane

It has been recognised that water soluble phospholipases are only able to attack their substrates, organized in a lipid mono- or bilayer, when the enzymes are able to interact with, or to penetrate into, that layer (Verger *et al.*, 1973). Studies involving monomolecular films of phospholipids, of which the composition mimics that of the outer membrane leaflet of the human erythrocyte, learned that each individual phospholipase has a characteristic penetration capacity that can be expressed by the maximal lateral surface pressure of such a film above which the enzyme is unable to attack and degrade its substrate. Using highly purified phospholipases of which the penetration capacity had been calibrated by the above method, it was estimated that under physiologic conditions of tonicity, temperature and pH, the surface pressure in the outer membrane leaflet of the human erythrocyte is approximately 33 dynes/cm (Demel *et al.*, 1975). Consequently, Naja naja phospholipase A_2, which can attack a phosphatidylcholine/sphingomyelin monomolecular film to a lateral surface pressure of up to 34.8 dynes/cm, can indeed degrade its substrates in the outer membrane leaflet of the intact human erythrocyte, in contrast to pig pancreatic phospholipase A_2 for which the monolayer technique indicated the limiting pressure to be 16.5 dynes/cm. It has recently been shown that the weak penetrating properties of this pancreatic phospholipase A_2 can be considerably improved by the covalent attachment of a long acyl chain to Lys[116] in the otherwise fully amidinated pancreatic phospholipase (AMPA). The enhancement in penetration capacity appeared to be proportional to the length of the acyl chain, at least up to palmitic acid (v.d. Wiele *et al.*, 1988). The palmitoyl derivative, Pal-116-AMPA, easily degrades its substrates in the outer monolayer of the intact human erythrocyte at a rate that is even appreciably faster than can be achieved with Naja naja phospholipase A_2 and, as summarized above, can therefore be used very efficiently in phospholipid localization studies. The lauric acid derivative, Lau-116-AMPA, on the other hand, which can penetrate a phosphatidylcholine monolayer packed at a pressure up to 27 dynes/cm, virtually fails to attack the (normal) human red cell.

In view of the above-mentioned decreased packing of lipids in the membrane of Plasmodium-infected red cells, we studied the possibility that Lau-116-AMPA might be able to discriminate between infected cells and non-infected ones, by attacking the former and leaving

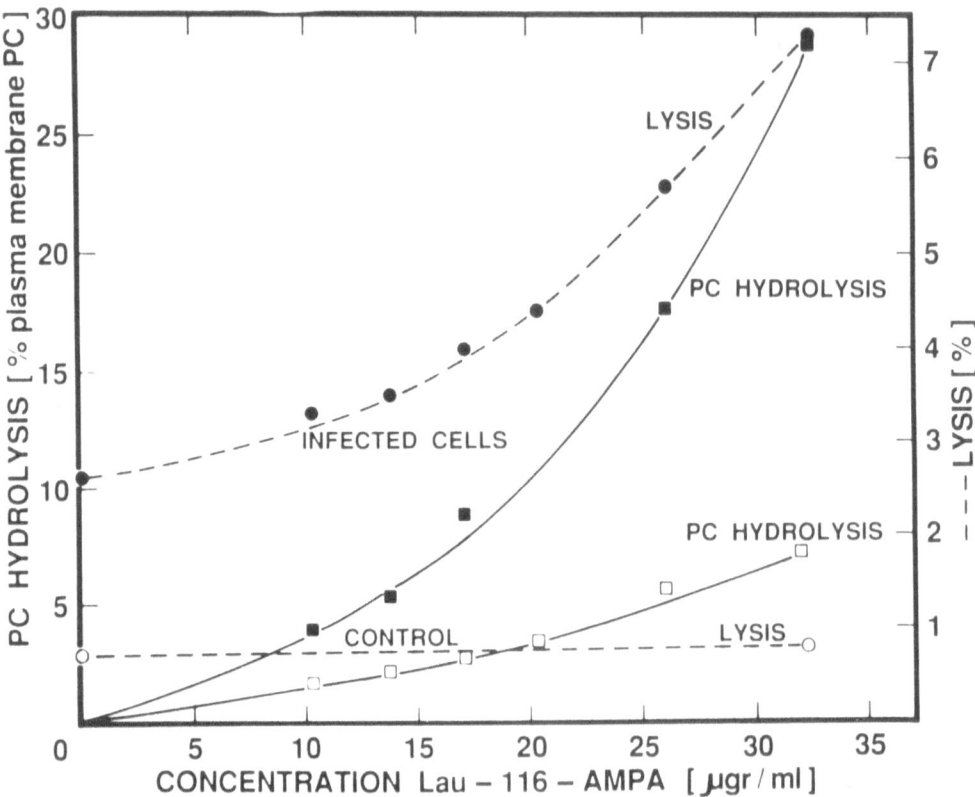

Figure 1: Action of Lau-116-AMPA on *P. knowlesi*-infected erythrocytes.
Haemolysis, measured by the release of haemoglobin and expressed as % of the total amount of haemoglobin and phosphatidylcholine hydrolysis, expressed as % of the amount of PC present in the plasma membrane, have been studied in control and infected cells which were incubated with the laurine modified phospholipase. Details of the experimental procedures are given elsewhere (Moll *et al.*, 1990b).

the latter undisturbed. Studies on erythrocytes obtained from *P. knowlesi*-infected rhesus monkeys (Macaca mulatta) showed that, in contrast to the uninfected cells, the 100% schizont-infected cells were highly susceptible to Lau-116-AMPA (Fig. 1). As little as 45 µm Lau-116-AMPA per 70 µl packed cells induced over 7% haemolysis within 1h in case of the infected cells, which was accompanied by the hydrolysis of 30% of the phosphatidylcholine present in the host cell plasma membrane. No hydrolysis of phosphatidylserine was observed under those conditions, whereas less than 5% (if any) of the phosphatidylethanolamine appeared to be degraded in the infected cells. Control experiments with pig pancreatic phospholipase A_2 of which all of the nine free amino groups had been blocked (v.d. Wiele *et al.*,1988), showed the complete failure of this weakly penetrating enzyme to attack the infected population of red cells.

As these studies clearly demonstrated that Plasmodium-infected erythrocytes are indeed specifically susceptible to Lau-116-AMPA, it became of interest to see what might happen to the parasite itself when this modified pancreatic phospholipase A_2 is present during its in vitro culturing. Figure 2 shows a dose-response curve of the effect of Lau-116-AMPA on the survival of P. falciparum cultured in a suspension of human erythrocytes. The indicated amounts of Lau-116-AMPA had been added to a 1.5-2.0% suspension of red cells (approx. 1% parasitaemia at zero time) in RPMI 1640 supplemented with 10% AB+ serum. Viability of the parasites was determined after 24 h. Similar experiments were performed using either Naja naja phospholipase A_2 or AMPA. In agreement with its higher penetration power, when compared to Lau-116-AMPA, N. naja phospholipase A_2 is much more effective in killing the parasites than the lauroyl derivative of the pancreatic enzyme is. It should be recalled, however, that in contrast to Lau-116-AMPA, the snake venom phospholipase A_2 also attacks the uninfected cells. Some inhibition of growth could also be achieved with AMPA, but a complete elimination of the parasites could not be reached at the highest concentration (7.3 µg/ml.) tested (Fig. 2)

Incubation of healthy human erythrocytes for 24h in the Plasmodium culture medium and in the presence of an amount of Lau-116-AMPA that causes a complete elimination of the parasite, does not cause more than 7% hydrolysis of phosphatidylcholine in those cells. Furthermore, it was observed from experiments in which either the total number of cells or the relative number of parasitized cell was varied, that the amount of Lau-116-AMPA that is needed to cause a 50% reduction in parasite survival was rather strictly correlated to the number of infected cells only (results not shown).

It is also worth noting that the lethal effect that Lau-116-AMPA exerts on the parasite, was not at all affected when the amount of AB+ serum present in the Plasmodium falciparum culture was varied from 5 to 15%, which precludes that serum (lipoproteins) play(s) an intermediary role in this process.

Taken together, our studies demonstrate that Lau-116-AMPA exclusively affects the infected cell and - equally important - that its action causes the death of the parasite. Although it

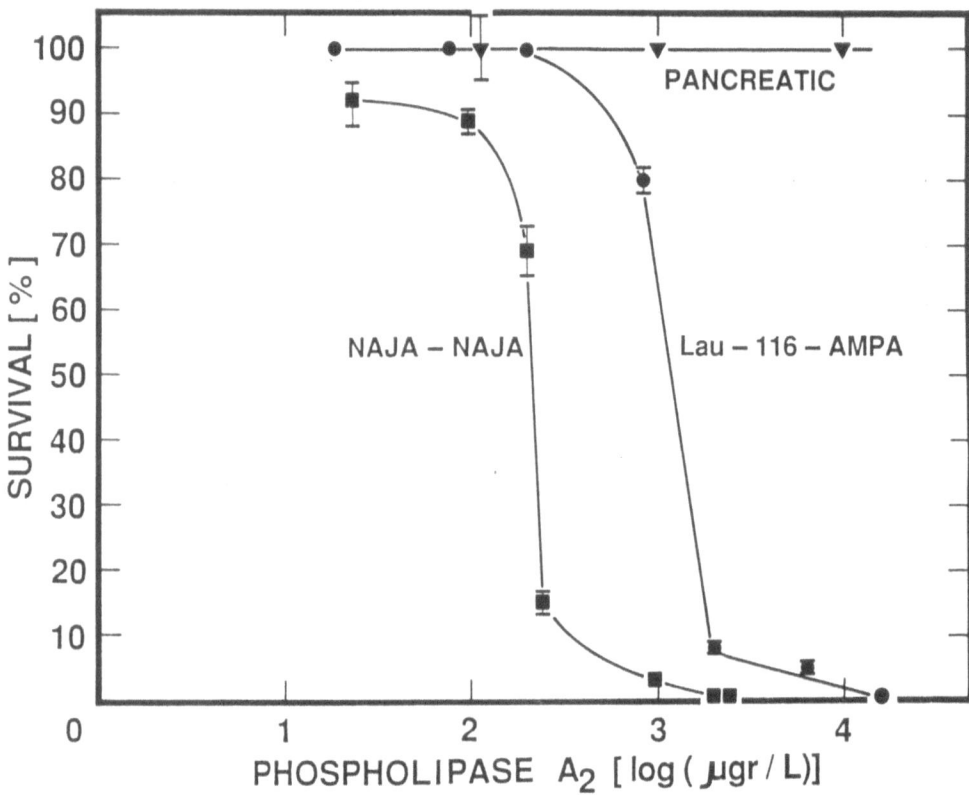

Figure 2: Survival of Plasmodium knowlesi following phospholipase treatment.
The various phospholipases are added, at the concentrations indicated, to growing
cells of Plasmodium knowlesi in an *in vitro* cultivation system. Survival, determined
by the incorporation of radiolabel from the precursor hypoxanthine into DNA is
expressed as percentage of the untreated control. Details and experimental
procedures are described elsewhere (Moll *et al.*,1990b).

may be tempting to speculate that this modified pancreatic phospholipase A_2 may open new
ways of strategy in the fight against malaria, it should be realised that its potential application
still has to await the answer on a number of questions and the solution of some pertinent
problems. For instance, it still has to be ascertained that Lau-116-AMPA is unable to affect
other healthy blood cells such as thrombocytes and leucocytes, similarly as it is unable of
attacking healthy erythrocytes. One of the major problems that have to be solved yet, is to
prevent a direct attack of the enzyme on the serum lipoproteins[10]. However, the essentially
absolute selectivity by which Lau-116-AMPA attacks only those erythrocytes that have been
infected by the Plasmodium parasite warrants further research to overcome those problems.

References

Beaumelle BD, Vial HJ, Bienvenue A (1988) Enhanced transbilayer mobility in malaria infected erythrocytes. J Cell Physiol 135 : 94-100

Blank ML, Robinson M, Fitzgerald V and Snyder F (1984) Novel quantitative method for determination of molecular species of phospholipids and diglycerides. J Chromatogr 298 : 473-482

Deguercy G, Schrevel J, Duportail G, Laustriat G and Kuhry JS (1986) Membrane fluidity changes in P.berghei-infected erythhrocytes, investigated with a specific plasma membrane fluorescent probe. Biochem Int 12 : 21-32

Demel RA, Geurts van Kessel WSM, Zwaal RFA, Roelofsen B and van Deenen LLM (1975)Relation between various phospholipase actions on human red cell membrane and the interfacial phospholipid pressure in monolayers.Biochim Biophys Acta 406:97-107.

Ginsburg H, Kutner S, Zangwil M and Cabantchik ZI (1986) Selective properties of pores induced in host erythrocyte membrane by Plasmodium falciparum infected cells. Effect of parasite maturation. Biochim Biophys Acta 861 : 194-196

Haldar K, de Amorim AF, Cross GAM, (1989) Transport of fluorescent phospholipid analogues from the erythrocyte membrane to the parasite in Plasmodium falciparum infected cells. J Cell Biol 108 : 2183-2192

Hommel M and Semoff S (1988) Expression and function of erythrocyte associated surface antigens in malaria. Biology of the Cell 64 : 183-203

Howard RJ (1988) Malarial proteins at the membrane of P.falciparum infected erythrocytes and their involvement in cytoadherence to endothelial cells. Prog Allergy, 41 : 98-147

Maguire PA, Prudhomme J and Sherman IW (1991) Alterations in the erythrocyte membrane phospholipid organization due to the intracellular growth of the human malaria parasite, Plasmodium falciparum. Parasitology,102:179-186

Joshi, P., Dutta, G.P. and Gupta, C.M. (1987) An intracellular simian malaria parasite (Plasmodium knowlesi) induces stage-dependent alterations in membrane phospholipid organization of its host erythrocyte.Biochem. J. 246:103-108

Joshi, P. and Gupta, C.M. (1988) Abnormal membrane phospholipid organization in Plasmodium falciparum infected human erythrocytes. Brit. J. Haematol.68:255-259

Moll GN, Vial HJ, Ancelin ML, Op den Kamp JAF, Roelofsen B and van Deenen LLM (1988) Phospholipid uptake by Plasmodium knowlesi infected erythrocytes. FEBS Lett 232 : 341-346

Moll GN, Vial HJ, Bevers EM, Ancelin ML, Roelofsen B, Comfurius P, Slotboom AJ, Zwaal RFA, Op den Kamp JAF and van Deenen LLM (1990a) Phospholipid asymmetry in the plasma membrane of malaria infected erythrocytes. Biochem. Cell. Biol. 68: 579-585

Moll GN, Vial HJ, Van der Wiele FC, Ancelin ML, Roelofsen B, Slotboom AJ, de Haas GH, van Deenen LLM and Op den Kamp JAF. (1990b). Selective elimination of malaria infected erythrocytes by a modified phospholipase A2 in vitro. Biochim Biophys Acta 1024:189-192.

Sherman J (1979) Biochemistry of Plasmodium . Microbiol Rev 43 : 453-495

Sherman, J.W. (1988) Parasitology 96:77-79.

Simões AP, Moll GN, Beamelle B, Vial HJ, Roelofsen B and Op den Kamp JAF (1990) Plasmodium knowlesi induces alterations in phosphatidylcholine and phosphatidylethanolamine molecular species composition of parasitized monkey erythrocytes.Biochim Biophys Acta 1022:135-145.

Simões AP,Roelofsen B and Op den Kamp JAF. (1991) Lipid compartmentalization in Plasmodium sp. parasitized erythrocytes. Parasitology Today.in press..

Tanabe K, Mikkelse RB and Wallach DFH (1982) Calcium transport of Plasmodium chabaudi infected erythrocytes. J Cell Biol 93:680-684

Taraschi TF, Parashar A, Hooks M and Rubin M (1986) Perturbation of red cell membrane structure during intracellular maturation of Plasmodium falciparum. Science 232: 102-104

Verger R, Mieras MCE and de Haas GH. (1973) Action of phospholipase A at interfaces. J.Biol.Chem. 248:4023-4034

Van der Schaft PH, Beaumelle B, Vial HJ, Roelofsen B, Op den Kamp JAF and van Deenen LLM (1987) Phospholipid organisation in monkey erytrocytes upon Plasmodium knowlesi infection. Biochim Biophys Acta 901 : 1-14

Van der Wiele, F.Chr., Atsma, W., Dijkman, R., Schreurs, A.M.M., Slotboom, A.J. and de Haas, G.H. (1988) Site specific epsilon-NH2 mono-acylation of pancreatic phospholipase A2. Biochemistry 27, 1683-1688

EFFECTS OF NONSUBSTRATE LIPIDS ON ACYLHYDROLASE ACTIVITIES OF LIPOPROTEIN AND HEPATIC LIPASES USING DEFINED EMULSION PARTICLES

A. Derksen, H.M. Laboda and D.M. Small
Department of Biophysics
Boston University School of Medicine
Housman Medical Research Center
80 East Concord Street
Boston MA 02118-2394
USA

Hydrolysis of triolein (TO) and egg phosphatidylcholine (PC) in emulsion particles by purified bovine milk lipoprotein lipase (LPL) and by rat hepatic lipase (HL) was studied as a function of the particle cholesterol (C) and/or ether-PC content. Methods included the formation and characterization of emulsion particles. A computer program calculated the particle surface and core lipid composition and average diameter. Negative stain electron microscopy verified morphology. Enzyme assays were designed to mimic physiological conditions. Independent of particles C content, LPL hydrolysis of the core was faster than the surface. With HL the surface was hydrolyzed much faster than the core. Increasing the particle C content from 0 to 4 mole% progressively stimulated TO hydrolysis and slowly inhibited PC hydrolysis by LPL, while the hydrolysis of both these lipids by HL was gradually inhibited. Ether-PC hardly affected TO hydrolysis by LPL, but HL was 60% inhibited. Thus non-substrate lipids in chylomicrons may aid its catalysis by LPL while impeding HL.

INTRODUCTION

Most investigations of the hydrolytic activities of lipolytic enzymes, such as Lipoprotein Lipase (LPL) and Hepatic Lipase (HL), use micelles, emulsions or monolayer systems composed of water-insoluble lipids (Olivecrona & Bengtsson, 1988;

NATO ASI Series, Vol. H 63
Dynamics of Membrane Assembly
Edited by J. A. F. Op den Kamp
© Springer-Verlag Berlin Heidelberg 1992

Jackson, 1983). Micelles and monolayers, however, do not precisely resemble the physiological substrates of these enzymes.
Advantages of using the emulsion system to study lipolytic
activity are: 1) emulsions have a large surface area to facilitate adsorption of all of the enzyme and cofactor molecules to
the lipid water interface, and 2) emulsions resemble the physiological substrates of the enzymes (i.e., chylomicrons). A
disadvantage is that it was difficult to characterize the
interface.

Miller and Small (1983) however, developed an emulsion
model for triglyceride-rich lipoproteins for which the composition of the surface and the core can be calculated (Miller &
Small, 1987). Briefly, when emulsion particles are made with
triolein, cholesterol and egg yolk phosphatidylcholine (EYPC),
the cholesterol can partition into the surface up to a molar
ratio with EYPC of 1:1 and even higher if supersaturated.
Triolein is soluble to about 3 mole% in an EYPC interface (Hamilton & Small, 1981). However, when cholesterol is increased
in the surface triolein is progressively pushed out and at
saturation triolein is virtually displaced from the emulsion
monolayer (Spooner & Small, 1987). When available, fatty acids
partition readily into the interface (Spooner et al., 1988;
Ekman et al., 1988).

The partition coefficient of protonated fatty acid between
surface and core is 7:1, whereas all of the ionized fatty acid
partitions into the interface. The apparent pKa of fatty acids
in the surface of emulsion particles and triglyceride-rich
lipoproteins is about 7.4, indicating that at physiologic pH
half of the fatty acid is ionized and half is protonated (Spooner et al., 1988). While cholesterol readily displaces triolein from the surface (Spooner & Small, 1987), it is unable to
displace fatty acids from the interface (Ekman et al., 1988).

Model emulsions can be used in several ways. First, to
study their interactions with pure apoproteins (Derksen &
Small, 1989; Derksen et al., 1989) and plasma (Derksen et al.,
1988). Second, the metabolism of lipid emulsion models can be
traced in rats (Redgrave & Maranhao, 1985; Maranhao et al.,
1986; Bennett Clark & Derksen, 1987; Bennett Clark et al.,
1991). Lastly, emulsion particles are used as model substrates

for appropriate lipolytic enzyme (Laboda et al., 1988; Bennett Clark & Laboda, 1991).

Here, we report the effects of non-hydrolyzable lipids such as cholesterol and ether-PC on the acyl hydrolase activities of lipoprotein and hepatic lipases towards triolein and EYPC using defined emulsion particles.

METHODOLOGY

Lipids. Nonradioactive lipids were used without further purification. All were >99% pure, as stated by the suppliers and confirmed by thin-layer chromatography (TLC) for lipid class purity and by gas-liquid chromatography (GLC) of FA methyl esters. Cholesterol, egg yolk phosphatidylcholine (EYPC), and 1-O-hexadecyl-2-oleoyl phosphatidylcholine (HOPC) were obtained from Sigma Chemical Co. (St. Louis, MO). Triolein was from Nu Chek Prep (Elysian, MN). Glycerol tri[9,10-^{3}H]oleate and 1-palmitoyl-2-[1-14^C]-oleoyl-L-3-phosphatidylcholine were from Research Products, Amersham Corp. (Arlington Heights, IL) were shown to be >99% pure by thin-layer chromatography of neutral lipid classes and liquid scintillation spectrometry.

Proteins. LPL was purified from bovine milk by the procedure of Socorro et al. (1985). HL was purified from rat liver perfusate by the method of Jensen & Bensadoun (1981). Both enzymes showed a single band on 10% SDS-PAGE with a mass of about 50 kD. Aliquots of a single, combined preparation of each enzyme were stored at −80°C in 50% aqueous glycerol and were used within minutes after thawing.

ApoC-II was isolated from human plasma VLDL using the method of Holmquist & Carlson (1977).

The purity was established by IEF electrophoresis and amino acid analysis. Aliquots of an aqueous solution were stored at −80°C. FA-free albumin (Fraction V) was from Sigma Chemical Co.

Preparation of Emulsions. Emulsions of different lipid compositions, namely ^{3}H-TO/^{14}C-UC/egg PC, ^{3}H-TO/UC/(^{14}C-POPC) egg PC, and ^{3}H-TO/^{14}C-UC/HOPC, were prepared according to the procedure of Derksen & Small (1989). The appropriate amounts of lipids were pipetted from stock solutions into 20 ml glass scintillation vials. The total amount of lipid was 30 mg, of which 84% was TO. The proportion of UC in the starting mixture was varied from 0 to 10 wt%, and of PC from 6 to 16 wt%. The lipid solutions were dried under N_2 and then in a vacuum desiccator overnight at 4°C. The lipids were resuspended in 10 ml of 0.15 M NaCl and were emulsified by continuous sonication for 10 min at approximately 30% of maximal power output. The solutions were transferred to polyallomer centrifuge tubes, overlayed with 1 ml of distilled water and centrifuged at 23,000 rpm for 11 min at 23°C in a Beckman SW-41 rotor. The top 0.6 ml of the emulsion was isolated with a tube slicer and was transferred to a plastic tube using a plastic syringe. The well of the slicer was rinsed with 300 μl of distilled water to complete the transfer of the creamy layer. Small aliquots were removed for isotopic and chemical lipid analysis. Emulsions (2-5 preparations of each emulsion type) were chemically and morphologically characterized by standard methods as described by Derksen et al. (1989). The compositions of the emulsions are shown in Table I.

TABLE I: Lipid Compositions of Emulsions Used in Enzyme Studies

Starting Mix	Lipid Composition of Isolated Emulsions[b]		
	TO	PC	Ch
EYPC-0 Ch(5)	89.7 ± 2.4	10.3 ± 2.4	– (0)
EYPC-1 Ch(5)	90.2 ± 0.9	9.1 ± 0.9	0.7 ± 0.1 (1.5)
EYPC-3 Ch(5)	90.4 ± 0.8	7.7 ± 0.8	1.9 ± 0.2 (4.1)
EYPC-10 Ch(3)	87.2 ± 0.6	5.7 ± 0.3	7.1 ± 0.2 (14.3)
HOPC-0 Ch(2)	92.1	7.9	– (0)
HOPC-1 Ch(2)	91.3	7.9	0.75 (1.6)
HOPC-3 Ch(2)	91.4	6.75	1.85 (4.0)
HOPC-10 Ch(1)	86.0	6.8	7.2 (14.5)

a) Starting mixtures were composed of egg yolk (EY) or 1-O-hexadecyl-2-oleoyl (HO) phosphatidylcholine (PC), the indicated weight percent cholesterol (Ch) and 25.2 mg triolein (TO) to contain 30 mg total lipid (# of emulsions prepared).

b) Values are the mean ± SEM and represent weight percent (and mole percent).

Enzyme Assays. Enzyme activities were monitored simultaneously, with each individual emulsion. Assays were performed in
duplicate for each preparation in round-bottom glass tubes.
Blank tubes contained all ingredients except enzyme. Single
batches of apoC-II, albumin, and heparin aqueous solutions were
used in all the assays.

Lipoprotein Lipase Assays. The components were added at
room temperature in the order NaCl, pH 7.4 potassium phosphate
buffer (KPB), heparin, emulsion, apoC-II, FA-free bovine serum
albumin (BSA). The final concentrations in the reaction mixture were NaCl 150 mM, heparin 2.65 μg/mL, apoC-II 16.7 μg/mL,
emulsion TG 3.33 μmole/mL, BSA 15 mg/mL, 10 mM pH 7.4 KPB.
Tubes were warmed to 37°C for 5 min after apoC-II was added and
before adding BSA. The reaction was then initiated by the
addition of LPL, freshly diluted in 10 mM pH 7.4 KPB, to yield
an enzyme/substrate ratio of 100 ng protein/μmole TG. The LPL
and apoC-II concentrations used provided about 1 and 300 molecules per emulsion particle, respectively, giving an apoC-
II/LPL ratio of 300. Incubation were routinely performed·for
20 min and, in certain instances, for up to 40 min (see Fig. 1)
with continuous shaking at 1 vibration/sec. The released
labeled free oleic acid was extracted from the substrate by the
liquid/liquid partition system of Belfrage & Vaughan (1969).
Liquid scintillation spectrometry was performed in Liquiscint
(National Diagnostics Corporation, Manville, NJ) to quantitate
FA released during the reactions.

The phospholipase activities of LPL on ^{3}H-TO/UC/egg PC
(P[^{14}C]OPC) emulsions were also determined. The assay conditions were the same as those used for the hydrolysis of TO in
an assay volume of 300 μl. The reaction was stopped by the
addition of 6 ml chloroform/methanol (2:1 v/v) followed by 1.2
ml of acid saline. The ^{14}C-lyso PC was isolated by TLC in
silica gel G using a developing solvent system of chloroform/
diethyl ether/water/acetic acid (65:25:4:1 v/v).

Hepatic Lipase. The reaction mixtures contained NaCl 100
mM, BSA 15 mg/mL and emulsion TG 3.33 μmole/mL in 10 mM pH 7.4
KPB. The reaction was initiated by addition of HL freshly
diluted in 10 mM pH 7.4 KPB, in a ratio of 100 ng protein to 1
μmole TG. Lipolysis followed at 37°C, routinely for 20 min and

was stopped as described above for LPL. Phospholipase of HL on each emulsion type was measured as described above for LPL.

Other Analytical Procedures. The lipid surface concentrations of the isolated emulsions were calculated by the phase analysis program of Miller & Small (1987), as modified by the method of Spooner & Small (1987). The protein concentrations of LPL and HL were determined by the method of Bensadoun & Weinstein (1970), using BSA as the standard. Electron micrographs of emulsions using a negative stain were prepared as described by Derksen & Small (1989).

RESULTS

Compositions of Isolated Emulsions. Tables I and II show the lipid compositions of the isolated emulsions in weight %, the surface composition in mole% and the amounts of lipid substrate located in the surface in nmoles per enzyme assay. The average diameters to the emulsion particles calculated from the lipid compositions ranged between 122-142 nm (EYPC) and from 144 to 160 nm for HOPC emulsions. Negative stain electron microscopy of these emulsions showed polydisperse populations of only large spherical particles with similar size distributions and mean diameters of 136 $\pm$ 13 nm (results not shown).

TABLE II: The Composition of the Surface and the Amounts of Lipid Substrate Located in the Surface[a]

Emulsion	Surface Composition[b]			Surface Substrate[c]	
mole % Ch	TO	PC	Ch	TO	PC
0	2.2	97.8	–	2.84	129.3
1.5	1.6	88.9	9.5	2.04	113.6
4.1	1.0	74.2	24.8	1.33	95.9
14.3	0.1	50.0	50.0	0.1	73.6

a) Calculated from the EYPC emulsion data listed in Table I.

b) Values represent the average and are expressed in mole%.

c) Substrate values represent nmoles per incubation containing 100 ng of enzyme which represent 1.2 pmole LPL (MW of dimer 83.4 KD) (Olivecrona & Bengtsson, 1987) and 0.5 pmole HL (MW of tetramer 200 KD) (Twu et al., 1984).

All assays contained equal amounts of TO and enzymes. When the emulsion-surface content of cholesterol is increased from 0 to more than 50 mole% the amounts of surface PC substrate is nearly halved from about 130 to less than 74 nmoles. At the same time the effective amounts of surface TO substrate were reduced from 2.9 to less than 0.1 nmoles.

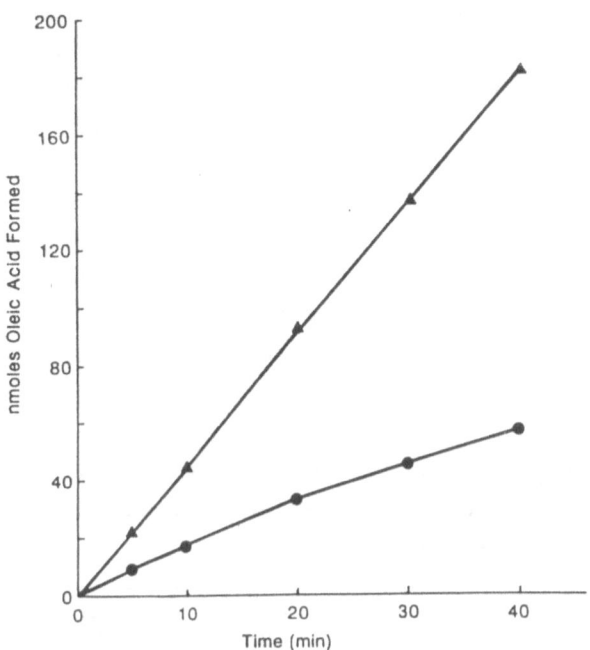

Fig. 1. Time course of the production of oleic acid from TO in an egg PC emulsion by LPL (Δ) and HL (·).

Triolein Hydrolysis by LPL and HL. The production of oleic acid was linear for at least 40 min for LPL and 20 min with HL (Fig.1). Similar time courses were observed with TO/HOPC emulsions. In addition to oleic acid, LPL and HL produced diolein and monoolein from TO. In all calculations of TO hydrolysis it was assumed that the enzymes produced 1 mole monoolein and 2 moles oleic acid from 1 mole TO.

Increasing Emulsion Cholesterol Affects the Hydrolysis of TO and EYPC by LPL. The hydrolysis of TO and EYPC in emulsions containing 0, 1.5, 4.1 and 14.3 mole% cholesterol by LPL and HL are listed in Tables III and IV respectively. In addition the hydrolysis of EYPC has been expressed as percent hydrolysis of the surface phospholipids and the hydrolysis of TO has been converted to the percent hydrolysis of core TO as well as the turnover of surface pools of TO. For TO hydrolysis the maximal activities of the two enzymes were similar, about 50 nmole TO/20 min/1.2 pmole lipase. In addition, these maximal enzyme activities occurred at different particle C contents. For HL the hydrolysis of TO remained maximal between 0 and 4 mole% C then the activity declined steadily with increasing C content and was >90% inhibited at 14 mole% C. For LPL maximal TO hydrolysis occurred at 4 mole% C while at 0 and 14 mole% C the enzyme activity was inhibited by at least 25 and 50% respectively.

For PC, hydrolysis remained maximal for both enzymes between 0-1.5 mole% C (LPL) and 0-4 mole% C (HL), then enzyme activities declined steadily with increasing C content and were 80% (LPL) and 90% (HL) inhibited at 14 mole% C. PC hydrolysis by HL was always about 3X faster than PC lipolysis by LPL when the particle C content was between 0 and 4 mole%. With LPL the percent hydrolysis of the core TO was always greater than the percent hydrolysis of the surface PC. In contrast, with HL the percent hydrolysis of surface PC was 2 to 3 times greater than the percent hydrolysis of core TO over the entire range of particle C contents.

TABLE III: The Effect of Emulsion Cholesterol on the Acylhydrolase Activity of LPL and the Percent Hydrolysis of the Emulsion Surface and Core.

Emulsion	Substrate Hydrolysis[a]		Percent Hydrolysis		Surface Turnover[b]
mole% CH	TO	PC	Surface PC	Core TO	TO Pools
0	36.8 (10.8)	4.3 (0.8)	3.4	3.7	13
1.5	41.3 (6.4)	3.7 (1.5)	3.2	4.1	20
4.1	49.5 (4.7)	3.4 (0.6)	3.5	5.0	37
14.3	21.8 (2.0)	0.9 (0.3)	1.2	2.2	>220

a) Values are the mean ± (SD) and represent nmoles.
b) Values represent number of surface pools of TO in 20 min per 1.2 pmole of enzyme.

Increasing the emulsion content of cholesterol from 0 to 14 mole% decreases the surface TO substrate pools from 2.9 nmoles to less than 0.1 nmole. At low cholesterol levels (0-4 mole%), the turnover of the TO pools in the surface increases several times for both enzymes to about 40 pools in 20 min. However, when the emulsion particles were saturated with cholesterol at

14 mole%, the turnover of the surface TO pools by LPL was increased another 5-fold to greater than 220 pools in 20 min. In
contrast, the turnover of the surface TO pools by HL leveled
off because the enzyme activity was 90% inhibited by particles
saturated with cholesterol.

TABLE IV: The Effect of Emulsion Cholesterol on the Acylhydrolase Activity of HL and the Percent Hydrolysis of the
Emulsion Surface and Core

Emulsion	Substrate Hydrolysis[a]		Percent Hydrolysis		Surface Turnover[b]
mole% CH	TO	PC	Surface PC	Core TO	TO Pools
0	49.2 (6.0)	12.9	10.0	4.9	17
1.5	49.2 (2.5)	10.7	9.4	4.9	24
4.1	47.5 (10.3)	12.5	13.0	4.8	36
14.3	4.8 (2.4)	0.8	1.1	0.5	>48

a) a and b as in Table III.

Effect of Ether-PC on the Hydrolysis of TO by LPL and HL.
The hydrolysis of TO in HOPC emulsions containing 0, 1.6, 4.0
and 14.5 mole% cholesterol by LPL and HL are listed in Table V.
At low cholesterol levels (0-4 mole%), TO hydrolysis by LPL
was 15 to 40% inhibited by HOPC as compared with EYPC (Table
III). In this range of cholesterol HL was strongly inhibited
(60-70%). At 14 mole% cholesterol both enzymes were equally
inactive towards TO with HOPC in the surface.

Table V: Hydrolysis of TO in HOPC/C Emulsions by LPL and HL.

Emulsion[a]	TO Hydrolysis[b]	
mole% C	LPL	HL
0	22	15
1.6	34	20
4.0	42	16
14.5	7	6

a) Emulsion composition as described in Table I.
b) Values are the mean of 2 experiments and represent nmoles
TO hydrolyzed in 20 min per 1.2 pmole LPL or HL.

DISCUSSION

Emulsions physically resemble lipoprotein substrates and can be chemically characterized (Table I). Lipid components on the surface and in the core can be calculated (Miller & Small, 1987; Table II). Lipoproteins are more complex in lipid composition than these emulsions and have various apoproteins added to their surfaces. The interaction of the lipids with themselves and also with proteins may influence lipolytic activity. Starting with a model system such as the emulsions described in this study, which can be controlled chemically, appears essential to gain a more detailed understanding of these interactions and their relation to the activities of lipolytic enzymes with their natural substrates.

The activities of the enzymes were assayed with an emulsion to enzyme ratio between 0.5 and 1 (for HL and LPL respectively), since it seems unlikely that an enzyme will work on more than one particle at a time. Most of the enzyme was assumed to be bound to the emulsion particle surface most of the time. Several observations support this assumption. First, LPL' has been shown to have high affinity in the nanomolar range for phospholipid surface (McLean & Jackson, 1985). Second, the phospholipid to enzyme ratios varied between 5×10^{-4} to 25×10^{-4} (Table II) which is about 15 to 75 times more phospholipid surface than is needed for the binding of apolipoproteins such as apoA1 and apoE3 (Derksen & Small, 1989). Finally, increasing the cholesterol in emulsions does not change the affinity for apoprotein binding to the surface of the particles (Derksen & Small, 1989).

Once the enzymes are on the surface of the emulsion particle, there always are saturating levels of phospholipid substrate available to ensure zero order reactions. For TO the surface concentration decreases 30-fold (Table II) when the particles become saturated with cholesterol, and the ratio of surface TO to LPL molecules for instance, falls to 80. This ratio is the lowest in this particular study but should be sufficient to saturate the enzyme.

Cholesterol affected the acylhydrolase activities of LPL towards TO and PC in different ways. The TO hydrolase showed optimal activity with emulsion particles containing 4 mole% C.

This corresponds to about 25 mole% cholesterol in the surface. This surface cholesterol content is like plasma VLDL and chylomicron remnants (Miller & Small, 1987). In contrast, the PC hydrolase activity of LPL was steadily inhibited by increasing particle cholesterol. Neither of the acylhydrolase activities of HL were affected by cholesterol in the physiological range from 0 to 4 mole% C but became strongly inhibited (>90%) when the emulsion particles were saturated with cholesterol. Since LPL hydrolyzed TO with nearly half maximal activity in particles saturated with cholesterol we concluded that LPL is less sensitive to high cholesterol than HL.

When the surface TO concentration decreased several fold with increasing particle content, the turnover of the surface TO pools increased several fold without much change in enzyme activity (Table III and IV). Thus the transfer of TO from the core to the surface was fast enough not to become rate-limiting during the 20 min incubation time.

The PC hydrolase activity of HL was 3 times greater than LPL. Because of this difference, the percent hydrolysis of the surface PC by HL was twice that of the core TO. After 20 min, with 10-13% of the surface PC deacylated, the hydrolysis of TO ceased to be linear (Fig. 1). In contrast, the percent hydrolysis of surface PC by LPL nearly kept pace with that of the core, and TO lipolysis was linear for over 1 hr. Thus, the shrinking of particle surface and core at similar rates are required to maintain optimal enzyme activity.

The relatively greater PC hydrolase activity of HL indicates that the catalytic site of this enzyme may have a higher affinity for phospholipid as compared to LPL. This higher affinity was also noticed with HOPC, which was able to inhibit the TO hydrolase of HL several fold more than that of LPL (Table V). Similarly, other non-substrate lipids like sphingomyelin may regulate the acylhydrolase activity of both these enzymes with natural substrates in vivo.

ACKNOWLEDGEMENTS

This work was supported by National Institutes of Health Research Grant HL-26335 and Training Grant HL-07291. We wish to thank Anne Plunkett for preparing the manuscript and Donald Gantz for carrying out electron microscopy of the emulsions.

REFERENCES

Belfrage P (1969) Simple liquid-liquid partition system for isolation of labeled oleic acid from mixtures with glycerides. J Lipid Res 234:466-468

Bennett Clark S, Derksen A (1987) Phosphatidylcholine composition of emulsions influences triacylglycerol lipolysis and clearance from plasma. Biochim Biophys Acta 920:37-46

Bennett Clark S, Derksen A, Small DM (1991) Plasma clearance of emulsified triolein in conscious rats: Effects of phosphatidylcholine species, cholesterol content and emulsion surface physical state. Experimental Physiol 76:39-52

Bennett Clark S, Laboda H (1991) Triolein-phosphatidylcholine-cholesterol emulsions as substrates for lipoprotein and hepatic lipases. Lipids 26:68-73

Bensadoun A, Weinstein D (1976) Assay of proteins in the presence of interfering materials. Anal Biochem 70:241-250

Derksen A, Small DM (1989) Interaction of apoA-1 and apoE-3 with triglyceride-phospholipid emulsion containing increasing cholesterol concentrations. Model of triglyceride-rich nascent and remnant lipoproteins. Biochemistry 28:900-905

Derksen A, Ekman S, Small DM (1989) Oleic acid allows more apoprotein A-1 to bind with higher affinity to large emulsion particles saturated with cholesterol. J Biol Chem 264:6935-6940

Derksen A, Bennett Clark S, Small DM (1989) Lipid composition of emulsions affects rapid protein binding: Flotation through plasma. 61st Scientific Sessions, American Heart Association Meetings, Washington. November 14-17, 1988. Arteriosclerosis 8:606a

Ekman S, Derksen A, Small DM (1988) The partitioning of fatty acid and cholesterol between core and surfaces of phosphatidylcholine-triolein emulsions at pH 7.4. Biochim Biophys Acta 959:343-348

Hamilton JA, Small DM (1981) Solubilization and localization of triolein in phosphatidylcholine bilayers: A ^{13}C NMR study. Proc Natl Acad Sci USA 78:6878-6882

Holmqvist L, Carlson K (1977) Selective extraction of human serum very low density apolipoproteins with organic solvents. Biochim Biophys Acta 493:400-409

Jackson RL (1983) Lipoprotein lipase and hepatic lipase. In: Boyer PD (ed) The Enzymes, vol XVI. Academic Press, New York, p 141-181

Jensen GL, Bensadoun A (1981) Purification, stabilization, and characterization of rat hepatic triglyceride lipase. Anal Biochem 113:246-252

Laboda HM, Bennett Clark S, Derksen A, Small DM (1988) Hydrolysis of triolein emulsions by hepatic lipase and lipoprotein lipase: Effect of cholesterol. 61st Scientific Sessions, American Heart Association Meetings, Washington. November 14-17, 1988. Arteriosclerosis 8:600a

Maranhao RC, Tercyak AM, Redgrave TG (1986) Effects of cholesterol content on the metabolism of protein-free emulsion models of lipoproteins. Biochim Biophys Acta 875:247-255

McLean LR, Jackson RL (1985) Interaction of lipoprotein lipase and apolipoprotein C-II with sonicated vesicles of 1,2-ditetradecylphosphatidylcholine: Comparison of binding constants. Biochemistry 24:4196-4201

Miller KW, Small DM (1987) Structure of triglyceride-rich lipo-
    proteins: An analysis of core and surface phases. In: Gotto
    AM (ed) New Comprehensive Biochemistry, vol 14, Plasma lipo-
    proteins. Elsevier Science, B.V., p 1
Olivecrona T, Bengtsson-Olivecrona G (1988) In: Borensztajn J
    (ed) Lipoprotein lipase. Evener Publishers, Inc., Chicago, p
    17-58
Redgrave TG, Maranhao RC (1985) Metabolism of protein-free
    lipid emulsion models of chylomicrons in rats. Biochim
    Biophys Acta 835:104-112
Socorro L, Green CC, Jackson RL (1985) Preparation of homogen-
    ous and stable form of bovine milk lipoprotein lipase. Prep
    Biochem 15:133-143
Spooner PJR, Small DM (1987) Effect of free cholesterol on
    incorporation of triolein in phospholipid bilayers. Biochem-
    istry 26:5820-5825
Spooner PJR, Bennett Clark S, Gantz DL, Hamilton JA, Small DM
    (1988) The ionization and distribution behavior of oleic
    acid in chylomicrons and chylomicron-like emulsion particles
    and the influence of serum albumin. J Biol Chem 263:1444-
    1453
Twu J-S, Garfinkel AS, Schotz MC (1984) Hepatic lipase. Purifi-
    cation and characterization. Biochim Biophys Acta 792:330-
    337

STUDIES ON THE FUNCTIONS OF CELL SURFACE GLYCOPROTEINS DURING EARLY DEVELOPMENT IN THE SEA URCHIN EMBRYO

K. Foltz, S.-P. Hwang, B. Kabakoff, R. Stears and W. J. Lennarz
Department of Biochemistry and Cell Biology
State University of New York at Stony Brook
Stony Brook, New York 11794-5215
USA

INTRODUCTION

At present there is a rather clear picture of the basic steps involved in assembly of cell surface glycoproteins containing N- and/or O-linked oligosaccharide chains. The earliest studies on the glycosylation of proteins utilized autoradiography and focused on the subcellular sites of protein synthesis and glycosylation. From these studies, it appeared that the assembly of the oligosaccharide chains of N-linked glycoproteins occurred in two phases: the initial phase was believed to take place in the endoplasmic reticulum, whereas the second occurred in the Golgi complex. These observations were followed by a series of important findings: the distinction between membrane-bound and free polysomes, the discovery of the signal peptide, the demonstration of the cotranslational glycosylation of ovalbumin, and the findings that, in accord with the signal hypothesis, newly translated and glycosylated proteins were sequestered within the lumen of the endoplasmic reticulum and that viral coat proteins were inserted with their carbohydrate chains facing the lumen. These findings were solidified by a number of in-depth biochemical studies on the mechanisms of assembly of the oligosaccharide chains of the glycoproteins. As a result of these studies one can draw the following general conclusions:

(1) All eukaryotic cells are capable of the synthesis of glycoproteins that are destined to become components of the plasma membrane. In addition, many cell types commit a significant portion of their protein biosynthetic activity to the synthesis of secreted and/or lysosome-packaged glycoproteins.

(2) The synthesis of membrane, secretory, or lysosomal glycoproteins is a highly segregated process that occurs with an intracellular membrane system composed of the endoplasmic reticulum, transfer vesicles, Golgi apparatus, and secretory vesicles. During their translation, glycosylation, and processing the glycoproteins are completely isolated from the cytoplasm and travel to the cell surface (in the case of membrane glycoproteins), to the extracellular environment (in the case of secretory glycoproteins), or the lysosomes (in the case of lysosomal enzymes) as part of, or within, these membrane compartments.

(3) The assembly of N-linked oligosaccharide chains occurs in the endoplasmic reticulum and involves the stepwise preassembly of the oligosaccharide chain on dolichyl phosphate followed by en bloc transfer of the oligosaccharyl unit to the growing polypeptide chain.

NATO ASI Series, Vol. H 63
Dynamics of Membrane Assembly
Edited by J. A. F. Op den Kamp
© Springer-Verlag Berlin Heidelberg 1992

(4) Subsequent modifications of the oligosaccharide chains on N-linked glycoproteins are initiated in the rough endoplasmic reticulum and completed in the Golgi apparatus.

(5) In contrast, the complete assembly of O-linked chains, starting with attachment of GalNAc to Ser or Thr residues in the polypeptide backbone occurs post-translationally in the Golgi complex.

(6) This process does not involve dolichyl phosphate or preassembly of the oligosaccharide chains. After addition of the GalNAc residue, single sugar residues are added stepwise to complete the O-linked oligosaccharide chain.

(7) Upon their completion in the Golgi complex the N- and/or O-linked glycoproteins destined for the cell surface are transported via vesicles that fuse with the cell surface, thereby bring these glycoproteins to their final locale in the plasma membrane.

Although a number of important questions remain about the mechanism of assembly of these glycoproteins, and the factors controlling their routing to the cell surface, an even greater challenge is to gain an understanding of their function(s). Indeed, it can be calculated that the energy requirements (in terms of ATP molecules) for assembly and addition of a single complex type oligosaccharide chain to a protein of 40,000 MW adds over 30% to the overall "cost" to build such a macromolecule. Given this fact, and the wide spread occurrence of oligosaccharide chains on proteins of diverse function, ranging from antibodies to zymogens, it seems likely that these chains serve important function(s).

One of the functions that has been demonstrated in a variety (but by no means all) glycoproteins is stabilization of the macromolecule to proteolysis. A second function that has been widely postulated is to serve as a specific recognition element. A few cases where the oligosaccharide chain has been demonstrated to serve as such a "tag" have been described: two examples are the recognition of galactose-terminal glycoproteins by the asialo glycoprotein receptor and the recognition of glycoproteins containing mannose-6-phosphate units by the so-called mannose-6-phosphate receptor. In general, however, little is known in detail about the function of glycoproteins in cellular recognition events. In view of this, a number of years ago we directed some of our efforts to studying this question. In this chapter, examples of two glycoproteins that function in important, but very dissimilar events in early embryonic development in the sea urchin embryo will be described.

AN EGG CELL SURFACE GLYCOCONJUGATE IS THE SPECIES SPECIFIC RECEPTOR FOR SPERM

It has been known for many decades that fertilization in sea urchin gametes is species specific. However, until relatively recently nothing was known about the molecular basis for this specific cellular interaction. In our initial studies, we established that the egg surface contained a molecule whose release was protease-sensitive and resulted in loss of ability of sperm to fertilize these eggs

(Schmell et al., 1977). Subsequently, biochemical studies revealed that the receptor had a very high apparent molecular weight and the composition of a proteoglycan. In addition, it was established that proteolytically-released fragments of the receptor inhibited fertilization in a species specific manner (Rossignol et al., 1984; Ruiz-Bravo & Lennarz, 1986; Ruiz-Bravo et al. 1986). Based on these observations we postulated that the receptor was a species specific, a complex proteoglycan-like molecule of high molecular weight that was associated with the cell surface. Evidence was also obtained supporting the hypothesis that the carbohydrate chains of the receptor serve as the adhesive element of the molecule, whereas the polypeptide chain defines the species specificity of the binding process (Ruiz-Bravo & Lennarz, 1986). To test this hypothesis we generated glycoprotein fragments of the receptor by limited proteolytic digestion of the egg cell surface (Ruiz-Bravo & Lennarz, 1986). Studies with various receptor preparations revealed that the presence of 30% of the polypeptide chain by weight is required to inhibit fertilization species specifically. The results of experiments with receptor fragments containing less polypeptide support the hypothesis that the species specificity of inhibition of fertilization observed is conferred by the polypeptide portion of the receptor molecule.

Given the loss of species specificity of extensively digested, soluble fragments of the receptor, we undertook to isolate homogenous fragments of the receptor that might retain species specificity. As shown in Fig. 1, although trypsin treatment of eggs yielded fragments that inhibited fertilization species specifically, these fragments were polydisperse in size and charge.

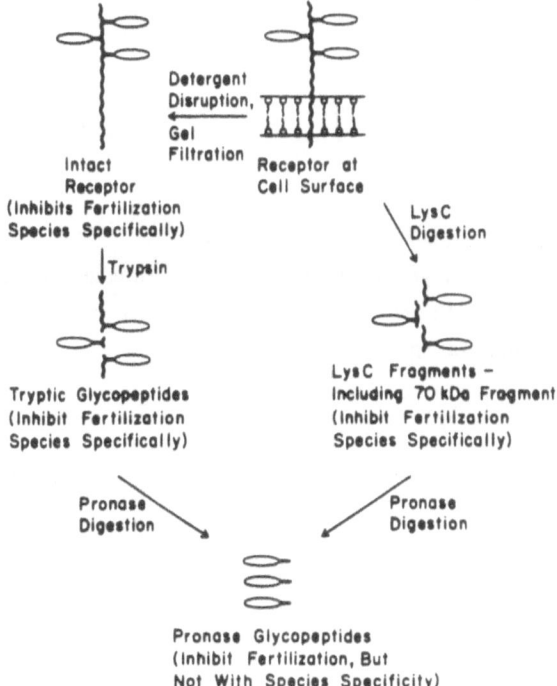

Figure 1. Summary of observations on the sperm receptor and fragments prepared from it.

Subsequently, we undertook studies with proteolytic enzymes of more restrictive specificity to release the receptor fragment (Foltz & Lennarz, 1990). It was found that treatment of S. purpuratus eggs with lysylendoprotease C abolished the ability of eggs to bind acrosome-reacted sperm and resulted in the release of proteolytic fragments that bound to sperm and showed inhibitory activity in a fertilization bioassay (see Fig. 1, right side). One of these fragments, presumed to be a fragment of the extracellular domain of the receptor, was purified to homogeneity and shown to be a 70kDa glycoprotein containing sulfated oligosaccharide chains. As shown in Fig. 2, addition of this glycoprotein fragment to a fertilization bioassay inhibited fertilization in a species specific manner.

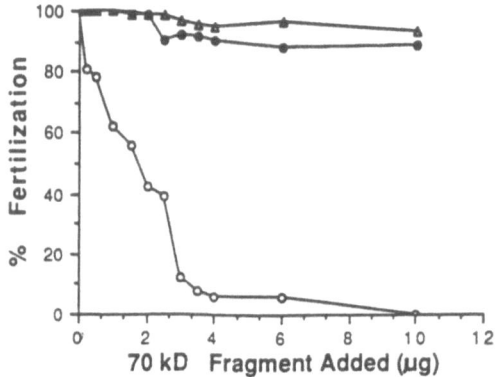

Figure 2. Species specificity of the purified 70kDa glycoproetins. Sperm were added to dejellied eggs of S. purpuratus (O–O), L. pictus (△–△), and S. drobachiensis (●–●) in the presence of increasing amounts of the purified 70kDa fragment isolated from S. purpuratus.

Further, when the 70kDa fragment was labeled with ^{125}I it could be shown that it bound to sperm, but only if the sperm had been induced to undergo the acrosome reaction (Fig. 3). As also shown in Fig. 3, this binding was species specific.

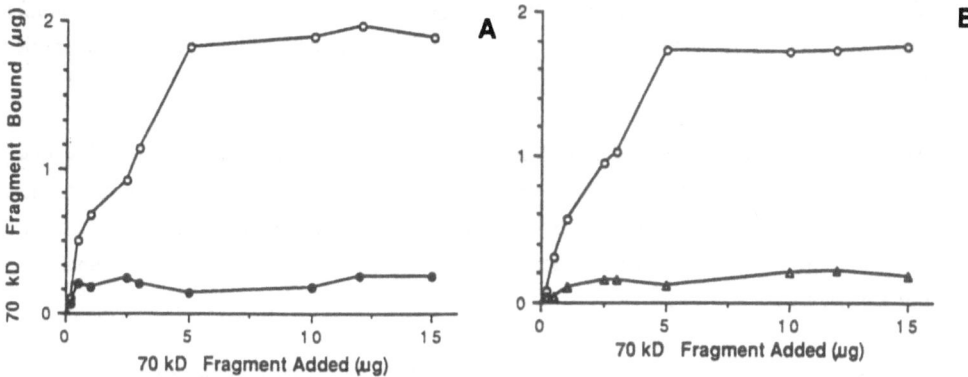

Figure 3. The 70kDa glycoprotein binds species specifically to acrosome-reacted sperm. Increasing amounts of purified, radiolabeled 70kDa fragment were incubated with nonacrosome-reacted (●—●) or acrosome-reacted (O—O) S. purpuratus sperm (A). In (B) acrosome-reacted S. purpuratus (O—O) or L. pictus sperm (△—△) were incubated with increasing amounts of labeled fragment.

Given the earlier findings mentioned above on the requirement of the polypeptide chain for retention of species specific inhibition, it was important to study the effect of degradation of the polypeptide chain of the 70kDa glycoprotein on its ability to inhibit fertilization. The results shown in Fig. 4 indicate that, as earlier found with the trypsin fragment, extensive degradation of the 70kDa glycoprotein fragment with Pronase yields a glycopeptide preparation that still inhibits fertilization, but other sea urchin species as well as in S. purpuratus.

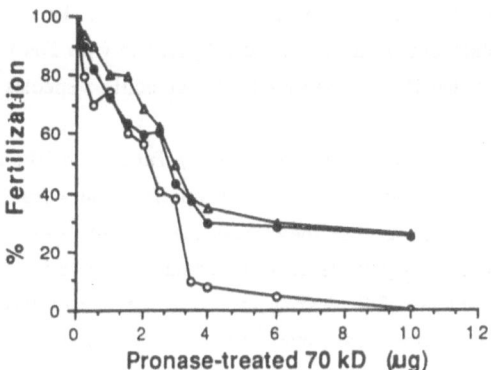

Figure 4. Pronase treatment of the 70kDa glycoproteins yield a glycopeptide that inhibits without species specific (same symbols as in Fig. 2.)

This finding provides further support that the oligosaccharide chains are important in binding to sperm, but their attachment to a polypeptide backbone in some way defines their species specificity.

To address structure-function questions pertaining to the molecular basis of gamete interaction, we have continued our studies of the egg receptor for sperm by: 1) preparing and characterizing an antiserum specific for the 70kDa extracellular fragment; 2) characterizing the intact receptor using the antiserum; 3) cloning and sequencing the receptor; and 4) analyzing receptor function.

With respect to 1) and 2), an anti-70kDa IgG has been prepared and shown to bind to the 70kDa fragment in lysC digests and to a 350-400kDa component, presumably the intact receptor, in cell surface preparations. The IgG and Fab fragments inhibit sperm binding. In addition, by indirect immunofluoresence studies using the anti-70kDa IgG and fluorescently tagged second antibody, it was established that binding occurs on the cell surface of egg prior to, but not after lysC digestion. This antibody binding is, as expected, species specific. Currently, immuno-cytochemistry studies are underway using antibody gold techniques to more precisely localize the receptor on the cell surface.

With respect to 3), the anti-70kDa IgG has been used to screen an oocyte/ovary λZAP expression library and immunoreactive clones have been isolated. Antibody purified from the expressed fusion proteins recognizes the 70kDa lysC fragment and binds to the egg surface. DNA sequencing of one such clone has provided the first sequence information on the receptor. The deduced amino acid sequence appears to be unique, not having been reported in the GeneBank. Thus far, except for the presence of numerous potential sites for O-linked oligosaccharides, and two potential N-linked sites, few noteworthy features are evident. This cDNA, encoding at least a portion of the extracellular domain, does not cross-hybridize with DNA from other sea urchin species, supporting the idea that the polypeptide backbone confers species specificity.

Finally, it is of particular interest with respect to 4), function of the receptor, that both the anti-70kDa IgG and its Fab fragment have a very novel effect on a fraction of the eggs - they induce at least partial activation because they cause the eggs to raise a fertilization envelope. Upon sperm binding, sea urchin eggs undergo an immediate increase in membrane potential as a fast block to polyspermy. Release of Ca^{2+} from internal stores causes cortical granule exocytosis and the elevation of the fertilization envelope. Thus, the fact that the anti-70kDa Fabs cause partial FE elevation link the receptor to the release of Ca^{2+}. Although various controls have confirmed the validity of these observations, the fact that only a subset of eggs (10-15%) respond to antibody treatment in this way makes it difficult to study this phenomenon. It does, however, raise the interesting possiblity that occupancy of the receptor, in this case by antibody rather than the sperm ligand, bindin, induces a signal that leads to activation.

Data from Jaffe and co-workers (reviewed in Turner & Jaffe, 1989) suggest that signal transduction via G proteins is involved in egg activation. Two working models for the receptor and its relationship to its 70kDa fragment are shown in Fig. 5. Hopefully, as sequencing and other structural studies provide more information, we will be able to determine which of these models is valid and how the receptor transduces a signal.

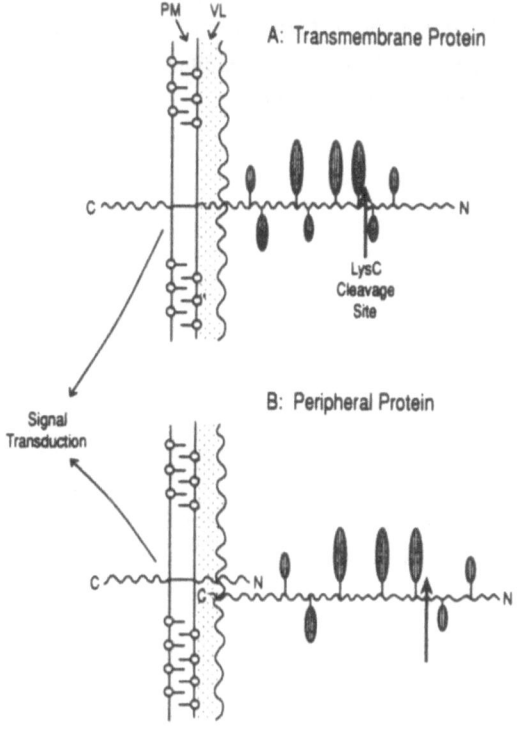

Figure 5. Alternative working models for the structure of the sperm receptor.

A GPI-ANCHORED CELL SURFACE GLYCOPROTEIN IS INVOLVED IN SKELETON FORMATION IN THE SEA URCHIN EMBRYO

After fertilization the sea urchin embryo undergoes a large number of cell divisions, resulting in formation first of a solid ball of cells (morula stage) and later a hollow ball made up of a single layer of cells (bastula stage). Upon hatching a subset of cells known as primary mesenchyme cells invade the hollow cavity (blastocoel) and invagination near the original site of these cells occurs. This process of gastrulation eventually leads to formation of a functional gut. However, before this process is complete, at the mid gastrula stage, the primary mesenchyme cells fuse with each other and begin assemblying a $CaCO_3$ - containing skeleton within the blastocoel.

Early studies from our laboratory on the effect of tunicamycin on embryonic development revealed that this drug blocked spicuogenesis when added at the late gastrula stage. More recently, we have studied this apparent requirement for N-linked glycoprotein synthesis during spicule formation in more detail (Kabakoff & Lennarz, 1990). The initial reason for this more focused effort was that after we utilized an in vitro cell culture system in which isolated primary mesenchyme cells differentiated into spicule-forming cells we found that addition of a serendipitously-isolated monoclonal antibody (MAb 1223) to such cultures blocked spiculogenesis assessed either microscopically or by $^{45}Ca^{2+}$ uptake (Carson et al., 1985). The inhibitory effect of this monclonal antibody (or an Fab derived from it) on $^{45}Ca^{2+}$ uptake into the spicule is shown in Fig. 6A. Also shown (Fig. 6B) is the fact that the inhibitory effect of MAb 1223 and acetazolamide (an inhibitor of carbonic anhydrase) are additive.

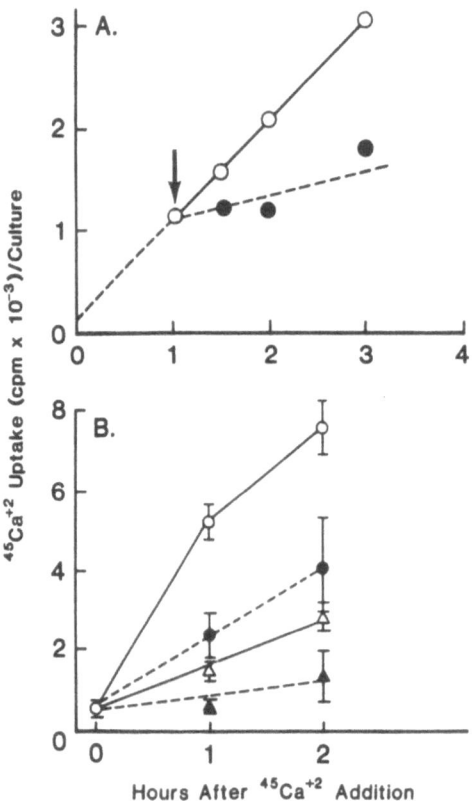

Figure 6A and 6B. Effects of 1223 antibody on calcium accumulation by specule-forming cell cultures. [^{45}Ca]Cl$_2$ was added at time zero, and accumulation was assessed at the indicated times by measureing the total cell-associated radioactivity. (A) Cultures were incubated in the absence of antibody for 2 days. On day 2 of culture, unattached cells were removed, and calcium accumulation into attached cells was assessed. [^{45}Ca]Cl$_2$ was added at time 0, and 1223 antibody was added to some cultures (●) after 1 hr (arrow). (B) Day 2 cultures consisting only of attached cells were pretreated for 10 min with inhibitor, with antibody, or with both. At time 0, [^{45}Ca]Cl$_2$ was added, and accumulation was measured at the time intervals indicated. (O) control (preimmune); (●) preimmune IgG and acetazolamide; (Δ) MAb 1223 and DMSO; (▲) 1223 and acetazolamide.

Immunofluorescence studies showed that the 1223 antigen was localized to primary mesenchyme cell. Immunoblot analysis revealed that one of the proteins containing the 1223 epitope was a 130kDa polypeptide whose level of expression correlated with the acquisition of the ability of primary mesenchyme cells to accumulate Ca^{2+} during spicule formation (Farach et al., 1987; Decker et al., 1988). These observations, coupled with subsequent studies establishing that the 130kDa protein was a glycoprotein, and that the MAb 1223 was directed toward an oligosaccharide chain on the glycoprotein (Farach-Carson et al., 1989), led us to study both the distribution of this epitope in the embryo and the nature of the oligosaccharide moiety in more detail. These studies have established that the epitope on the 130kDa glycoprotein recognized by mAb 1223 is a complex, N-linked oligosaccharide chain that appears in association with the cell surface and the Golgi complex of primary mesenchymal cells just prior to spiculogenesis (Decker et al., 1988; Decker et al., 1987). While this work was in progress, Raff and co-workers independently characterized the 130kDa protein by cloning and sequencing it (Parr et al., 1990). Their studies established the presence of six potential N-glycosylation sites, at least two potential O-glycosylation sites and two glycine rich domains. Further, on the basis of sequence analysis and studies with PI-specific phospholipase C they concluded that the 130kDa 1223 antigen contained a GPI anchor. A diagram of the primary sequence of the 130kDa protein is shown in Fig. 7.

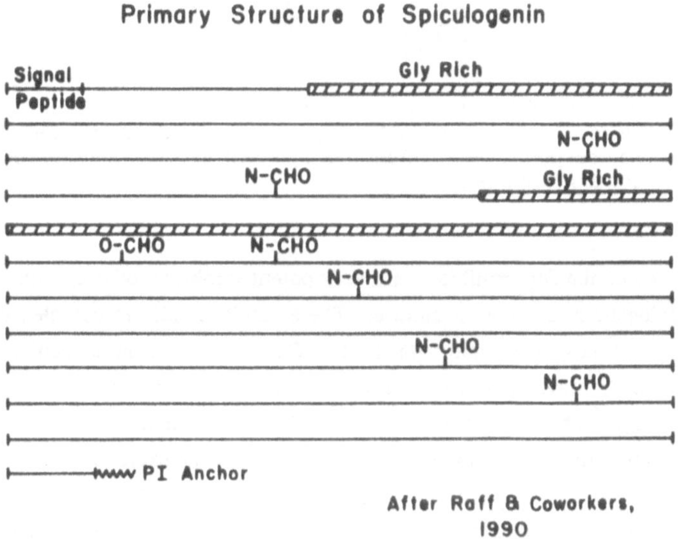

Figure 7. Deduced primary structure of the 130kDa glycoprotein (Parr et al, 1990).

We have confirmed and extended this observation by carrying out _in_ _vitro_ metabolic labeling studies in isolated primary mesenchyme cells with components of GPI anchors (fatty acids and ethanolamine) and by treating the protein with phospholipases. The results of these studies reveal that indeed the 130kDa 1223 antigen does contain a GPI anchor and that primary mesenchyme cells contain two other proteins that also contain both GPI anchors and N-linked oligosaccharide chains that react with MAb 1223.

Given the apparently essential role of the 130kDa glycoprotein in a step in spiculogenesis, it seemed likely that inhibitors of the processing of oligosaccharide chains would block spicule formation. Recently we have carried out both biochemical and morphological studies with such inhibitors using both intact embryos and primary mesenchyme cells in culture. The results shown in Fig. 8 with four different inhibitors of glycoproteins processing clearly reveal that all of these inhibit expression of the expression of mature 1223 epitope.

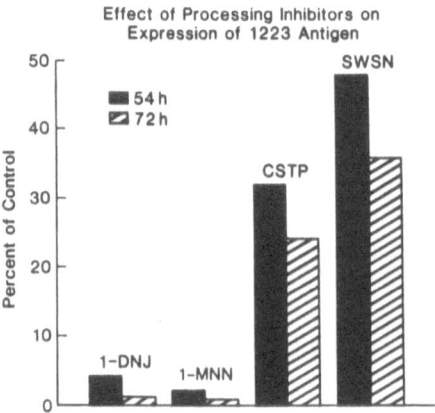

Figure 8. Quantitation of the effect of processing inhibitors on the level of 1223 epitope expression. The levels of the 1223 epitope associated with the 130kDa protein at two time points were expressed as a percentage relative to control levels for the corresponding time point.

As shown, two of the four inhibitors are very potent inhibitors of processing and these two also inhibit spiculogenesis by cells in culture. These results implicate complex oligosaccharide chains in general, and those associated with the 130kDa glycoprotein in particular, in the process of spiculogenesis in the developing sea urchin embryo. Currently we are trying to define the structure of the oligosaccharide chains of the 130kDa glycoprotein in more detail in order to understand how they control Ca^{2+} uptake and deposition into the skeleton.

Because little is known about the Ca^{2+} uptake and deposition processes that occur in these spicule-forming cells, we have initiated studies utilizing Ca^{2+} channel blockers and drugs that affect the secretory pathway. The rational for this approach is based on the following observations:

(1) Ca^{2+} entry from the extracellular medium must be tightly controlled since the concentration of Ca^{2+} in the culture media is very high (10 mM).

(2) Reflective laser confocal microscopy has revealed the presence of deposits with the reflective properties of $CaCO_3$ in vesicles in the cytosol of these cells.

(3) The $CaCO_3$ matrix of the spicule is known to contain several spicule matrix proteins (Sucov et al.; Katah-Fukui et al., 1991), some of which may be identical to glycoconjugates detected by electron microscopy (Decker et al., 1987; Decker & Lennarz, 1988).

185

(4) The 1223 antigen is found in the Golgi apparatus as well as on the cell surface.

(5) The available evidence indicates that although it is enwreathed in membrane, the spicule *per se* is extracellular (Decker et al., 1987; Decker & Lennarz, 1988).

Recently, we have found that nifedipine, a Ca^{2+} channel blockes, blocked spicule formation, but only after long-term treatment with the drug. Studies with brefeldin A, which blocks traffic from the ER to the Golgi apparatus, and with monensin, which disrupts the Golgi apparatus, show that these drugs have an immediate, marked inhibitory effect on $^{45}Ca^{2+}$ incorporation into spicules. The results of these inhibitor studies, in the context of the above observations, strongly support the working hypothesis that the Ca^{2+} that ultimately is secreted as $CaCO_3$ originates in the extracellular medium and subsequently enters the cell and moves from the cytoplasm to the ER, where it is concentrated. Subsequently it is routed through the Golgi complex and finally deposited extracellularly as pseudocrystalline $CaCO_3$. Presumably, at some point subsequent to entry in the ER, the Ca^{2+} is converted to $CaCO_3$, perhaps while it interacts with proteins destined for the spicule matrix and at this time it assumes its reflective, pseudocrystalline characteristics. The diagram in Fig. 9 summarizes these findings.

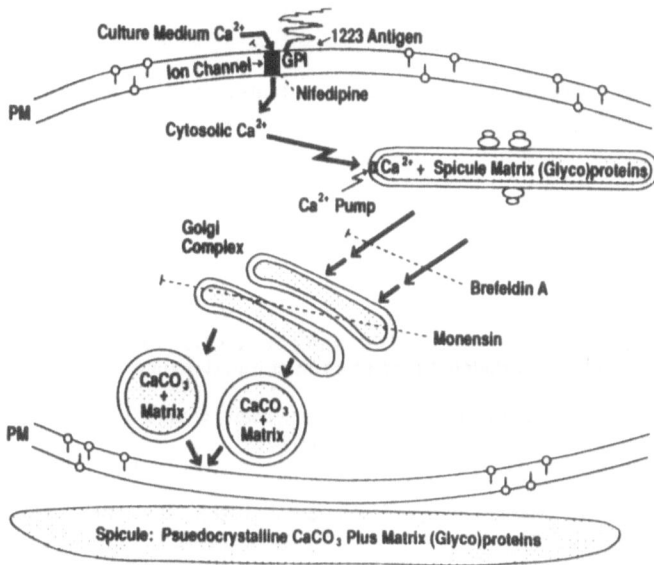

Figure 9. Summary of observation on the effects of inhibitors on spiculogenesis in primary mesenchymal cell cultures.

In the context of the above observations, it is postulated that nifedipine only has an effect after long-term treatment because the Ca^{2+} stores in the ER are large. In contrast the other drugs have an immediate effect because they disrupt the membrane traffic pathway leading to the conversion of Ca^{2+} to $CaCO_3$ and its subsequent secretion. Further studies will focus on integrating our emerging knowledge of this pathway for routing of $CaCO_3$ into the spicule and the precise role the 1223 antigen in assembly of this novel, paracrystalline structure.

CONCLUSION

The above studies represent two examples in which oligosaccharide chains of glycoproteins have been implicated to play a functional role in a developmental process. In the case of the sperm receptor, it is clear that binding of the oligosaccharide chains of the egg receptor to bindin on the sperm occurs in the absence of the polypeptide backbone. However, the structure(s) of the oligosaccharide chain and the precise mechanism of the binding remain to be established. With respect to spiculogenesis, several lines of indirect evidence, such as the use of inhibitors of oligosaccharide chain assembly and inhibitors of oligosaccharide chain processing, coupled with the fact that a monoclonal antibody (or its Fab) to the oligosaccharide chain blocks calcification, leads one to conclude that this moiety, attached to proteins on the surface of primary mesenchyme cells, plays a function role in this process. As in the case of the sperm receptor, the exact nature of this role, and the structure of the oligosaccharide chains, are important unanswered questions.

ACKNOWLEDGEMENT

The work in WJL's laboratory was supported by grants from the National Institutes of Health (HD21483 and HD18590).

REFERENCES

Carson DD, Farach MC, Earles DS, Decker GL and Lennarz WJ (1985) A monoclonal antibody inhibits calcium entry and skeleton formation in cultured embryonic cells of the sea urchin. Cell 41:639-648

Decker GL, Morrill JB and Lennarz WJ (1987) Mesenchyme cells and spicules during biomineralization in vitro. Development 101:297-312

Decker GL and Lennarz WJ (1988) Skeletogenesis in the sea urchin embryo. Development 103:231-247

Decker GL, Valdizan MC, Wessel GM and Lennarz WJ (1988) Developmental distribution of a cell surface glycoprotein in the sea urchin Strongylocentrotus purpuratus. Develop. Biol. 129:339-349

Farach MC, Valdizan M, Park HR, Decker GL and Lennarz WJ (1987) Developmental expression of a cell surface protein involved in calcium uptake and skeleton formation in sea urchin embryos. Develop. Biol. 122:320-331

Farach-Carson MC, Carson DD, Collier JL, Lennarz WJ (1989) Transcription of the spec 1-like gene of Lytechinus is selectively inhibited in response to disruption of the extracellular matrix. Development 106:355-365

Foltz KR and Lennarz WJ (1990) Purification and characterization of an extracellular fragment of the sea urchin egg receptor for sperm. J. Cell Biol. 111:2951-2960

Kabakoff B and Lennarz WJ (1990) Inhibition of glycoprotein processing blocks assembly of spicules during development of the sea urchin embryo. J. Cell Biol. 111:291-400

Katoh-Fukui Y et al. (1991) The corrected structure of the SM50 spicule matrix protein of Strongylocentrotus purpuratus. Develop. Biol. 145:201-202

Parr BA, Parks AL and Raff RA (1990) Promoter structure and protein sequence of msp130, a lipid-anchored sea urchin glycoprotein. J. Biol. Chem. 265:1408-1413

Rossignol DP, Earles BJ, Decker GL and Lennarz WJ (1984) Characterization of the sperm receptor on the surface of eggs of Stronglylocentrotus purpurtus. Develop. Biol. 104:308-321

Ruiz-Bravo N, Earles D and Lennarz WJ (1986) Identification and partial characterization of sperm receptor associated with the newly formed fertilization envelope from sea urchin eggs. Develop. Biol. 117:204-208

Ruiz-Bravo N and Lennarz WJ (1986) Isolation and characterization of proteolytic-fragments of the sea urchin sperm receptor that retain species specificity. Develop. Biol. 118:202-208

Schmell E, Earles BJ, Breaux C and Lennarz WJ (1977) Identification of a sperm receptor on the surface of the eggs of the sea urchin Arbacia punctulata. J. Cell. Biol. 72:35-46

Sucov HM, Benson S, Robinson JJ, Britten RJ, Wilt F and Davidson EH (1987) A lineage-specific gene encoding a major matrix protein of the sea urchin embryo spicule. Develop. Biol. 120:507-519

Turner R and Jaffe L (1989) G-proteins and the regulation of oocytes maturation and fertilization. In The Cell Biology of Fertilization (Schatten H and Schatten G, eds.) pp 299-317, Academic Press, New York

CHARACTERIZATION OF POLYPRENYLATION OF *DROSOPHILA* NUCLEAR LAMINS*

Janine M. Susan, Bruce Kabakoff,
Paul A. Fisher and William J. Lennarz
Dept. of Biochemistry and Cell Biology
SUNY at Stony Brook
450 Life Sciences
Stony Brook, NY 11794

INTRODUCTION

In the late 1970's studies on the structures of peptides involved in fungi mating responses resulted in discovery of a novel post-translational modification; the covalent attachment of isoprene groups (Kamiya, et al., 1978).   More recently, similar lipid modifications have been identified by the incorporation of an isoprene precursor, mevalonic acid into numerous proteins in a variety of higher eukaryotes (Schmidt, et al., 1984; Bruenger and Rilling, 1986; Maltese and Sheridan, 1987; Beck, et al., 1988).   Most isoprenylated proteins contain a carboxy-terminal CaaX sequence con-sisting of a cysteine residue followed by two aliphatic amino acids and any amino acid.   The isoprenyl groups are bound to the cysteine via a thioether linkage (Rilling, et al., 1989).   In most proteins studied the three terminal residues are subsequently removed by proteolysis and the modified cysteine is carboxymethylated (Anderegg, et al., 1988).   Chem-ical characterization shows that different isoprenoid species may be attached to different proteins; farnesol (C15) and geranylgeraniol (C20) have been identified in covalent linkages to proteins (Maltese, 1990).   The significance of these differences is not yet known but recent studies indicate that the precise carboxy-terminal sequence may define the type of polyprenyl group attached to cysteine.   Proteins ending in methionine or serine are preferentially modified with a farnesyl moiety, whereas proteins ending in leucine are preferentially modified with geranyl-geraniol (Seabra, et al., 1991).   Proteins ending in GGCC or CXXX have also

*Research in the laboratory of William J. Lennarz supported by NIH Grant 33184.

been shown to incorporate mevalonic acid (Kinsella and Maltese, 1991; Kinsella, et al., 1991) but the structure of the polyprenyl group is not yet known.

Several classes of proteins involved in the regulation of the cell cycle and signal transduction have been identified as incorporating [3H]-mevalonate, including ras proteins, ras-related proteins, heterotrimeric GTP-binding proteins, and nuclear lamins (Maltese, 1990). Here we examine the incorporation of [3H]-mevalonate into polypeptides of the *Drosophila* cell line Schneider 2 (S2). Insects are auxotrophic for cholesterol since they lack a key enzyme in this branch of the isoprenoid pathway that synthesizes squalene, the precursor of cholesterol, from farnesyl pyrophosphate. The absence of this major product greatly facilitates the detection of incorporation of radiolabel into the other minor products of this pathway and makes S2 cells an ideal system for studying the synthesis of isoprenylated proteins.

This report contains a preliminary chromatographic characterization of the isoprenoid species of the interphase forms of the *Drosophila* nuclear lamins. The nuclear lamins are intermediate filament proteins which form a fibrous network lining the inner face of the nuclear mem-brane. They are thought to function in stabilizing this membrane and in anchoring of nuclear pore complexes and interphase chromosomes (Burke, 1990). Isoprenylation has been shown to be important in the association of the lamins with the nuclear membrane (Krohne, et al., 1989; Kitten and Nigg, 1991). It is yet not known if isoprenylation plays a role in targeting lamins to the nucleus, in assembling of the network, or in the disassembly and reassembly of the nucleus at different stages of mitosis. It is also not known whether both farnesol and geranylgeraniol would act similarly or if the natural occuring species is critical for proper functioning of the nuclear lamins.

MATERIALS AND METHODS

Metabolic labeling of S2 cells with [3H]-mevalonate.
Schneider Line 2 (S2) cells were grown at 25° C in Schneider's media (Gibco) containing 8% FCS (Gibco) and Penicillin-Streptomycin (Gibco). During log growth phase the cells were pelleted by centrifugation and resuspended in fresh media at a concentration of 1 x 10^7 cells/ml. These cells were then incubated 16 hrs. in the presence of 8 μCi/ml [3H]-mev-alonate (NEN). After incubation the cells were pelleted by centrifugation,

washed three times with phosphate-buffered-saline, pH 7.4 (PBS) and sonicated in 500 µl 10 mM Tris, 10 mM CaCl$_2$, 70 mM NaCl, pH 8.0 (HET buffer) containing protease inhibitors. Protein concentration was determined by the method of Lowry (Lowry, et al., 1951) and 99 µg of total cellular protein was electrophoresed utilizing a 10% SDS gel under reducing conditions. The gel was stained with Coomassie Blue, treated with En3Hance (NEN), and exposed to x-ray film.

Immunoprecipitation of D$_{m1}$ and D$_{m2}$ with antibodies against *Drosophila* nuclear lamins. S2 cells were labeled with [3H]-mevalonate as described above. After incubation the cells were pelleted by centrifugation, washed three times with PBS, and solubilized by boiling in 500 µl 10% SDS/20 mM DTT. A 400 µl aliquot was incubated with 50 µl 1 M iodoacetate in darkness. After 30 min., 50 µl 1 M DTT was added and the final volume increased to 2 ml with dH$_2$O. Proteins were precipitated with TCA and the pellet was redissolved in 400 µl IPA buffer (20 mM Tris, 5 mM EDTA, 150 mM NaCl, 0.2% SDS) and 2 M Tris base was added to neutralize pH. Triton X-100 was added to a final concentration of 1%. Anti-lamin antibody (donated by the laboratory of Paul Fisher) bound to protein A-sepharose beads was incubated with the TCA precipitated sample at 37^0 C for 2 hrs. The beads were pelleted by centrifugation and washed three times with IPA buffer containing 1% Triton X-100. The pelleted beads were boiled in 20 µl sample buffer and electrophoresed through a 7.5% SDS mini gel under reducing conditions. The gel was stained with Coomassie Blue, treated with En3Hance, and exposed to x-ray film.

Chemical cleavage of the [3H]-mevalonate-labeled isoprenoid from immunoprecipitated *Drosophila* nuclear lamins. The procedure utilized for the cleavage of the isoprenoid moiety from the nuclear lamin was as described by Maltese and coworkers (Maltese and Erdman, 1989). After immunoprecipitation the pelleted anti-lamin protein A-sepharose beads were resuspended in 1.4 ml of 3% formic acid. After the addition of 550 µl of CH$_3$I (Aldrich), the sample was rotated in darkness for 24 hrs. at 37^0 C. The sample was spiked with 8.8 µg of non-radioactive all-trans farnesol (Aldrich) to serve as an internal standard and to monitor recovery of chloroform extractable isoprenoids. The CH$_3$I was evaporated under a stream of nitrogen and the sample volume was returned to 1.4 ml with H$_2$O. The pH was adjusted to 10 with 10 N NaOH and the sample was again rotated in darkness for 20 hrs. at 37⁰ C. A CH₃OH/CHCl₃ extraction was done by adding 2 ml of methanol followed by 4 ml of chloroform and the phases separated by centrifugation. Aliquots of both phases were counted in a scintillation counter. The radioactivity partitioned into the CHCl₃ phase which was collected and evaporated to dryness under a stream of nitrogen.

Reverse-phase chromatography of the chloroform-extractable [³H]-mevalonate-labeled isoprenoid material. The cleavage product was dissolved in methanol and loaded onto a Zorbax ODS reverse phase column (0.46 x 25 cm) (DuPont). The sample was eluted with 20% H_2O/CH_3OH at 2 ml/min for 15 min and 1 ml fractions were collected. The solvent was changed to 100% CH_3OH and chromatography was continued for an additional 15 min. The fractions were monitored by their absorbance at 210 nm and by counting aliquots in a scintillation counter.

Straight-phase chromatography of the chloroform-extractable [³H]-mevalonate-labeled isoprenoid material. The cleavage product was dissolved in 1% isopropanol/hexane and loaded onto a Zorbax Sil straight phase column (0.46 x 25 cm) (DuPont). The sample was eluted with 1% isopropanol/hexane at 2 ml/min for 15 min and 1 ml fractions were collected. The fractions were monitored at 210 nm and by counting aliquots in a scintillation counter.

RESULTS AND DISCUSSION

The study of the isoprenylation of proteins has relied mainly on metabolic labeling of tissue culture cells with [³H]-mevalonate, a precursor in the isoprenoid biosynthetic pathway. Although this has been shown to be an effective approach there are some disadvantages with mammalian systems because isoprenylated proteins are a minor product of the isoprenoid pathway. When using mammalian cell lines it is a standard procedure to include mevinolin, an inhibitor of endogenous mevalonate synthesis, in metabolic labeling experiments in order to minimize isotopic dilution of the exogenous labeled mevalonic acid and thereby maximize incorporation of radioactivity into the protein end products. Even so it is clear from most published studies that 30-40 days may be required to obtain a reasonable signal on a fluorogram of the labeled proteins.

In order to circumvent these problems we have used *Drosophila* S2 cells for similar metabolic labeling studies. Insects lack the enzyme which converts farnesyl-PP to squalene and therefore they cannot synthesize cholesterol. Our results indicated that lack of this branch of the pathway resulted in incorporation of a larger percentage of the radioactive precursor, [3H]-mevalonate, into the protein end products, thereby facilitating detection of these proteins. More than 20 polypeptides incorporated the [3H]-mevalonate (Fig. 1), including at least 5 polypeptides of molecular weight greater than 35kD and numerous polypeptides of less than 30kD. The SDS-PAGE gel was exposed to film for 4 days, illustrating that the time needed to obtain a fluorogram of the labeled proteins is greatly reduced compared to that needed in other cell systems under the same conditions.

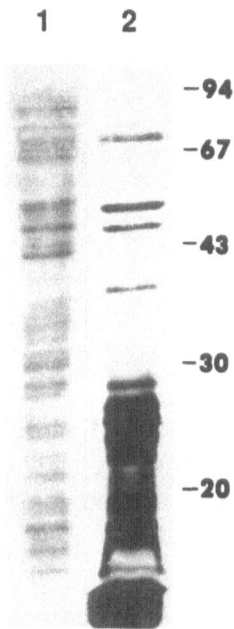

Figure1.   SDS-PAGE protein profile and fluorogram of [3H]-mevalonate-labeled S2 cells. Ninety-nine micrograms of total cellular protein was electrophoresed through a 10% SDS gel under reducing conditions. Lane 1 shows the Coomassie Blue-stained protein profile of the S2 cell lysate. Lane 2 is the corresponding fluorogram after exposure to x-ray film for 4 days.

Two of the labeled polypeptides were identified as the interphase forms of the *Drosophila* nuclear lamins. D_{m1} and D_{m2}, of 74kD and 76kD respectively, incorporated [3H]-mevalonate and were immunoprecipitated from an S2 cell extract with a *Drosophila* anti-lamin antibody (Fig. 2). The deduced amino acid sequences indicate that D_{m1} and D_{m2} contain the carboxy-terminal sequence, CAIM (Gruenbaum, et al., 1988). This conforms to the CaaX sequence found in many proteins that have been shown to be isoprenylated.

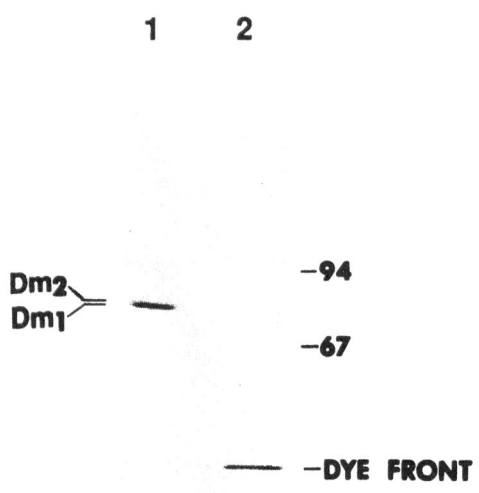

Figure 2.   Immunoprecipitation of Dm₁ and Dm₂ with *Drosophila* anti-lamin antibody.   A fluorogram of a 7.5% SDS mini gel is shown; lane 1 is the proteins bound to the anti-lamin antibody and lane 2 is the proteins in the unbound supernatant.

Thus, it appeared that the incorporation of [3H]-mevalonate into D_{m1} and D_{m2} involved the formation of a thioether bond between isoprenyl groups and the sulfur of the cysteine. Thioether bonds can be cleaved chemically by forming the S-methylsulfonium derivative via treatment with methyliodide. The sulfonium salt is subsequently cleaved with strong base to release the free isoprenoid. Consequently, this method was used to remove the isoprenoid from immunopurified D_{m1} and D_{m2} in order to identify the isoprenoid groups released. Possible products include the isoprenoids, geranylgeraniol and farnesol, and rearrangement products such as nerolidol, a tertiary alcohol, or bisabolol, a cyclic compound.

Greater than 95 % of the radioactivity was recovered by extraction with chloroform. The chloroform-extractable material was subjected to reverse phase C18 HPLC eluting with 20% H_2O/CH_3OH. It was predicted that farnesol is the isoprenoid modification of *Drosophila* nuclear lamins because of its CaaX sequence, CAIM, however only 20% of the radioactive material co-chromatographed with an all-trans farnesol standard. Sixty-five percent of the remaining labeled species was more hydrophobic than farnesol and eluted from the column with 100% CH_3OH. In an attempt to characterize this hydrophobic species, normal phase chromatography of the CH_3I cleavage product was also performed using a silica column and eluting with 1% isopropanol/hexane. Similar to the reverse phase chromatographic system, 25% of the labeled material eluted with the farnesol standard in 21 ml of solvent. The remaining 75% eluted as a single peak of radioactivity in 4 ml of solvent, whereas the nerolidol standard eluted in 7.8 ml and the geranylgeraniol standard eluted in 18.8 ml. This unidentified species constituted the major CH_3I cleavage product of immunoprecipitated *Drosophila* nuclear lamins and migrated as a more hydrophobic species than any of the isoprenoid standards chromatographed under these conditions. Other possibilities for the identity of this material may include rearrangement products of the methylation reaction other than nerolidol or condensation products. Clearly, further study on this potential new adduct is warranted.

REFERENCES

Anderegg RJ, Betz R, Carr SA, Crabb JW, and Duntze W (1988) Structure of the Saccharomyces cerevisiae mating hormone a-factor: identification of S-farnesyl cysteine as a structural component. J. Biol. Chem. 263: 18236-18240.

Beck LA, Hosick TJ, and Sinensky M (1988) Incorporation of a product of mevalonic acid metabolism into proteins of Chinese Hamster Ovary cell nuclei. J. Cell Biol. 107: 1307-1316.

Bruenger E and Rilling HC (1986) Prenylated proteins from kidney. Biochem. Biophys. Res. Comm. 139: 209- 214.

Burke B. (1990) The nuclear envelope and nuclear transport. Curr. Op. Cell Biol. 2: 514-520.

Gruenbaum Y, Landesman Y, Drees B, Bare JW, Saumweber H, Paddy MR,

Sedat JW, Smith DE, Benton BM, and Fisher PA (1988) Drosophila nuclear lamin precursor D_{mo} is translated from either of two developmentally regulated mRNA species apparently encoded by a single gene. J. Cell Biol. 106: 585-596.

Kamiya Y, Sakurai A, Tamura S, Takahashi N, Abe K, Tsuchiya E, Fukui S, Kitada C, and Fujino M (1978) Structure of rhodotorucine A, a novel lipopeptide, inducing mating tube formation in Rhodosporidium toruloides. Biochem. Biophys. Res. Comm. 83: 1077-1083.

Kinsella BT, Erdman RA, and Maltese WA (1991) Carboxyl-terminal isoprenylation of ras -related GTP-binding proteins encoded by rac 1, rac 2, and ral A. J. Biol. Chem. 266: 9786-9794.

Kinsella BT and Maltese WA (1991) rab GTP-binding proteins implicated in vesicular transport are isoprenylated in vitro at cysteines within a novel carboxyl-terminal motif. J. Biol. Chem. 266: 8540-8544.

Kitten GT and Nigg EA (1991) The CaaX motif is required for isoprenylation, carboxyl methylation, and nuclear membrane association. J. Cell Biol. 113: 13-23.

Krohne G, Waizenegger I, and Hoger TH (1989) The conserved carboxy-terminal cysteine of nuclear lamins is essential for lamin association with the nuclear envelope. J. Cell Biol. 109: 2003-2011.

Lowry OH, Rosebrough NJ, Farr AL, and Randall RJ (1951) Protein measurement with the folin phenol reagent. J. Biol. Chem. 193: 265-275.

Maltese WA (1990) Posttranslational modification of proteins by isoprenoids in mammalian cells. FASEB J. 4: 3319-3328.

Maltese WA and Erdman RA (1989) Characterization of isoprenoid involved in the post-translational modification of mammalian cell proteins. J. Biol. Chem. 264: 18168-18172.

Maltese WA and Sheridan KM (1987) Isoprenylated proteins in cultured cells: subcellular distribution and changes related to altered morphology and growth arrest induced by mevalonate deprivation. J. Cell. Phys. 133: 471-481.

Rilling HC, Bruenger E, Epstein WW, and Kandutsch AA (1989) Prenylated proteins: demonstration of a thioether linkage to cysteine of proteins. Biochem. Biophys. Res. Comm. 163: 143-148.

Schmidt RA, Schneider CJ, and Glomset JA (1984) Evidence for post-translational incorporation of a product of mevalonic acid into Swiss 3T3 cell proteins. J. Biol. Chem. 259: 10175-10180.

Seabra MC, Reiss Y, Casey PJ, Brown MS, and Goldstein JL (1991) Protein farnesyltransferase and geranylgeranyltransferase share a common α subunit. Cell 65: 429-434.

LIPID MODIFICATIONS OF PROTEINS

Milton J. Schlesinger
Department of Molecular Microbiology, Box 8230
Washington University School of Medicine
660 South Euclid Ave
St.Louis, MO 63110 USA

A protein's structure and function is determined by both its primary amino acid sequence and by covalent modifications of these amino acids. Among the very large number of such alterations, which can vary from simple inorganic substituents to complex arrays of oligosaccharides, are a group of lipids. The latter have been detected in many proteins that associate with cellular membranes and in a variety of virus structural polypeptides. Thus far, four distinct types of lipid modifications have been identified in eukaryotic cell and virus proteins (Fig. 1).

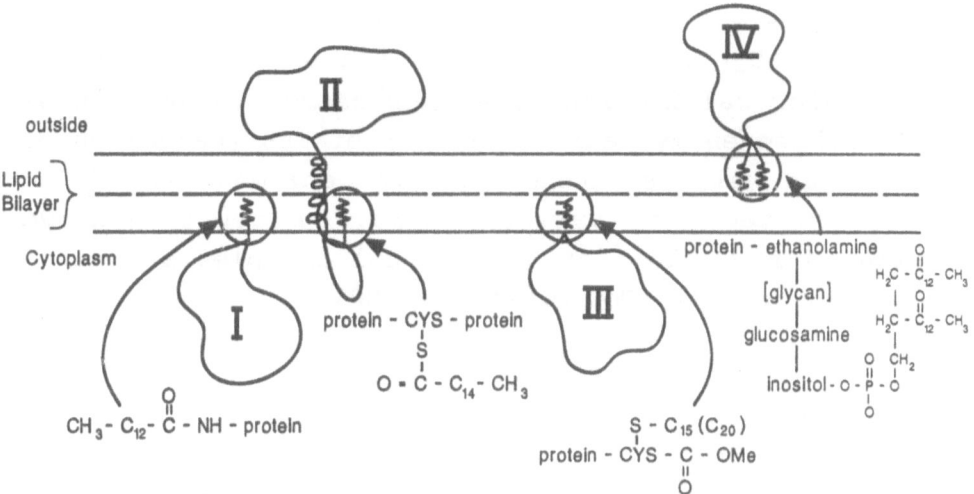

Figure 1. Fatty acylation of eukaryotic cell proteins

NATO ASI Series, Vol. H 63
Dynamics of Membrane Assembly
Edited by J. A. F. Op den Kamp
© Springer-Verlag Berlin Heidelberg 1992

They consist of (1) acylation of the amino-terminus of the polypeptide by myristic acid (C14:0), (2) acylation of cysteines by long chain fatty acids-predominantly palmitic acid (C16:0), (3) isoprenylation -predominantly farnesyl or geranylgeranyl - via thioether linkage to cysteine and (4) substitution of the carboxyl terminal sequence with a glycophospho-inositol lipid, refered to as glypiation. Another kind of lipid modification that is found only in prokaryotic cells alters the proteins's amino terminal sequence by both thioester acylation and phospholipid amidation at a cysteine. The latter will not be discussed here; however, this kind of lipid-protein binding is described in a report of an in vitro lipid processing system (Tokunaga et al 1982). Protein modification by glycophosphoinositol glycan is described in another chapter of this monograph. In this chapter I will summarize information about the first three of these lipid modifications.

MYRISTOLYLATION

The covalent addition of myristic acid via an amide bond to an NH_2-terminal glycine of a protein was first reported for calcineurin B, a phosphatase (Aitken et al 1982). Subsequently, animal virus proteins (Henderson et al 1983) as well as other cellular enzymes were found to be acylated with myristic acid. A representative listing is in Table 1 and additional examples are noted in a recent review (Towler et al 1988).

Table 1. Myristoylated Proteins

<u>Cellular Proteins</u>   <u>Virus Proteins</u>

cAMP-protein kinase   Retrovirus matrix
Src-tyrosine kinase   Picorna virus capsid
$GTP_{o\alpha}$ -binding subunit   Reovirus $\mu 1$ peptide
Cytochrome b5 reductase   Papova virus capsid
Phosphatases   Vaccinia"late" prot.
          HIV *nef* reg.protein

A major "break-through" in elucidating the detailed biochemistry of protein myristoylation came from the discovery that a preparation of yeast extract could covalently bind myristoylCoA to a synthetic octapeptide in vitro (Towler and Glaser 1986). This assay allowed for the purification of the acylating enzyme, N-myristoyltransferase, from *S. cerevisiae*. Properties of this enzyme have recently been reviewed (Gordon et al 1991) and are summarized in Table 2.

Table 2. Properties of N-Myristoyltransferase

1. Highly selective for C14:0
   Utilizes the CoA-SH form of the fatty acid.
2. Absolute requirement for NH_2-terminal glycine
   as the acceptor. Serine at position 5 of
   the polypeptide and a basic residue at positions
   7 & 8 lead to optimal acylation rates.
3. A non-membranal monomeric 53kDa protein
   not an -SH or metalloprotein.
4. In yeast, encoded by a single gene of 455
   amino acids.
   Essential for yeast cell viability.
5. Not in prokaryotic cells.

A vast amount of information regarding substrate analogues of fatty acids and peptides and the detailed kinetic mechanism of the enzymatic reaction has been obtained for the yeast enzyme (Gordon et al 1991). Mutated forms of the enzyme are also being studied.

Myristoylation occurs while the polypeptide is still nascent on non-membrane bound polysomes but most of the acylated proteins ultimately associate with cellular membranes. A 32 kD plasma membrane receptor for the myristoyl group of the pp60 src kinase has been identified (Resh 1989), and the transport of this kinase to the membrane involves a complex with a 90 kD protein identical to the heat-shock protein of this size (Oppermann et al 1981).

The function of the myristoyl group has been examined for a number of proteins by using "reverse genetics", a technique involving site-mutagenesis that removes the glycine at the amino terminus of the protein. Several examples are cited in Table 3 and they show that a protein lacking the fatty acid is functionally altered with regard to its membrane-associated activities.

Table 3. Site-mutagenized N-Myristoylated Proteins

Protein	Altered Activity
pp60 src kinase	no targeting to membrane
	no cell transformation
	no phosphorylation of 115 kD
GTPα binding protein	lower affinity for trimers
HIV Gag 55 kD	no virus assembly
poliovirus VP4	no virus assembly; poor proteolytic processing

The lack of the myristoyl group in membrane-associated tyrosine kinases alters the pattern of phosphorylation (Linder and Burr 1988). A myristoyl-minus Rous sarcoma virus pp60 fails to transform cells in culture and produces in chickens small tumors that regress. Mutationally-altered GTPα-binding proteins that are defective in myristoylation fail to associate properly with other form of G proteins (Linder 1991). Non-myristoylated retrovirus, picornavirus and polyoma virus structural proteins fail to assemble properly and virions are defective in entry into cells (Schultz and Rein 1989; Gottlinger et al 1989; Bryant and Ratner 1990; Marc et al 1990; Krausslich et al 1990; Krauzewicz et al 1990). A crystal structure of the poliovirus reveals a packing of the myristoyl groups at interfaces of the capsid proteins in the icosahedral lattice (Chow et al 1987). Portions of the hydrocarbon chain interact with related regions from other myristate tails and other portions interact with amino acid residues. Thus, the myristoyl groups stabilize the virus structure and, in addition, facilitate the early stages of virus assembly during formation of pentameric intermediates.

Analogues of myristic acid that affect the hydrophobic character of the lipid can perturb membrane association of myrisotylated proteins (Johnson et al 1990), and this type of analysis as well as the genetic studies clearly implicate the myristoyl group as a determinate in effective protein-membrane interactions. Other data, noted above, suggest that the myristoyl group functions as a membrane targeting signal for some proteins.

PALMITOYLATION

The covalent addition of palmitic acid to cysteine residues in proteins is much more widespread than N-myristoylation and, as noted with the latter, both virus and cellular proteins are substrates for fatty acylation (Table 4; Towler et al 1988). Almost all of the proteins listed interact much more strongly with lipid bilayers than the myristoylated proteins. For both cellular and viral proteins, the palmitoyl groups are found on those cysteine residues localized to cytoplasmic domains of transmembranal proteins.

Table 4. Palmitoylated Proteins[+]

Cellular Proteins	Virus Proteins
myelin proteolipid	Sindbis, Semliki Forest (E1,E2 and 6K)
HLA B/DR antigens	Vesicular stomatitis G
Insulin receptor	Influenza hemagglutinin
Ii,Ia α & β antigens	Mumps HN
IgE receptor	Newcastle disease F
acetylcholine recep.	Respiratory syncyctial F
TGF-α receptor	Herpes glycoprotein gE
Transferrin receptor	Rous sarcoma gp37
β-adrenergic receptor	LaCrosse G1
ankyrin, vinculin	

[+]Stoichiometries vary from 1 to 4 fatty acids per polypeptide chain.

The brain myelin proteolipid protein was the first cellular protein to be identified as containing bound fatty acid (Folch and Lees 1951), but definitive evidence for covalent palmitoylation of a membrane

associated protein was first shown for the transmembranal glycoproteins of Sindbis and vesicular
stomatitis viruses (Schmidt et al 1979). In the latter
examples, virus-infected cells given ^{3}H-labeled palmitic
acid effectively incorporated the label into the
glycoprotein in a bond that was stable to a variety of
protein denaturants. In addition proteolytic fragments
that retained the lipid label could be isolated and,
importantly, the radiolabeled lipid could be released
from the protein by mild hydrolysis and recovered as the
intact fatty acid. The attachment site was inferred to
be cysteine based on the sensitivity of the protein-
bound lipid to hydrolysis by neutral hydroxylamine but
definitive assignment came from analyses of site-
directed mutations, first obtained by Rose et al (1984)
with the G glycoprotein of vesicular stomatitis virus.

The enzymatic pathway for palmitoylation of
proteins has not been definitively elucidated although a
palmitoyl CoA transferase that attaches long chain fatty
acids to virus glycoproteins has been partially
purified from liver microsomes (Berger and Schmidt
1984). No specific polypeptide sequences other than a
cysteine positioned near the membrane on the cytoplasmic
domain of the protein have been identified as the
acceptor. Furthermore, palmitoylation can occur at
different sites in a cell including the plasma membrane
and a putative transitional organelle between the rough
endoplasmic reticulum and the cis-Golgi compartment
(Bonatti et al 1989).

The function of the palmitoyl group has been
difficult to define and this fatty acid may serve
different roles depending on the protein. For example,
replacement by site-directed mutagenesis of cysteines in
the Sindbis virus E2 cytoplasmic domain and in the
hydrophobic 6K virus protein dramatically alters both
the rate of virus assembly and the type of virus

structures released (Gaedigk-Nitschko et al 1990; Gaedigk-Nitschko and Schlesinger 1991). Instead of small single-cored enveloped viruses, the mutationally altered strains produced large multicored membrane-enclosed infectious particles. These results suggest that the fatty acids play a role in restructuring the lipid bilayer during virus assembly. Removal of fatty acids from the influenza virus hemagglutinin by hydroxylaminolysis inhibited the capacity of this protein to fuse membranes (Lambrecht and Schmidt 1988). A recent study by Naeve and Williams (1991) that utilized site-directed mutagenesis to remove three hemagglutinin cysteines showed both loss of fatty acids and defects in membrane fusion. However, a report by Veit et al (1991) detected no effect on fusogenesis by a hemagglutinin that had its cysteinyl-acylation sites removed by mutagenesis.

For the β-adrenergic receptor, mutagenesis of a cysteine that is normally palmitoylated blocks signal transduction between receptor and adenyl cyclase (Linder et al 1990) via G proteins, which are also modified by either myristoyl groups or farnesyl groups(see below). Ovchinnikov (1988) has suggested that the fatty acid anchors a part of the cytoplasmic domain of the receptor to form a fourth cytoplasmic loop in a structure analogous to that of fatty acylated rhodopsin.

Removal of two fatty acylation sites in the human transferrin receptor did not affect total recycling of receptor but led to an increased rate of endocytosis when the receptor was expressed in CHO cells (Alvarez et al 1990)suggesting that the palmitoyl groups normally affect interactions between receptor and adaptor proteins in membrane coated pits.

Based on the various studies cited here, it has been postulated that the palmitoyl groups provide membrane anchors that restrict the conformation of cytoplasmic domains of these proteins. In some cases, the fatty acid together with the polypeptide as the head group may substitute for normal phopholipids in the inner leaflet of the bilayer. Replacement of the latter may be necessary for some membrane activities such as envelopment of bilayers around virus nucleocapsids or vesicle formation during intracellular organelle transport.

ISOPRENYLATION

An unusual kind of lipid modification to protein is the covalent attachment of polyisoprene units in thioether linkage to cysteines (Gibbs 1991). An isoprenylated protein, the fungal mating factor rhodotorucine A, was first reported by Kamiya et al (1978) but the significance of this kind of lipid in protein structure was recognized only recently when a number of key regulatory cell proteins (Table 5) were found to be post-translationally altered by isoprenyl groups.

Table 5. Isoprenylated Proteins

Protein	Carboxyl-sequence	Prenylation
Ras2 (yeast)	G.C.C.I.I.S	farnesyl
Rap1 (human)	K.S.C.L.L.L	geranylgeranyl
Transducin γ	G.G.C.V.I.S	farnesyl
Go,Gi,Gs γ	F.F.C.A.I.L	geranylgeranyl
Lamin A	Q.N.C.S.I.M	farnesyl
cGMP phosphodi-diesterase	K.T.C.L.M.L	?

Several other yeast G-binding proteins are reported to be prenylated but the type of isoprene unit has not yet been characterized (Powers 1991). Thus far, only two isoprenes have been found linked to proteins, the 15 carbon farnesyl group and the related 20-carbon geranylgeranyl moiety. The latter is derived from geranyl pyrophosphate in a side reaction from the central metabolic pathway that converts mevalonate to cholesterol. Farnesyl pyrophosphate is also an intermediate in this pathway and side reactions of this compound lead to dolichol, ubiquinones and farnesylated proteins (Maltese 1990).

Two distinct multimeric enzymes prenylate proteins but they appear to share a common α-subunit of 49 kD that binds the prenyl-group (Reiss 1991). The β-subunits recognize different carboxyl terminal protein sequences (cf Table 5); the β-subunit of the farnesylating enzyme is 46 kD that binds the Ras protein substate (Seabra et al 1991). Prenylation is accompanied by removal of the C-terminal amino acids and the carboxylmethylation of the newly formed cysteine as the terminus. Prenylation of G-binding proteins is essential for protein-protein interactions (Fukada et al 1989) and for membrane targeting and anchoring (Hancock et al 1989; Jackson et al 1990). A *S.cerevisiae* gene, DR1-RAM1, encodes a polypeptide subunit of a prenyltransferase (Schafer et al 1990), and the yeast Bet2 gene, which is required for transport of proteins from RER to Golgi apparatus, is identical to DR1-RAM1 (Rossi et al 1991).

It is of interest that G-binding proteins, which are so crucial to transmembrane signaling in all eukaryotic cells, are covalently modified with all three types of lipids discussed here- myristic acid, palmitic acid and geranylgeranyl groups. The putative insertion of these lipophilic side chains should affect membrane stability as well as facilitate binding of the proteins to the membrane.

REFERENCES

Aitken A, Cohen P, Santikarn S, Williams DH, Calder AG, Smith A, Klee CB (1982) Identification of the NH2-terminal blocking group of calcineurin B as myristic acid. FEBS Letters 150:314-318

Alvarez E, Girones N, Davis RJ (1990) Inhibition of the receptor-mediate endocytosis of diferric transferrin is associated with the covalent modification of the transferrin receptor with palmitic acid. J Biol Chem 265:16644-16655

Berger M, Schmidt MFG (1984) Cell free fatty acid acylation of Semliki Forest viral polypeptides with microsomal membranes from eukaryotic cells. J Biol Chem 259:7245-7252

Bonatti S, Migliaccio G, Simons K (1989) Palmitoylation of viral membrane glycoproteins takes place after exit from the endoplasmic reticulum. J Biol Chem 264:12590-12595

Bryant M, Ratner L (1990) Myristoylation-dependent replication and assembly of human immunodeficiency virus 1. Proc Natl Acad Sci USA 87:523-527

Chow M, Newman JFE, Filman D, Hogle JM, Rowlands DJ, Brown F (1987) Myristylation of picornavirus capsid protein VP4 and its structural significance. Nature 327:482-486

Folch-Pi J, Lees M (1951) Proteolipids a new type of tissue lipoproteins: their isolation from brains. J Biol Chem 191:807-817

Fraser, CM (1989) Site-directed mutagenesis of the b-adrenergic receptors: indentification of conserced cysteine residues that independently affect ligand binding and receptor activation. J Biol Chem 264:9266-9270

Fukada Y, Takao T, Ohguro H, Yoshizaea T, Akino T, Shimonishi Y (1990) Farnesylated γ subunit of photoreceptor G protein indispensable for GTP-binding. Nature 346:658-660

Gaedigk-Nitschko K, Ding M, Aach-Levy M, Schlesinger MJ (1990) Site-directed mutations in the Sindbis virus 6K protein reveal sites for fatty acylation and the underacylated protein affects virus release and virion structure. Virology 175:282-291

Gaedigk-Nitschko K, Schlesinger MJ (1991) Site-directed mutations in Sindbis virus E2 glycoprotein's cytoplasmic domain and the 6K protein lead to similar defects in virus assembly and budding. Virology 183:in press

Gibbs JB (1991) Ras C-terminal processing enzymes- new drug targets? Cell 65:1-4

Gordon, JI, Duronio RJ, Rucnick DA, Adams SP, Gokel GW (1991) Protein N-Myristoylation. J Biol Chem 266:8647-8650

Gottlinger HG, Sodroski JG, Haseltine WA (1989) Role of capsid precursor processing and myristoylation in morphogenesis and infectivity of human immunodeficiency virus type 1. Proc Natl Acad Sci USA 86:5781-5785

Hancock JF, Magee AI, Childs JE, Marshall CJ (1989) All *ras* proteins are polyisoprenylated but only some are palmitoylated. Cell 57:1167-1177

Henderson LE, Krutzsch HC, Oroszlan S (1983) Myristyl amino-terminal acylation of murine retrovirus proteins: an unusual post-translational protein modification. Proc Natl Acad Sci USA. 80:339-343

Johnson RD, Cox AD, Solski PA, Devadas B, Adams SP, Leimbruber RM, Heuckeroth RO, Buss JE, Gordon JI (1990) Functional analysis of protein N-myristoylation: metabolic labeling studies using three oxygen-substituted analoys of myristic acid and cultured mammalian cells provide evidence for protein-sequence-specific incorporation and analog-specific redistribution. Proc Natl Acad Sci USA 87:8511-8515

Kamiya Y, Sakurai A, Tamura S, Takahasi N, Tsuchiya E, Abe K, Fukui S (1979) Structure of rhodoturicine A, a peptidyl factor inducing mating tube formation in *Rhodosporidium torulaides*. Arig Biol Chem 43:363-369

Krausslich H-G, Holscher C, Reuer Q, Harber J, Wimmer E. (1990) Myristoylation of the poliovirus polyprotein is required for proteolytic processing of the capsi and for viral infectivity. J Virol 64:2433-2436

Krauzewicz N, Streuli CH, Stuart-Smith N, Jones MD, Wallace S, Griffin BE (1990) Myristylated polyomavirus VP2: role in the life cycle of the virus. J Virol 64:4414-4420.

Lambrecht B, Schmidt MFG (1986) Membrane fusion induced by influenza virus hemagglutinin requires protein bound fatty acids. FEBS Lett 202:127-132

Linder ME, Burr JG (1988) Nonmyristoylated p60 v-src fails to phosphorylate proteins of 115-120 kDa in

chicken embryo fibroblasts. Proc Natl Acad Sci USA
85:2608-2612

Linder ME, Iok-Hous P, Duronio RJ, Gordon JI, Sternweiss
PC, Gilman AG (1991) Lipid modifications of G protein
subunits: myristoylation of Goα increases its affinity
for βγ. J Biol Chem 266:4654-4659

Marc D, Masson G, Girard M, van der Werf S (1990)
Lack of myristoylation of poliovirtus capsid
polypeptide VP0 prevents the formation of virions or
results in the assembly of noninfectious virus
particles. J Virol 64:4099-4107

Maltese WA (1990) Posttranslational modification of
proteins by isoprenoids in mammalian cells. The FASEB
J 4:3319-3328

Naeve CE, Williams D (1990) Fatty acids on the
A/Japan/305/57 influenza virus hemagglutinin have a
role in membrane fusion. EMBO J 8:2661-2668

O'Dowd BF, Hnatowich M, Caron MG, Lefkowitz RJ Bouvier
M (1989) Palmitoylation of the human β2-adrenergic
receptor: mutation of cys341 in the carboxyl tail
leads to an uncoupled nonpalmitoylated form of the
receptor. J Biol Chem 264:7564-7569

Oppermann H, Levinson W, Bishop JM (1981) A cellular
protein that associates with a transforming protein of
Rous sarcoma virus is also a heat-shock protein. Proc
Natl Acad Sci USA 78:1067-1071

Ovchinnikov YA, Abdulaev NG, Bogachuk AS (1988) Two
adjacent cysteine residues in the C-terminal
cytoplasmic fragment of bovine rhodopsin are
palmitylated. FEBS letters 230:1-5

Powers S (1991) Protein prenylation: a modification
that sticks. Curr Biology 2:114-116

Reiss Y, Seabra MC, Armstrong SA, Slaughter CA,
Goldstein JL, Brown MS (1991) Nonidentical subunits of
p21 H-ras farnesyl transferase. J Biol Chem 266:10672-
10677

Resh MD (1989) Specific and saturable binding of pp60
v-src to plasma membranes: evidence for a myristyl-src
receptor. Cell 58: 281-286

Rose JK, Adams GA, Gallione CJ (1984) The presence of
cysteine in the cytoplasmic domain of the vesicular
stomatitis virus glycoprotein is required for
aplmitate addition. Proc Natl Acad Sci USA 81:2050-
2054

Rossi G, Jiang Y, Newman AP, Ferro-Novick A (1991) Dependence of Ypt1 and Sec4 membrane attachment on Bet2. Nature 351:158-161

Schafer NR, Trueblood CE, Yang C-C, Mayer MP, Rosenberg S, Poulter CD, Kim S-H, Rine J (1990) Enzymatic coupling of cholesterol intermediates to a mating pheromone precursor and to the Ras protein. Science 249:1133-1139

Schmidt MFG, Bracha M, Schlesinger MJ (1979) Evidence for covalent attachment of fatty acids to Sindbis virus glycoproteins. Proc Natl Acad Sci USA 76:1687-1691

Schultz AM, Rein A (1989) Unmyristylated Moloney murine leukemia virus Pr65^{gag} is excluded from virus assembly and maturation events. J Virol 63:2370-2373.

Seabra MC, Reiss Y, Casey PJ, Brown MS, Goldstein JL (1991) Protein farnesyltransferase and geranylgeranyl

Tokunaga M, Tokunaga H, Wu HC (1982) Post-translational modification and processing of *Escherichia coli* prolipoprotein in vitro. Proc Natl Acad Sci USA 79:2255-2259

Towler DA, Adams SP, Eubanks SR, Towery DS, Jackson-Machelski E, Glaser L, Gordon JI (1987) Purification and characterization of yeast myristoyl CoA:protein N-myristoyltransferase. Proc Natl Acad Sci USA 84:2708-2712

Towler DA, Gordon JI, Adams SP, Glaser L (1988) The biology and enzymology of eukaryotic protein acylation. Annu Rev Biochem 57: 69-99

Veit M, Kretzschmar E, Kuroda K, Garten W, Schmidt MFG, Klenk H-D, Rott R (1991) Site-specific mutagenesis identifies three cysteine residues in the cytoplasmic tail as acylation sites of influenza virus hemagglutinin. J Virol 65:2491-2500

MODEL STUDIES ON MEMBRANE INSERTION AND TRANSLOCATION OF PEPTIDES

A.I.P.M. De Kroon[1], J. De Gier[1] and B. De Kruijff[1,2]
[1]Division Biochemistry of Membranes,
Centre for Biomembranes and Lipid Enzymology, and
[2]Institute of Molecular Biology and Medical Biotechnology,
University of Utrecht, Utrecht, NL

Introduction

Lipid-peptide/protein interactions determine the structure and function of biomembranes. For example, these interactions play an important role in the activity of membrane-bound enzymes, and in the transport of solutes across membranes (McElhaney, 1982; Carruthers & Melchior, 1986). Furthermore, they are involved in the regulation of membrane fusion (Hong et al., 1987). The action of many toxic peptides depends on lipid-peptide interactions (Neville & Hudson, 1986; Bernheim & Rudy, 1986). There is growing experimental evidence that the interaction of peptides or proteins with the lipid phase of biological membranes is also playing a role in processes such as membrane insertion and translocation of precursor proteins. A number of biophysical studies employing model membranes, have demonstrated that signal peptides and mitochondrial presequences by virtue of their amphipathic character, are very well suited to interact directly with the membrane lipids (for review see Briggs & Gierasch, 1986; Batenburg et al., 1988; Epand et al., 1986; Roise et al., 1986; Tamm, 1986). Moreover, *in vivo* and *in vitro* studies on protein translocation in *E. coli* strongly point to a role for acidic phospholipids in these processes (De Vrije et al., 1988; Lill et al., 1990), consistent with the hypothesis of a direct signal peptide-phospholipid interaction, which emerged from the biophysical studies. Similarly, in the research area of the receptor binding of peptide hormones and neuropeptides there are numerous biophysical and theoretical studies supporting the view that an initial adsorption of these peptides to the membrane lipids catalyzes their eventual specific binding to a receptor protein (for review see Sargent et al., 1988). The proposed mechanism of the catalysis is 2-fold: (1) the membrane lipids provide a matrix for fast two-dimensional diffusion, (2) the membrane lipids can modulate the peptide structure thus facilitating binding to the receptor. As yet, direct unequivocal proof for the involvement of peptide-lipid interactions in these

NATO ASI Series, Vol. H 63
Dynamics of Membrane Assembly
Edited by J. A. F. Op den Kamp
© Springer-Verlag Berlin Heidelberg 1992

processes is lacking.

The study of peptide-lipid model systems provides information on the possibilities and limitations of the putative function of peptide-lipid interactions in the above biological processes. The present paper summarizes the results obtained in well defined systems consisting of synthetic model peptides and phospholipid vesicles. Emphasis will be laid on the question whether a peptide has the ability to translocate across a phospholipid bilayer. The main parameters affecting peptide-lipid interactions that will be addressed, are the peptide's charge and hydrophobicity, the membrane surface charge, and in addition, the presence of transmembrane ion gradients that give rise to pH gradients and membrane potentials. The synthetic peptides playing the leading parts in this review are intended to be a general simple model for the biological peptides introduced above and have been listed in Fig. 1.

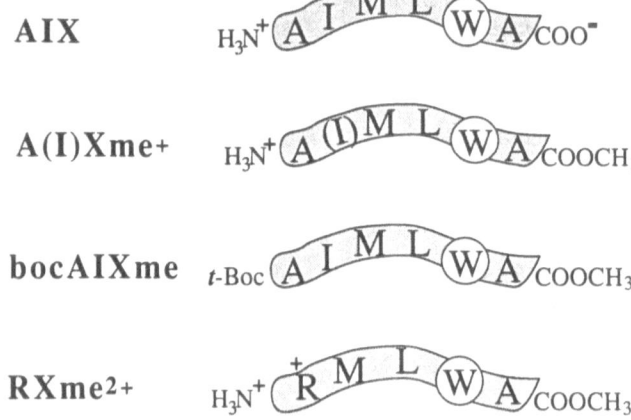

Fig. 1. Schematic representation of the model peptides and their nomenclature. The sequence MLWA (one letter code) is abbreviated as X; me denotes methylation of the carboxyl terminus; boc indicates the attachment of a *tert.*-butyloxycarbonyl group to the amino terminus.

They are derivatives of the sequence H-Ala-Met-Leu-Trp-Ala-OH, which vary in net charge and hydrophobicity. The design of the peptide sequence was guided by the requirement of a general hydrophobic character to ensure a basic level of affinity for model membranes. The peptide length was kept limited in order to avoid the possibility of a membrane spanning conformation. The tryptophan residue served as an intrinsic fluorescent reporter group. The sequence was inspired by that of the peptide hormone pentagastrin of which the interaction with model membranes has been extensively investigated (Surewicz & Epand, 1984; 1986).

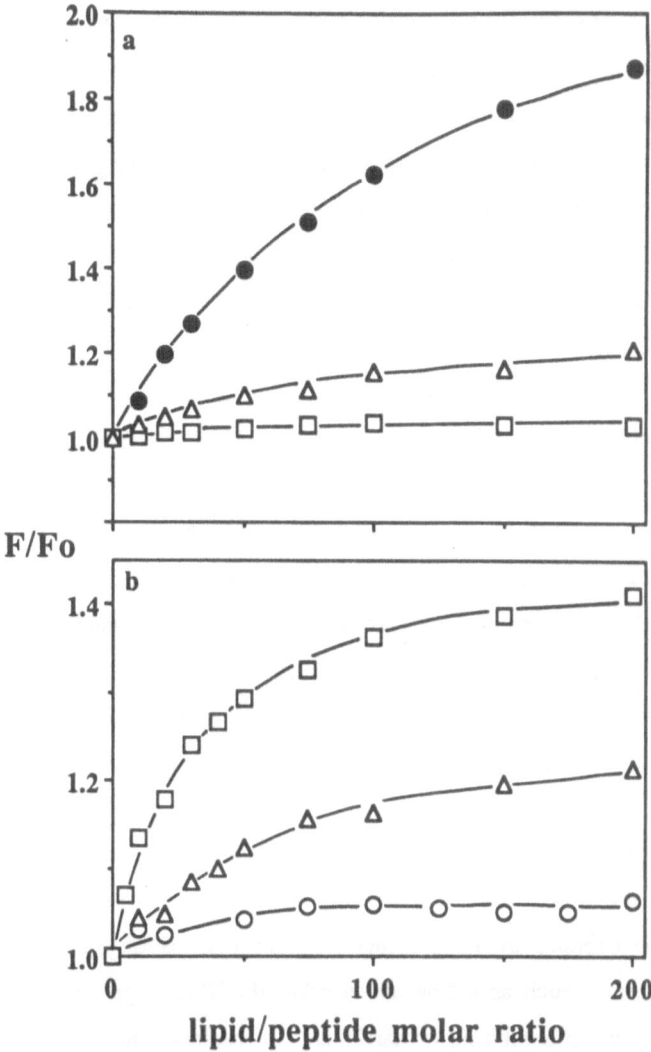

Fig. 2. Changes in tryptophan fluorescence emission intensity at 340 nm in titrations of the peptides AIX (O), AIXme[+] (Δ), bocAIXme (●), and RXme[2+] ([]) with small unilamellar vesicles consisting of eggPC (panel a) and of beef heart cardiolipin (panel b). For further details see De Kroon et al. (1990a), from which the data were taken.

Membrane affinity and topology of the peptides

The measurement of tryptophan fluorescence is a useful method to detect and characterize lipid-peptide interactions. The increase in fluorescence emission intensity and the blue shift of the wavelength of maximum emission which occur upon entry of the tryptophan residue into an environment with a smaller dielectric constant (Cowgill, 1967), serve as a first indication for peptide-lipid interaction. Figure 2 shows the changes in fluorescence intensity occurring in titrations of several of the peptides with small unilamellar vesicles (SUV) consisting either of the zwitterionic phospholipid phosphatidylcholine (PC) or of the acidic phospholipid cardiolipin (CL). Quantitative analysis of these titration curves yields the affinity parameter K_Dn, where K_D represents the dissociation constant of the peptide-lipid complex and n the number of lipid molecules constituting one peptide "binding site" (Bashford et al., 1979). K_Dn values allow a comparison of the lipid affinities of the different peptides for PC or for CL vesicles. However, they do not allow a distinction between electrostatic and hydrophobic energy contributions to the mechanism of binding (cf. Beschiaschvili & Seelig, 1990).

Figure 2a illustrates that the affinity for PC bilayers increases by increasing the peptide's hydrophobicity (decreasing the number of charged groups), with the uncharged bocAIXme showing the largest extent of binding (K_Dn = 0.40 mM). Electrostatic attraction primarily determines the order of affinity for CL bilayers with the divalent peptide RXme^{2+} (K_Dn = 0.07 mM) binding stronger than the monovalent AIXme$^+$ (K_Dn = 0.22 mM) as shown in Fig. 2b. The fluorescence increases are accompanied by a modest decrease in the wavelength of maximum emission which does not exceed a total value of 4 nm for the model peptides studied, at a lipid/peptide molar ratio of 200. A comparison with blue shift data obtained for membrane spanning alamethicin analogs which contain a tryptophan residue at different positions (Voges et al., 1987), indicates that the tryptophan residue of the bound model peptides becomes localized near the lipid-water interface.

An additional criterion for the lipid affinity of the peptides is provided by the residual accessibility of the peptides in the presence of lipid vesicles to aqueous quenchers of tryptophan fluorescence, such as iodide and acrylamide. This is particularly useful when comparing peptides which exhibit no or hardly any fluorescence increase upon addition of vesicles, due either to a very low membrane affinity, or to the relative position of the tryptophan residue within the peptide molecule (cf. De Kroon et al., 1990a).

The quenching of tryptophan fluorescence by membrane incorporated quenchers present at different depths provides information on the topology of the membrane inserted peptide. Brominated PCs, which have been used for this purpose, have the advantage over other membraneous quenchers, such as spin-labeled or anthranoyl-phospholipids, that they are less membrane perturbing (Roseman et al., 1978; Lytz et al., 1984). In fact the physical properties of phospholipids with a bromine attached to one of the acyl chains have been reported to closely resemble those of phospholipids with unsaturated acyl chains (East & Lee, 1982). X-ray diffraction analysis has shown that the bromines in brominated PC bilayers are

215

well localized in the hydrocarbon region, reflecting their acyl chain position (McIntosh & Holloway, 1987). The mechanism of quenching by BrPCs is not known. Recent evidence suggests that quenching by brominated lipids does not require contact between the tryptophan and the bromines. Yet there appears to be a distance dependence of the quenching like in energy transfer, however over a much shorter range (6-7 Å; Bolen & Holloway, 1990).

Figure 3 shows quenching profiles obtained for the peptides RXme^{2+} and AIXme$^+$ bound to CL vesicles and for bocAIXme bound to PC vesicles. The overall extent of the fluorescence quenching appears to reflect the peptide's affinity for the lipid vesicles: the fluorescence of RXme^{2+} is more efficiently quenched than that of AIXme$^+$ by BrPC incorporated in CL vesicles. The shapes of the quenching profiles show that the peptides attain an interfacial localization with the tryptophan of the neutral peptide bocAIXme penetrating slightly deeper into the bilayer than that of the charged peptides (Fig. 3), in agreement with the latter being anchored at the interface by their charged moiety. When compared to the accurate determination of the topology of a membrane spanning model peptide containing a tryptophan by BrPC fluorescence quenching (Bolen & Holloway, 1990), the shapes of the quenching profiles obtained for the amphiphilic model peptides look rather flat. This lack of resolution is most likely due to the disordering effect exerted by the peptides on the acyl chains (see section 3).

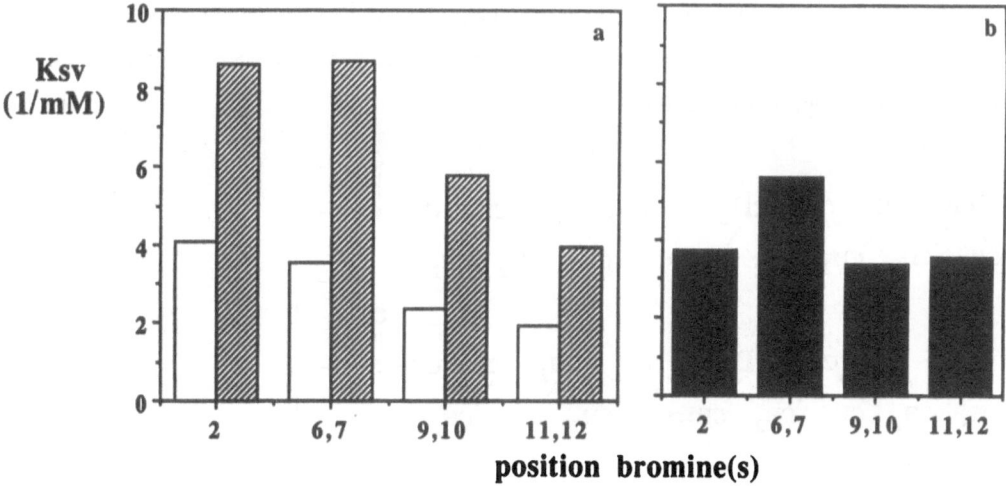

Fig. 3. Quenching profiles of AIXme$^+$ (open columns) and RXme^{2+} (dashed columns) interacting with CL SUV (panel a), and bocAIXme interacting with PC SUV (panel b). The lipid/peptide molar ratio is 100. Quenching constants, K_{sv}, were determined from the slope of Stern-Volmer plots as described (De Kroon et al., 1990a). The quenching by 2-BrPC is underestimated because the bromine content of this phospholipid is less than 1 bromine per PC, whereas the other brominated PC's contain 2 bromines per molecule (De Kroon et al., 1990a).

The conclusions reached with respect to membrane affinity and membrane topology of the peptides are in agreement with previous studies in which the interactions of similar sets of small peptides with model membranes were investigated (Dufourcq et al., 1981; Surewicz & Epand, 1986; Jain et al., 1985; Jacobs & White, 1986, 1989). With regard to the membrane topology of tryptophan residues present in peptides and proteins, it seems that this amino acid has a general tendency to localize in or near the lipid-water interface (Jacobs & White, 1989; De Kroon et al., 1990a, 1991c). Therefore the interfacial localization found for the peptides not containing any charged groups could be due to properties of the tryptophan residue.

Consequences of peptide insertion for structural and dynamic properties of the phospholipid bilayer

In this section the effects of the peptides that were shown to localize in the membrane-water interface, on properties of the phospholipid molecules will be summarized. ^{2}H NMR on membranes consisting of deuterated phospholipids is a convenient, non-perturbing method to measure these effects. The main parameter derived from ^{2}H NMR spectra is the quadrupolar splitting ($\Delta\upsilon_q$), which is the distance between the two maximum resonance positions. $\Delta\upsilon_q$ is proportional to the deuterium order parameter S_{CD} and provides information about the average orientation and the fluctuations of the C-^{2}H bond (Seelig, 1977). Jacobs & White (1987) observed that positively charged tripeptides exert a disordening effect along the entire acyl chain of acyl chain perdeuterated DMPC in lipid-peptide dispersions prepared by hydrating lipid-peptide mixed films.

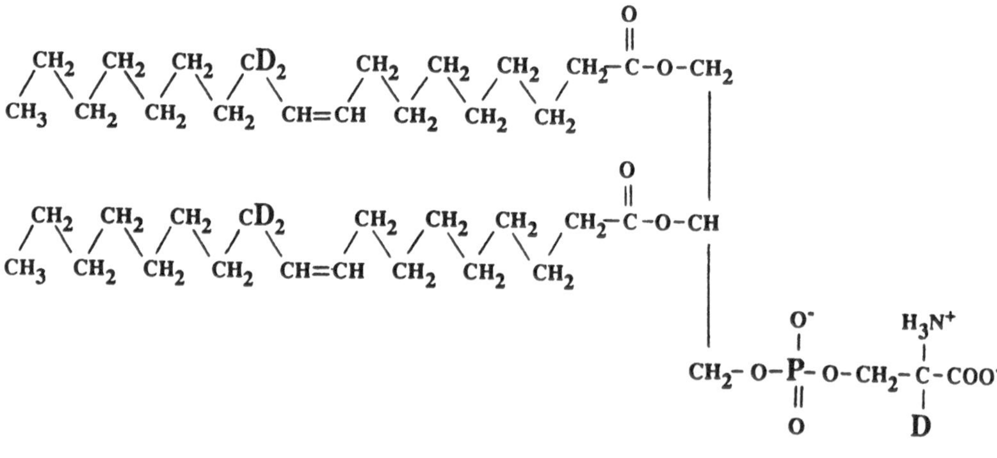

Fig. 4. Position of the deuterons in headgroup ([2-^{2}H]DOPS) and acyl chain ([11,11-^{2}H$_2$]DOPS) deuterated DOPS.

Titration of preexisting lipid bilayers with peptides requires that the peptides are water soluble at the concentrations needed in the NMR experiments (up to ~ 7.5 mM). The model peptides AXme⁺ and RXme²⁺ meet this requirement whereas AIX, AIXme⁺ and bocAIXme form aggregates at these concentrations and are not amenable in these experiments. Since AXme⁺ and RXme²⁺ show the highest affinity for acidic phospholipids, titrations were carried out on liposomes consisting of DOPS which was specifically deuterium labeled either in the headgroup (Browning & Seelig, 1979; De Kroon et al., 1990b), or in the acyl chains (Farren et al., 1984), yielding [2-²H]DOPS and [11,11-²H₂]DOPS), respectively (see Fig. 4 for chemical structure). This allowed the separate analysis of the effects exerted at two levels of the bilayer.

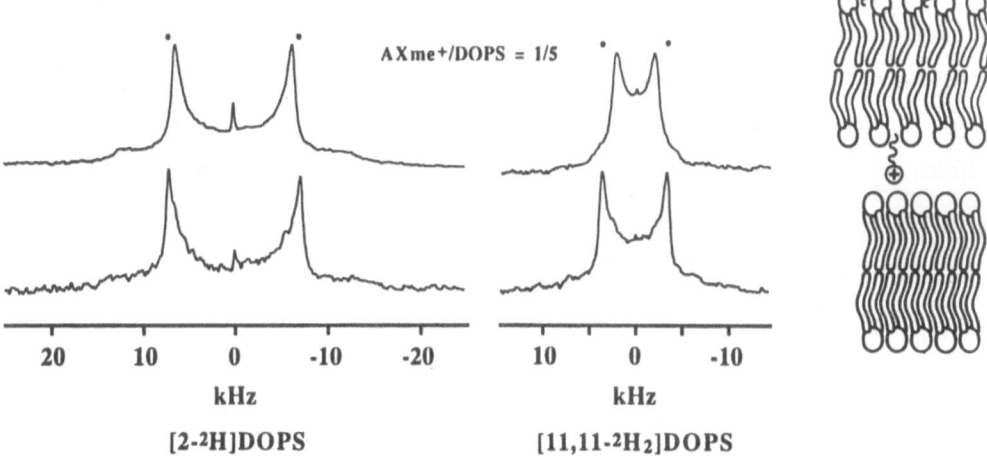

$[2\text{-}^2\text{H}]\text{DOPS}$ $\qquad\qquad$ $[11,11\text{-}^2\text{H}_2]\text{DOPS}$

Fig. 5. ²H NMR spectra of [2-²H]DOPS and [11,11-²H₂]DOPS with and without the peptide AXme⁺ added to a 1/5 molar ratio with respect to DOPS. Samples were equilibrated by 10 cycles of freeze-thawing. The dots in the upper spectra mark the quadrupolar splittings of pure DOPS. For details see De Kroon et al. (1991a).

DOPS in a bilayer organization gives rise to axially symmetric ²H NMR spectra with single characteristic quadrupole splittings of 14.2 and 7.3 kHz, respectively, for [2-²H]DOPS and [11,11-²H₂]DOPS. Upon addition of the peptide AXme⁺ and equilibration by freeze-thawing the bilayer organization is retained, however the values of $\Delta\upsilon_q$ decrease. At a ratio of 0.15 mol AXme⁺ per mol DOPS bound, $\Delta\upsilon_q$ values of ~ 12.4 and 4 kHz were read for headgroup and acyl chain deuterated DOPS, respectively (Fig. 5). One explanation for these smaller $\Delta\upsilon_q$ values is a decrease in the order of the entire DOPS molecule. However, the relative change in $\Delta\upsilon_q$ is smaller for the headgroup deuteron than for the acyl chain deuterons, which suggests that different mechanisms are involved. The latter view is supported by the data depicted in Fig. 6, where the change in $\Delta\upsilon_q$ is plotted against the number of positive charges bound per DOPS, and which compares a titration with AXme$^+$ to a pH titration. The pH dependence of the $\Delta\upsilon_q$ was converted into the protonation degree dependence by using an apparent pK_a value of the PS carboxyl group of 4.5 (MacDonald et al., 1976; De Kroon et al., 1990b). The linear decrease of the $\Delta\upsilon_q$ value of [2-^{2}H]DOPS in response to binding of the peptide, is virtually identical to that obtained upon increasing the protonation degree. The small divergence is most likely due to the 2°C temperature difference between the experiments. This result indicates that the effect of the peptide on the $\Delta\upsilon_q$ of the headgroup deuteron can entirely be accounted for by a partial surface charge neutralization. In contrast, for [11,11-^{2}H$_2$]DOPS counter-directional changes in $\Delta\upsilon_q$ are observed: whereas neutralization of surface charges gives rise to a slight increase, binding of AXme$^+$ results in a strong reduction of the $\Delta\upsilon_q$ value. Binding of AXme$^+$ to [11,11-^{2}H$_2$]DOPC also causes a decrease of the $\Delta\upsilon_q$, however, much higher peptide concentrations were required to observe significant effects, as expected from the low affinity of the peptide for PC bilayers.

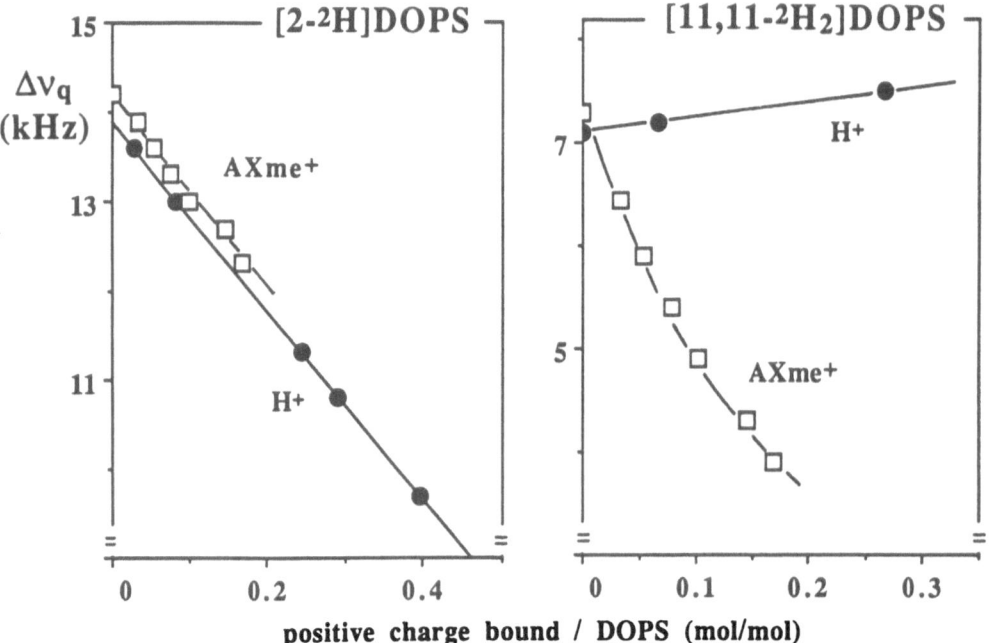

Fig. 6. Variation of the quadrupolar splitting ($\Delta\upsilon_q$) of [2-^{2}H]DOPS and [11,11-^{2}H$_2$]DOPS with the number of positive charges bound per DOPS molecule. The $\Delta\upsilon_q$ values were derived from a titration with the peptide AXme$^+$ ([]) performed at 18°C (De Kroon et al., 1991a), and from a pH titation with HCl (●) carried out at 20°C (De Kroon et al., 1990b).

The NMR data are consistent with the mode of peptide insertion deduced from the fluorescence measurements, which is schematically depicted in Fig. 5. By virtue of their interfacial localization the peptides cause a spacing of the phospholipid headgroups, which gives rise to an increased motional freedom of the acyl chains. The decreased acyl chain order probably contributes to the flat shape of the BrPC quenching profiles in the previous section. The property of perturbing the phospholipid acyl chain order appears to be confined to peptides and proteins that localize in the lipid-water interface, since neither superficially membrane-adsorbed molecules such as polylysine or cytochrome c (Roux et al., 1988; Devaux et al., 1986), nor membrane incorporated peptides such as gramicidin (Chupin et al., 1987) or the membrane spanning peptide $K_2GL_{20}K_2A$ (Roux et al., 1989) display this behaviour.

The peptide-induced linear decrease of the Δv_q of [2-^{2}H]DOPS is probably due to a surface charge dependent change in the average orientation of the headgroup, rather than to a change in headgroup flexibility. Consequently the [2-^{2}H]DOPS data fit in with the concept of phospholipid headgroups as sensors of electric charge which was proposed by Seelig et al. (1987). A variety of charged amphiphilic compounds was shown to give rise to counter-directional linear changes of the Δv_q values of the deuterons at the α- and β-position of the PC headgroup, leading to the conclusion that the dipolar phosphocholine headgroup in PC bilayers changes its orientation with respect to the plane of the bilayer upon introduction of charges at the membrane surface. Strong dipolar fields are generated by the changed headgroup orientation, which may be of physiological relevance. For the phosphoserine headgroup a similar conformational change induced by membrane bound charges has been proposed, based on ^{2}H NMR studies on the interaction of a transmembrane peptide (Roux et al., 1989), of melittin (Dempsey et al., 1989), and of metallic cations (Roux & Bloom, 1990) with deuterated PS bilayers.

Titrations of deuterated DOPS with RXme^{2+} yielded identical changes in Δv_q per mol of peptide bound as compared to AXme$^+$, not only for the acyl chain deuterons but also for the headgroup deuteron (De Kroon et al., 1991a). The latter finding is remarkable since the formal charge of RXme^{2+} is twice that of AXme$^+$ and according to the electrometer concept one would expect RXme^{2+} to give rise to a larger decrease of the headgroup Δv_q. Instead the PS headgroups experience an effective charge of only +1 per RXme^{2+} bound. The reason for this discrepancy is unknown at present, but it may result from the positioning of the charges relative to the plane of the headgroups. In this respect the small increase of the Δv_q of [2-^{2}H]DOPS observed in response to the binding of polylysine (Roux et al., 1988; De Kroon et al., 1991a) is interesting. Differences between formal and effective charge have been observed in a number of peptide-lipid systems (Seelig & Macdonald, 1989; Kuchinka & Seelig, 1989; Stankowski & Schwartz, 1990).

Peptide translocation and ion gradients

The ^{2}H NMR approach in addition provides insight into the ability of the peptides AXme$^+$ and RXme^{2+} to cross the DOPS bilayer. When the freeze-thaw procedure to equilibrate the peptide-lipid mixtures is omitted in the peptide-DOPS titrations, RXme^{2+} gives rise to a two component ^{2}H spectrum, of which the outer quadrupolar splitting corresponds to that of the pure lipid (Fig. 7b). In contrast, for the peptide AXme$^+$ only one Δv_q is observed under these conditions (Fig. 7a). This apparent difference in lipid accessibility to RXme^{2+} and AXme$^+$ has been interpreted as a difference in translocation competence between the two peptides (De Kroon et al., 1991a). This interpretation is supported by the different kinetics of the tryptophan fluorescence increase observed upon addition of vesicles to these peptides (De Kroon et al., 1991a; cf. Jain et al., 1985). The study of vesicle systems which exhibit ion gradients giving rise to a membrane potential and/or a pH gradient, provided the unambiguous proof that peptides with a single ionizable group are membrane permeable, as described below.

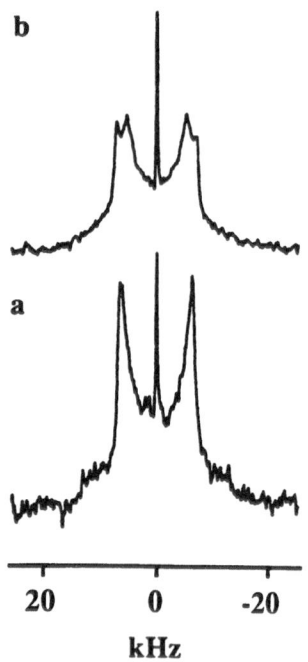

Fig. 7. ^{2}H NMR spectra of [2-^{2}H]DOPS in the presence of AXme$^+$ (a) and RXme^{2+} (b) at molar ratios of 1 peptide added per 5 DOPS molecules. In this experiment the equilibration of the samples by freeze-thawing was omitted. For details see De Kroon et al. (1991a).

A membrane potential is often involved in membrane insertion and translocation of proteins in biological systems. The import of most precursor proteins into mitochondria depends on the presence of a membrane potential across the inner mitochondrial membrane (Pfanner & Neupert, 1985). In the export of proteins from Gram-negative bacteria, there is an important role for the proton motive force (Bakker & Randall, 1984; Schiebel et al., 1991), although in this case it appears not to be essential for protein translocation (see e.g. Yamada et al., 1989). Furthermore membrane potential has been implicated in the entry of several toxins into cells, e.g. of diphtheria toxin (Hudson et al., 1988), and in the formation of colicin channels in membranes (Merrill & Cramer, 1990).

In the above processes the membrane potential may act by influencing the interaction of the protein/peptide with the membrane lipids. The black lipid membrane experimental set up which monitors the formation of voltage-dependent channels is often used to study peptide-lipid interaction under the influence of a membrane potential (see e.g. Tosteson & Tosteson, 1981; Menestrina et al., 1986). In order to allow for a direct comparison with the data described above, we will here focus on how peptide-lipid interaction is affected by a membrane potential applied to phospholipid vesicles. The membrane potential (negative inside) was generated by adding the K^+ ionophore valinomycin to large unilamellar vesicles (LUV) exhibiting a K^+_{in}/Na^+_{out} ion gradient.

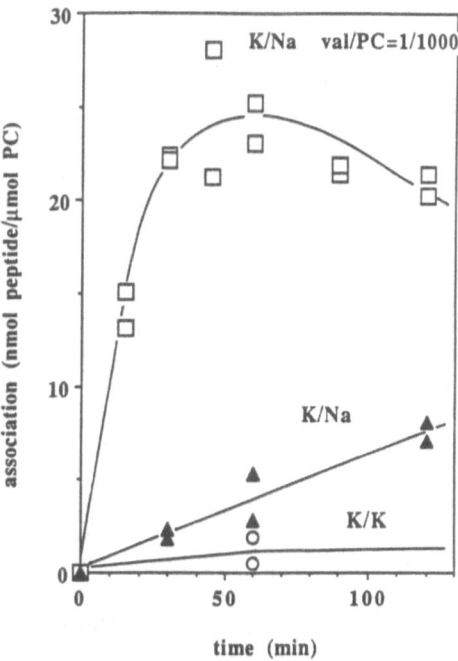

Fig. 8. Time course of the ion gradient-induced association of AlXme$^+$ with eggPC LUV assayed by the minicolumn gel filtration method. The vesicles were prepared in 150 mM K_2SO_4, 20 mM Hepes, pH 7.0 (K^+_{in}/K^+_{out}). A K^+_{in}/Na^+_{out} ([]) ion gradient was generated by passing the vesicles through a Sephadex G50 column eluted with 150 mM Na_2SO_4, 20 mM Hepes pH 7.0. Samples of 150 ml withdrawn from a 1 mM PC vesicle suspension containing 0.1 mM peptide were passed through a syringe containing 1.5 ml Sephadex G50 by 2 min centrifugation at 300 x g (De Kroon et al., 1989). The filtrates were collected and analyzed for phospholipid phosphorus and bound peptide. Valinomycin was added at t = 0 to a 10^{-3} molar ratio with respect to PC. ▲ without valinomycin added, O without a K^+ gradient. For details see De Kroon et al. (1991b).

For this purpose LUV were prepared in K_2SO_4 containing buffer and subsequently the external medium was exchanged for Na_2SO_4 containing buffer by gel filtration. When applied to PC vesicles made by an extrusion technique, very stable ion gradients were obtained (cf. Blok et al., 1974; Hope et al., 1985). The so-called minicolumn gel filtration method (Penefsky, 1977; Hope et al., 1985) was found to be a suitable method to monitor peptide-vesicle association.

Figure 8 shows that the association of the peptide AIXme$^+$ with PC LUV is dramatically enhanced upon application of a valinomycin-induced K$^+$ diffusion potential. In the absence of valinomycin, with the K$^+$ gradient present, a gradually increasing peptide association is observed as a result of the building up of a diffusion potential due to the difference in membrane permeability for K$^+$ and Na$^+$ ions. Measurements of the membrane potential ($\Delta\Psi$), which were carried out in parallel using the radioactively labeled probe tetraphenylphosphonium (TPP$^+$), demonstrated that the vesicle association of AIXme$^+$ results in a slowly proceeding dissipation of $\Delta\Psi$ (De Kroon et al., 1991b). The latter is also apparent from the gradual release of peptide after 60 min of incubation (Fig. 8).

To elucidate the vesicle localization of the associated peptide, advantage was taken of the fluorescence properties of the intrinsic tryptophan residue. As is shown in Fig. 9, the K$^+$ diffusion potential-induced peptide-vesicle association is paralleled by a fluorescence increase, indicative for the entry of the tryptophan residue into an apolar environment. Figure 9 shows two additional features of the peptide-vesicle association.

Fig. 9. Time course of the ion gradient-induced tryptophan fluorescence change at 340 nm of 2 mM AIXme$^+$ added to 300 mM PC LUV exhibiting a K$^+_{in}$/Na$^+_{out}$ gradient without ionophores ([]), with valinomycin (■), and with both valinomycin and FCCP (Δ), added at t = 0. Valinomycin, FCCP, and NFG were added to molar ratios of 10^{-3}, 1.3×10^{-4}, and 2.5×10^{-3} with respect to PC; buffer conditions as in Fig. 8. Taken from De Kroon et al. (1991b).

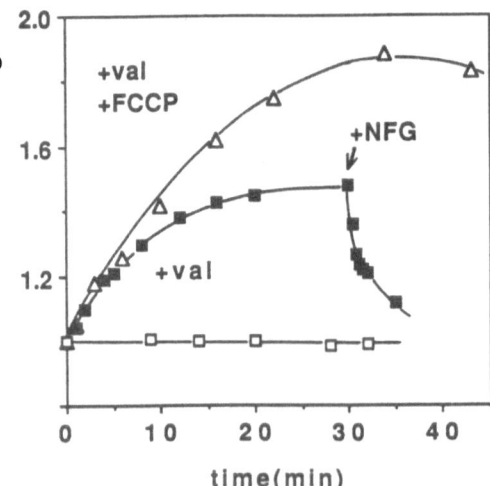

Firstly, the association is completely reversible, which is apparent from the return of the fluorescence intensity to its initial level, upon dissipation of the ion gradients by the addition of the non-fluorescent channel forming tryptophan-N-formylated gramicidin (NFG). Secondly, with both valinomycin and the protonophore FCCP present, the fluorescence increase is accelerated and it reaches a higher value. Under these conditions, FCCP enhances the $\Delta\Psi$-induced pH gradient from ~1.3 to ~3 pH units as was shown using the ΔpH probe methylamine (De Kroon et al., 1991b; cf. Nichols et al., 1980; Redelmeier et al., 1989). The latter result indicates that the pH gradient is involved in the peptide-vesicle association. In fact, under conditions of a high buffering capacity which prevent the development of any measurable pH gradient, the fluorescence response of AlXme$^+$ to the K$^+$ diffusion potential was strongly reduced at pH 7 and virtually absent at pH 6, indicating that the pH gradient is the major, if not the only driving force for the enhanced peptide-vesicle association (De Kroon et al., 1991b). Mechanistically, this implies that the deprotonated form of the peptide distributes across the bilayer according to the pH gradient, as was found for other lipophilic weak bases (Nichols & Deamer, 1976; Bally et al., 1988). Consequently, a localization of the peptide at the inside of the vesicles is predicted.

This prediction was verified in experiments employing the brominated phospholipids introduced in section 2. Fluorescence quenching measurements revealed an interfacial localization for the tryptophan residue of AlXme$^+$ in PC LUV exhibiting a K$^+$ diffusion potential, similar to that observed for this peptide when bound to CL SUV (Fig. 3a; De Kroon et al., 1991b). In order to distinguish between a localization at the inner or at the outer leaflet of the bilayer, LUV with an asymmetric distribution of BrPC were prepared by phosphatidylcholine transfer protein (PCTP) mediated PC exchange (De Kroon et al., 1991b). Asymmetric BrPC containing vesicles have been successfully used previously in determining the membrane topology of cytochrome b5 which was found to insert into the membrane outer leaflet only (Everett et al., 1986). Briefly, repeated incubation of LUV consisting entirely of di(9,10-dibromostearoyl)PC with a 20-fold molar excess of egg PC SUV in the presence of PCTP resulted in the exchange of at least 80% of the LUV outer leaflet BrPC for eggPC. The asymmetric LUV were recovered from the incubation by high speed centrifugation. The selective introduction of BrPC into the outer leaflet of the LUV up to ~10%, was accomplished by incubating eggPC LUV with SUV containing BrPC in the presence of PCTP.

As is shown in Fig. 10 the fluorescence of AlXme$^+$ is quenched when a K$^+$ diffusion potential is applied to the asymmetric vesicles with BrPC in the inner leaflet.

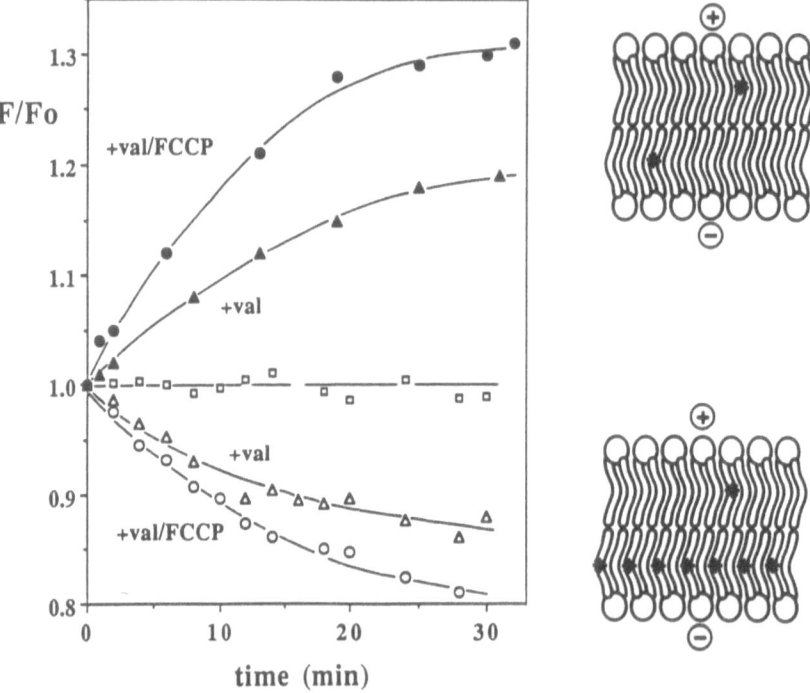

Fig. 10. Fluorescence change of AIXme$^+$ in reponse to LUV with an asymmetric transbilayer distribution of 100% Br$_4$PC in the inner and ≤ 20% Br$_4$PC (≥ 80% PC) in the outer leaflet of the membrane, in the absence of an ion gradient ([]), in the presence of a K$^+_{in}$/Na$^+_{out}$ gradient after the addition of valinomycin (Δ), and after addition of both valinomycin and FCCP (O). The corresponding closed symbols represent identical conditions applied to symmetric LUV with 20% Br$_4$PC incorporated. Valinomycin was added to a 10^{-4} molar ratio with respect to PC, other concentrations as in Fig. 9. Control experiments showed that the membrane potential stability of the asymmetric vesicles is indistinguishable from that of symmetric vesicles containing up to 20% BrPC, and that the bilayer asymmetry is not perturbed by the peptide uptake (De Kroon et al., 1991b). The schematic drawings depict the (a)symmetry of the vesicles used with the asterisk representing the bromines.

The control experiment employing vesicles containing 20% BrPC evenly distributed over both leaflets still shows a fluorescence increase, indicating that at least part of the peptide's fluorescence is quenched by bromines in the inner leaflet of the bilayer. Conversely, using the vesicles with BrPC exclusively present in the outer leaflet, a fluorescence increase is observed that does not significantly differ from that obtained when using eggPC LUV (Fig. 11).

It is concluded, that the peptide AIXme$^+$ accumulates at the interface of the inner leaflet of the bilayer in response to a K$^+$ diffusion potential. This process is primarily determined by the pH gradient evoked by ΔΨ, however, a direct stimulatory effect by the ΔΨ cannot be excluded (De Kroon et al., 1991b). In the following section the effect of a membrane potential on the vesicle-interaction of bioactive and model peptides containing more than one ionizable group will be described.

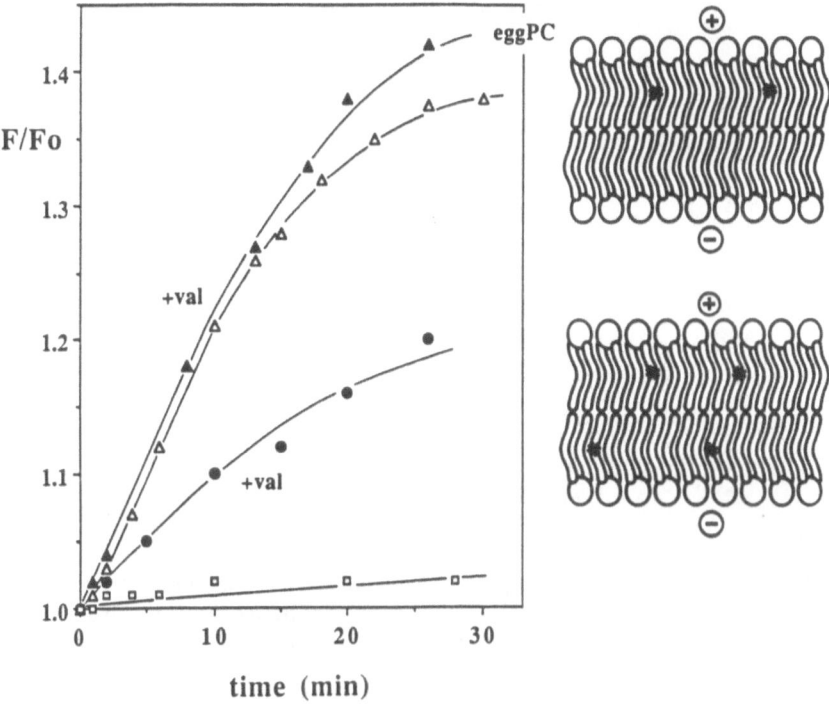

Fig. 11. Fluorescence change of AlXme⁺ upon addition to LUV containing ~10% 6,7-Br$_2$PC exclusively in the outer leaflet of the bilayer and experiencing a K^+_{in}/Na^+_{out} gradient without ([]) and with valinomycin added at t = 0 to a 2 x 10⁻⁴ molar ratio with respect to PC (Δ). For comparison, the fluorescence response recorded for PC LUV (▲) and for 12.5% 6,7-Br$_2$PC containing symmetric LUV (●) exhibiting a K^+ diffusion potential is shown. For details see De Kroon et al. (1991b). The schematic diagrams show the BrPC transbilayer distribution of the vesicles used.

Membrane potential and biologically active peptides

Mastoparans are toxic tetradecapeptides from wasp venom that, e.g., stimulate the degranulation of mast cells (Hirai et al., 1979), via a mechanism that probably involves the activation of G-proteins (Higashijima et al., 1988, 1990). Mastoparans have a strong binding affinity for (model) membranes (Uzu et al., 1985), and adopt an α-helix conformation when bound to phospholipid vesicles (Higashijima et al., 1983; Wakamatsu et al., 1983). Mastoparan X is a naturally occurring mastoparan analogue which contains a tryptophan residue and which has a net charge of 4+ (Hirai et al., 1979).

226

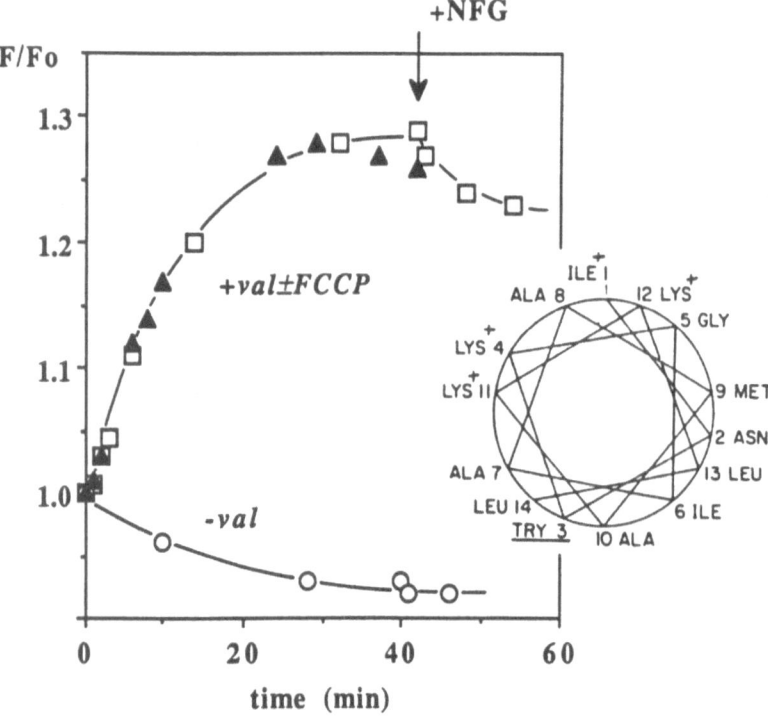

Fig. 12. Time course of the K^+ diffusion potential-induced tryptophan fluorescence change at 340 nm of 2 mM mastoparan X added to 300 mM PC LUV experiencing a K^+_{in}/Na^+_{out} gradient without (O), with valinomycin ([]), and with both valinomycin and FCCP added at t = 0 (▲). Conditions as in Fig. 9. The helical wheel projection of mastoparan X is also depicted. Data taken from De Kroon et al. (1991c).

Figure 12 shows that mastoparan X exhibits a tryptophan fluorescence increase similar to that of AlXme$^+$ (Fig. 9), upon applying a K^+ diffusion potential to PC LUV. Yet there are some differences: the mastoparan X fluorescence increase is not affected by FCCP, and no enhanced mastoparan X-vesicle association is detectable in the minicolumn gel filtration assay (De Kroon et al., 1991c). These results rule out the possibility of a large scale accumulation of mastoparan X inside the vesicles, as was found for the model hexapeptide. Centrifugation experiments revealed that in the presence of a membrane potential, significantly more mastoparan X binds to the vesicles than in its absence (De Kroon et al., 1991c).

Membrane potential has been proposed to electrophoretically transfer mastoparan across the plasma membrane, thus enabling it to activate the intracellularly localized G-proteins (Higashijima et al., 1988). The data summarized here argue against a large scale electrophoresis of mastoparan across a lipid bilayer, but do not exclude that a small subpopulation of the peptide molecules present, does get across the membrane in response to

ΔΨ. The formation of voltage dependent channels in black lipid membranes by mastoparan (Mellor & Sansom, 1990), might in the light of the vesicle system data, also be due to only a small fraction of the mastoparan molecules present.

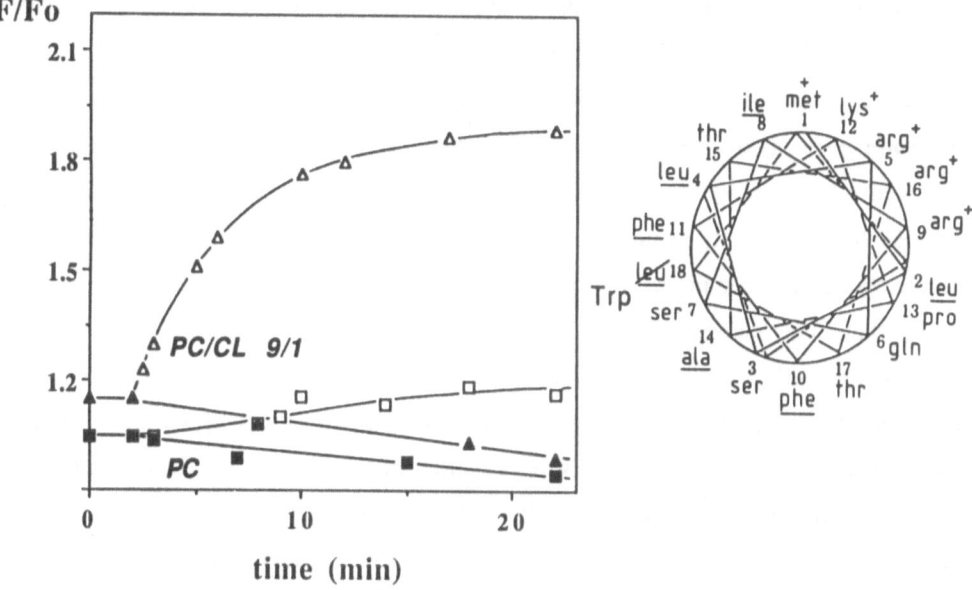

Fig. 13. Tryptophan fluorescence change at 340 nm of the tryptophan containing presequence of cytochrome oxidase subunit IV upon addition of LUV consisting of PC ($\square$) and PC/CL 9/1 (Δ), and exhibiting a K^+_{in}/Na^+_{out} gradient in the absence (closed symbols) and presence of valinomycin which was added to a 10^{-4} valinomycin/phospholipid molar ratio (open symbols). The fluorescence intensity F is related to F_o, the fluorescence intensity of the peptide in the absence of vesicles. The helical wheel projection of the presequence showing the position of the tryptophan residue is depicted. For details see De Kroon et al. (1991c).

The presequence of the mitochondrial precursor protein cytochrome oxidase subunit IV has been reported to induce the release of phospholipid vesicle contents, which is enhanced by a membrane potential, negative inside (Roise et al., 1986). The synthesis of this presequence with a tryptophan residue replacing the leucine at position 18, makes the peptide amenable for tryptophan fluorescence experiments, which allow the direct measurement of its response to a membrane potential. Figure 13 shows that the fluorescence increase of the tryptophan containing presequence upon applying a membrane potential to PC vesicles, is fairly modest compared to that of mastoparan X (Fig. 12). However after incorporation of 10% CL into the vesicles, the rate and the extent of the fluorescence increase are dramatically enhanced (Fig. 13). Interestingly, this CL content more or less corresponds to the acidic phospholipid content determined for the mitochondrial outer membrane (Hovius et al., 1990).

The rate of the $\Delta\Psi$-induced, ΔpH independent fluorescence increase of mastoparan X was likewise enhanced by raising the CL content of the LUV (De Kroon et al., 1991c). The introduction of a negative surface charge increases the affinity of the positively charged peptides for the membrane, which, apparently, increases their susceptibility to $\Delta\Psi$.

Interestingly, this also holds true for the model peptide RXme^{2+}. When the CL content of the LUV is raised to 50%, a small but significant $\Delta\Psi$-induced, ΔpH insensitive increase in resonance energy transfer of the peptide's tryptophan to vesicle incorporated dansyl-PE is observed (De Kroon et al., 1991b). Similarly, the regulatory peptides dynorphin A and ACTH1-24 which hardly or not associate with PC bilayers, but which do bind strongly to negatively charged membranes, only start to show a $\Delta\Psi$ response at a vesicle CL content of 25% and 50%, respectively (De Kroon et al., 1991c). Obviously, a peptide's susceptibility to $\Delta\Psi$ and its a priori affinity for the lipid vesicles are correlated. The $\Delta\Psi$-induced, ΔpH independent, effects on the peptides' fluorescence properties probably originate from an increased binding to the vesicles, as was demonstrated for mastoparan X and dynorphin A (Leenhouts & De Kroon, unpublished), with possible contributions from a changed conformation or orientation of the peptide in the membrane bound state.

Another factor that may determine whether or not a peptide is $\Delta\Psi$ sensitive, is its secondary structure. Mastoparan X has a very strong tendency to change from an unordered structure in aqueous solution into an α-helical conformation upon binding to a membrane (Wakamatsu et al., 1983; De Kroon et al., 1991c). The presequence of cytochrome oxidase subunit IV adopts an α-helix conformation in negatively charged bilayers, but it assumes an unordered structure in bilayers consisting of PC (Tamm & Bartoldus, 1990). The presequence of pre-ornithine transcarbamylase binds to phospholipid vesicles but the binding is not affected by a K$^+$ diffusion potential (Skerjanc et al., 1987). Interestingly, this presequence preferentially adopts a β-pleated structure when membrane bound (Epand et al., 1986). For the peptides ACTH1-24 the formation of α-helical structure in the presence of lipids has been reported (Gremlich et al., 1983), for dynorphin A it is controversial (Maroun & Mattice, 1981; Schwyzer, 1986; Lancaster et al., 1991).

Taken together, these and other data tentatively suggest a parallel between the α-helicity of a peptide when membrane bound, and its susceptibility to $\Delta\Psi$, and they point to the possibility of $\Delta\Psi$ acting on the dipole moment of the α-helix (Hol, 1985), rather than on its charged groups (cf. Tosteson et al., 1990).

An alternative mechanism by which a membrane potential might influence peptide-lipid interaction takes into consideration the effect $\Delta\Psi$ conceivably could have on the conformation of the membrane lipids. A $\Delta\Psi$-induced change in the orientation of the phospholipid headgroup dipoles as proposed by Seelig et al. (1987), could possibly lead to an altered interaction with peptides.

Acknowledgements

We gratefully acknowledge the contributions of R. Van 't Hof, J.A. Killian, G. De Korte-Kool, M.W. Soekarjo, J.W. Timmermans, and B. Vogt to the research described in this article. This work was carried out under the auspices of the Netherlands Foundation of Biophysics with financial support from the Netherlands Organization for Scientific Research (NWO).

References

Bashford CL, Chance B, Smith JC and Yoshida T (1979) Biophys J 25: 63-85
Bakker EP and Randall LL (1984) EMBO J 3: 895-900
Bally MB, Mayer LD, Loughrey H, Redelmeier T, Madden TD, Wong K, Harrigan PR, Hope MJ and Cullis PR (1988) Chem Phys Lipids 47: 97-108
Batenburg AM, Demel, RA, Verkleij AJ and De Kruijff B (1988) Biochemistry 27: 5678-5685
Bernheimer AW and Rudy B (1986) Biochim Biophys Acta 864: 123-141
Beschiaschvili G and Seelig J (1990) Biochemistry 29: 10995-11000
Blok MC, De Gier J and Van Deenen LLM (1974) Biochim Biophys Acta 367: 202-209
Bolen EJ and Holloway PW (1990) Biochemistry 29: 9638-9643
Briggs MS and Gierasch LM (1986) Adv Protein Chem 38: 109-180
Browning J and Seelig J (1979) Chem Phys Lipids 24: 103-118
Carruthers A and Melchior DL (1986) TIBS 11: 331-335
Chupin V, Killian JA and De Kruijff B (1987) Biophys J 51: 395-405
Cowgill RW (1967) Biochim Biophys Acta 133: 6-18
De Kroon AIPM, De Gier J and De Kruijff B (1989) Biochim Biophys Acta 981: 371-373
De Kroon AIPM, Soekarjo MW, De Gier J and De Kruijff B (1990a) Biochemistry 29: 8229-8240
De Kroon AIPM, Timmermans JW, Killian JA and De Kruijff B (1990b) Chem Phys Lipids 54: 33-42
De Kroon AIPM, Killian JA, De Gier J and De Kruijff B (1991a) Biochemistry 30: 1155-1162
De Kroon AIPM, Vogt B, Van 't Hof R, De Kruijff B and De Gier J (1991b) Biophys J 60: in press
De Kroon AIPM, De Gier J and De Kruijff B (1991c) Biochim Biophys Acta: in press
Dempsey C, Bitbol M and Watts A (1989) Biochemistry 28: 6590-6596
Devaux PF, Hoatson GL, Favre E, Fellman P, Farren B, MacKay AL and Bloom M (1986) Biochemistry 25: 3804-3812
De Vrije T, De Swart RL, Dowhan W, Tommassen J and De Kruijff B (1988) Nature 334: 173-175

Dufourcq J, Faucon JF, Maget-Dana R, Pileni MP and Hélène C (1981) Biochim Biophys Acta 649: 67-75

East JM and Lee AG (1982) Biochemistry 21: 4144-4151

Epand RM, Hui SW, Argan C, Gillespie LL and Shore GC (1986) J Biol Chem 261: 10017-10020

Everett J, Zlotnick A, Tennyson J and Holloway PW (1986) J Biol Chem 261: 6725-6729

Farren SB, Sommerman E and Cullis PR (1984) Chem Phys Lipids 34: 279-286

Gremlich HU, Fringeli UP and Schwyzer R (1983) Biochemistry 22: 4257-4264

Higashijima T, Wakamatsu K, Takamitsu M, Fujino M, Nakajima T and Miyazawa T (1983) FEBS Lett 152: 227-230

Higashijima T, Uzu S, Nakajima T and Ross EM (1988) J Biol Chem 263: 6491-6494

Higashijima T, Burnier J and Ross EM (1990) J Biol Chem 265: 14176-14186

Hirai Y, Yasuhara T, Yoshida H, Nakajima T, Fujino M and Kitada C (1979) Chem Pharm Bull 27: 1942-1944

Hol WGM (1985) Prog Biophys Molec Biol 45: 149-195

Hong K, Düzgünes N, Meers PR and Papahadjopoulos D (1987) in Cell Fusion: 269-285, Plenum Press New York

Hope MJ, Bally MB, Webb G and, Cullis PR (1985) Biochim Biophys Acta 812: 55-65

Hovius R, Lambrechts H, Nicolay K and De Kruijff B (1990) Biochim Biophys Acta 1021: 217-226

Hudson TH, Scharff J, Kimak MAG and Neville DM Jr (1988) J Biol Chem 263: 4773-4781

Jacobs RE and White SH (1986) Biochemistry 25: 2605-2612

Jacobs RE and White SH (1987) Biochemistry 26: 6127-6134

Jacobs RE and White SH (1989) Biochemistry 28: 3421-3437

Jain MK, Rogers J, Simpson L and Gierasch LM (1985) Biochim Biophys Acta 816: 153-162

Kuchinka E and Seelig J (1989) Biochemistry 28: 4216-4221

Lancaster CRD, Mishra PK, Hughes DW, St. Pierre SA, Bothner-By AA and Epand RM (1991) Biochemistry 30: 4715-4726

Lill R, Dowhan W and Wickner W (1990) Cell 60: 271-280

Lytz RK, Reinert JC, Church SE and Wickman HH (1984) Chem Phys Lipids 35: 63-76

MacDonald RC, Simon SA and Baer E (1976) Biochemistry 15: 885-891

Maroun R and Mattice WL (1981) Biochem Biophys Res Commun 103: 442-446

McElhaney RN (1982) Curr Top Membr Transp 17: 317-380

McIntosh TJ and Holloway PW (1987) Biochemistry 26: 1783-1788

Mellor IR and Sansom MSP (1990) Proc R Soc Lond B239: 383-400

Menestrina G, Voges KP, Jung G and Boheim G (1986) J Membrane Biol 93: 111-132

Merrill AR and Cramer WA (1990) Biochemistry 29: 8529-8534

Neville Jr DM and Hudson TH (1986) Ann Rev Biochem 55: 195-224

Nichols JW and Deamer DW (1976) Biochim Biophys Acta 455: 269-271

Nichols JW, Hill MW, Bangham AD and Deamer DW (1980) Biochim Biophys Acta 596: 393-403

Penefsky HS (1977) J Biol Chem 252: 2891-2899

Pfanner N and Neupert W (1985) EMBO J 4: 2819-2825

Redelmeier TE, Mayer LD, Wong KF, Bally MB and Cullis PR (1989) Biophys J 56: 385-393

Roise D, Horvath SJ, Tomich JM, Richards JH and Schatz G (1986) EMBO J 5: 1327-1334

Roseman MA, Lentz BR, Sears B, Gibbs D and Thompson TE (1978) Chem Phys Lipids 21: 205-222

Roux M and Bloom M (1990) Biochemistry 29: 7077-7089

Roux M, Neumann JH, Bloom M and Devaux PF (1988) Eur Biophys J 16: 267-274

Roux M, Neumann JH, Hodges RS, Devaux PF and Bloom M (1989) Biochemistry 28: 2313-2321

Sargent DF, Bean JW and Schwyzer R (1988) Biophys. Chem. 31: 183-193

Schiebel E, Driessen AJM, Hartl HU and Wickner W (1991) Cell 64: 927-938

Schwyzer R (1986) Biochemistry 25: 4281-4286

Seelig A and Macdonald PM (1989) Biochemistry 28: 2490-2496

Seelig J (1977) Q Rev Biophys 10: 353-418

Seelig J, MacDonald PM and Scherer PG (1987) Biochemistry 26: 7535-7541

Skerjanc IS, Shore GC and Silvius JR (1987) EMBO J 6: 3117-3123

Stankowski S and Schwarz G (1990) Biochim Biophys Acta 1025: 164-172

Surewicz WK and Epand RM (1984) Biochemistry 23: 6072-6077

Surewicz WK and Epand RM (1986) Biochim Biophys Acta 856: 290-300

Tamm LK (1986) Biochemistry 25: 7470-7476

Tamm LK and Bartoldus I (1990) FEBS Lett 272: 29-33

Tosteson MT and Tosteson DC (1981) Biophys J 36: 109-116

Tosteson MT, Alvarez O, Hubbell W, Bieganski RM, Altenbach C, Caporales LH, Levy JJ, Nutt RF, Rosenblatt M and Tosteson DC (1990) Biophys J 58:1367-1375

Uzu S, Nakajima T, Saito K, Wakamatsu K, Miyazawa T and Fujino M (1985) in Peptide Chemistry (Ed. Y. Kiso) pp: 229-234, Osaka Protein Research Foundation

Voges KP, Jung G and Sawyer WH (1987) Biochim Biophys Acta 896: 64-76

Wakamatsu K, Higashijima T, Fujino M, Nakajima T and Miyazawa T (1983) FEBS Lett 162: 123-126

Yamada H, Matsuyama S, Tokuda H and Mizushima S (1989) J Biol Chem 264: 18577-18581

RECONSTITUTION OF PROTEIN TRANSLOCATION ACTIVITY INTO LIPID VESICLES

Henrik Fridén, Gay L. Bush, Daniel Niclas, Adam J. Savitz and
David I. Meyer
Department of Biological Chemistry
UCLA School of Medicine and the
Molecular Biology Institute
Los Angeles, CA 90024
USA

Translocating a protein across a membrane, or getting one
assembled into it, is one of the most complex--yet routine--
activities that cells are called upon to perform. It is a form
of transport that transports transporters, uses receptors to
assemble other receptors, and represents one of the most ancient
forms of signal transduction. It is one of the best
characterized intracellular traffic routes, and most
importantly, the mechanism of its action is on the verge of
becoming accessible to study by functional reassembly from
isolated components. This exciting prospect raises several
questions that bear on everything from the approaches and
systems being used, to the final interpretation of the results.

The translocation of secretory proteins into the lumen of the
endoplasmic reticulum was the first such reaction to be
faithfully duplicated in vitro. Since then, many other protein
translocation reactions have been studied in the test tube and
biochemical dissections carried out. We now have considerable
information on the components mediating translocation of
proteins from the cytosol into mitochondria, chloroplasts, the
nucleus and into the bacterial periplasm. As the endoplasmic
reticulum membrane and the bacterial plasma membrane are (in
contrast to those of mitochondria and chloroplasts) composed of
single membranes, the chances of success in reconstituting a
functional unit from component parts is the greatest. It is
therefore not unexpected that the greatest progress in this area
has been achieved with such single membrane systems.

Both the bacterial plasma membrane and the endoplasmic reticulum are multifunctional structures. These membranes are the site of many cellular processes in addition to protein translocation. This suggests that only a fraction of their proteinaceous components are involved in translocation. Thus the first major question facing the investigator involves discriminating the participants from the by-standers in the translocation reaction. Since the number of endoplasmic reticulum-specific proteins is probably in the 50-150 range (Hortsch et al. 1987), this is no mean feat. Strategies must be devised that permit the selective isolation of the relevant proteins in an active state. Once this has been accomplished, and translocation has been replicated by proteins correctly assembled in an artificial membrane system, the major goal will be to define the role played by the individual components in the process. It is only at that point that a complete understanding of the process of protein translocation across a membrane will be acquired.

It is the purpose of this article to consider some of the approaches that are available in dealing with these problems. Certain of them have already provided us with valuable information regarding putative players in the game. Just as important, however, is that they also have given insights into the feasibility and limitations of a particular approach.

Step One: Can a solubilized membrane be functionally reconstituted? The most basic biochemical approach begins with a starting material that comprises some crude and easily obtained active fraction. In this case, it is a rough microsomal preparation from a mammalian secretory tissue, specifically canine pancreas. Reconstitution of a transmembrane, unidirectional transport step relies, by definition, on being able to reconstitute the lipid bilayer across which transport is to occur. This is easily accomplished when detergents are removed from solutions of phospholipids. As one can imagine, obtaining the correct quantity and orientation of proteins in such an artificial membrane, when detergent is removed from solubilized microsomes, is less likely to occur with such ease. Operationally, this means that the levels of translocation activity obtained from a reconstituted membrane will be considerably lower than from even the crudest of microsomal preparations.

Despite this handicap, workers in several laboratories have reported success in reconstituting translocation from partially (Yu et al., 1989; Zimmerman and Walter, 1989), and even completely solubilized microsomes (Nicchita and Blobel, 1990; Nicchita et al., 1991). In the case of rough ER-derived membranes, the difference between partial and complete solubilization resides in the type of detergent employed. As rough ER-specific proteins tend to interact with each other through electrostatic interactions, ionic detergents are required to solubilize them. High salt concentrations will accomplish the same task when non-ionic detergents are employed. Using the ionic detergent sodium cholate, microsomes have been completely solubilized, and dilution or removal of the detergent has yielded proteoliposomes capable of translocating secretory proteins (Nicchita and Blobel, 1990; Nicchita et al., 1991). Non-ionic detergents seem to leave rough ER-specific proteins in large aggregates that retain their functionality when the detergent is removed and proteoliposomes form (Yu et al., 1989; Zimmerman and Walter, 1989). This latter approach has the advantage that complexes may be isolated which contain most or all of the membrane components that participate in translocation. The former approach, although representing a true solubilization, will depend upon an as yet undefined strategy for the specific recovery of the components of the translocation machinery, or it will rely on their ability to "self-assemble". The concept of self-assembly may turn out to play an important role in such efforts (Nicchita et al., 1991).

The hope that the components of the translocation machinery form functional complexes and/or self-assemble is easily appreciated when one considers that translocation is a process requiring the participation and correct interaction of numerous membrane proteins. With this in mind, it is understandable that initially translocation might best be studied in a reconstituted system "one step at a time". This would involve being able to first determine individual subreactions of the translocation process in intact membranes. Such previously characterized subreactions include the release of an SRP-mediated translation arrest by docking protein (Meyer et al., 1982; Gilmore et al., 1982), the binding of preproteins to membrane receptors (Sanz and Meyer, 1989), and the binding of ribosomes to ribosome-free (stripped) rough microsomes (Savitz and Meyer, 1990). This type of approach will most certainly be continued to include blocking the translocation process at particuluar stages by withholding required cofactors, or by using truncated preproteins as substrates, or through the in vitro use of membranes derived from yeast mutants defective in translocation. Through such studies candidate proteins, that will be included in the roster of participants to be tested by incorporation into proteoliposomes, will be identified.

Step 2: Deciding which proteins to incorporate into liposomes. The process of translocation, for the sake of convenience, can be divided into three major steps. The first two steps, the recognition of those nascent proteins that are to become incorporated or translocated across the membrane, and their targeting to and docking with the ER membrane occur in the easily-accessible cytosolic compartment and have been well characterized by conventional biochemical means. It is the third step, that of the translocation of the nascent chain across the lipid bilayer, whose analysis requires the solubilization and reconstitution of a functional membrane. Deciding, then, which membrane proteins to include in the mixtures of lipid and detergent used to make proteoliposomes is a major focus of concern.

The three approaches from which viable candidates have been obtained include purification on the basis of a functional assay, chemical cross-linking of preproteins and membrane components proximal to them during their translocation, and identification of the gene products that are defective in translocation mutants. The first membrane component to be isolated by a functional assay was the docking protein, or SRP receptor (Meyer et al., 1982; Gilmore et al., 1982). It is integrally associated with the rough ER membrane, with most of its 69.5 kDa having a cytosolic disposition (Lauffer et al., 1985). After the recognition of a nascent secretory or membrane protein by the signal recognition particle (SRP), the docking protein mediates the interaction of the translation complex with the membrane by virtue of its affinity for the SRP (Meyer et al., 1982; Gilmore et al., 1982). Binding and hydrolysis of GTP have been shown to mediate this interaction (Connolly et al., 1991). The participation of docking protein in the translocation process was confirmed by its ability to release an SRP-induced translation arrest observed in a wheat germ cell-free translation system. To date no data have appeared that confirms the role of this component in translocation in vivo.

Another membrane component to be isolated using a functional assay is the ribosome receptor (Savitz and Meyer, 1990). This 180 kDa integral membrane protein was isolated by the ability of a derived proteolytic fragment to inhibit ribosome binding to intact stripped rough microsomes. Ultimately, the purified receptor was incorporated into liposomes which could then be demonstrated to have ribosome binding capabilities approximating those of intact membranes. This represents the first case of reconstituting an individual subreaction of the translocation process into proteoliposomes, thereby defining an individual participant in the reaction. Subsequent studies on docking protein have shown that an SRP-mediated translation arrest can be relieved by proteoliposomes containing, among other things, docking protein.

Chemical cross-linking has been successfully used to identify three proteins that may be playing a role in translocation (Reviewed in Rapoport, 1990). It is now quite certain that the 54 kDa subunit of SRP is involved in signal sequence recognition. An identical approach, that of supplementing cell-

free translations of secretory proteins with photoreactive
lysine analogs charged onto lysyl t-RNA and allowing their
translocation to occur in vitro, has demonstrated that two
integral membrane glycoproteins of 25 kDa and 35 kDa are
proximal to the translocating nascent chain. Antibodies to the
35 kDa species have been shown to have an inhibitory effect on
translocation. Because the original work showed that
photoreactive lysines in the signal sequence were cross-linked
to these proteins their function was postulated to be as
subunits of a membrane-bound signal sequence receptors (SSR).
The proximity of the SSR to translocating proteins was
determined to be maintained through later stages in
translocation, as photoreactive lysines located in the mature
portion of the nascent secretory protein were cross-linked to
the SSR proteins as well. Thus the current thinking is that
these membrane components might function as members of a
postulated pore complex through which nascent chains travel
during their vectorial transfer from the cytosol to the lumen of
the ER (Blobel and Dobberstein, 1975). Proteoliposomes capable
of carrying out translocation reactions should therefore also
include SSR proteins.

Genetic approaches in manipulable organisms such as *S.
cervisiae* have led to the identification of three membrane
components that deserve further attention (Reviewed in Schatz,
1991). Known by the genetic designation of the mutant
phenotype, *sec61*, *sec62* and *sec63* (the latter having also been
identified in other labs as *ptl1 and npl1*) are all defective in
protein translocation. Additional mutations that show
translocation defective phenotypes are *sec65*, the yeast homolog
of an SRP subunit (see contribution by C. Stirling in this
volume), and *kar2*, the yeast equivalent of the mammalian ER
lumenal protein known as BiP. Although membranes derived from
kar2 yeast are incapable of carrying out translocation *in vitro*,
mammalian membranes from which BiP has been depleted retain this
ability (Bulleid and Freedman, 1990). Thus a requirement for
the incorporation of BiP into functional proteoliposomes is not
yet definite.

Cloning of the *SEC61, SEC62 and SEC63* genes has suggested their products to be integral membrane proteins (Deshaies and Schekman, 1989; Sadler, et al., 1989). Antibodies that react with Sec62p and Sec63p have been used to verify that these proteins are 32 kDa and 75 kDa residents of the ER respectively (Friden, H., Bush, G.L. and Meyer,D.I., unpublished). Sec61p, based on its amino acid sequence (R. Schekman, personal communication), is a very hydrophobic 53 kDa membrane protein, sharing a very limited homology to the bacterial SecY protein, a component of the *E. coli* translocation machinery. To date, no mammalian homologs of these *SEC* genes have been identified. This points out the fact that reconstitution of translocation in proteoliposomes derived from yeast microsomes will and should be a separate endeavor from that using components of mammalian origin. It also raises the question as to the overall similarity between translocation in bacteria, yeast and mammalian cells.

Step 3: Isolating active components individually or as complexes. Despite the fact that several membrane proteins have been identified and the steps involved in their isolation perfected, it would be surprising if this handful of molecules could reconstitute translocation when integrated into liposomes. The approach of building liposomes containing individually-purified components is worthwhile when one is reasonably certain that a minimum number of substituents have been identified. This is presently working out nicely in bacteria, where extensive genetic studies suggest that a limited number of membrane proteins, 6 to be exact, is involved. The same might be possible in yeast (Deshaies et al., 1991). In this instance, however, identification of putative participants have come from genetics (SEC61-63,65) as well as from isolation of genes encoding homologs of mammalian membrane proteins (e.g., docking protein; P. Walter, personal communication). It is not yet clear that the SEC genes and the SRP/docking protein homologs interact with one another in a singular translocation reaction, and thus inclusion of all of them into the same liposome might not lead to reconstitution. More extensive biochemical analyses, coupled with reverse genetic studies are needed to establish involvement of the mammalian homologs in the secretion process in yeast. This is justified, since it is not at all clear that bacterial homologs (Römisch et al., 1989; Bernstein, et al., 1989) of SRP subunits and docking protein participate in the translocation of the most commonly studied preproteins (Bassford et al., 1991).

For the moment, the preferred approach is the isolation of "translocation complexes" from ER-derived membranes, whether from mammalian or yeast cells. In each case, certain proteins have been identified that can be used as handles to pull out such complexes. For example, the docking protein, or the ribosome receptor have large cytosolic domains. Previous work has shown that these two proteins, as well as other putative participants such as ribophorins, the signal peptidase complex and the SSR proteins are not solubilized by non-ionic detergents at low salt concentrations (Hortsch et al., 1987; Yu et al., 1989; Zimmerman and Walter, 1989). This insoluble material has already been shown to be capable of mediating protein translocation when incorporated into liposomes. The only task that remains is to more precisely separate out any proteins that are not required for this process. This could be accomplished by binding the insoluble complexes to anti-docking protein antibodies which are immobilized on beads or other suitable matrices. Washing at increasing salt concentrations, in the presence of detergent, will give rise to pools of ER membrane proteins that are either loosely, or more tightly associated with the docking protein. Similar studies could be carried out using anti-ribosome receptor antibodies, and so forth. When protein profiles from such elutions are compared, components most tightly associated with the various known components will be obvious. Furthermore, fractions can be combined for incorporation into liposomes and tested for activity beginning with the known protein plus the most tightly associated fraction, and then adding the next most tightly associated fraction and so on until active proteoliposomes are obtained.

REFERENCES:

Bassford P, Beckwith J, Ito K, Kumamoto C, Mizushima S, Oliver D, Randall L, Silhavy T, Tai PC and Wickner B (1991) The primary pathway of protein export in E. coli. Cell 65: 367-368

Berstein HD, Poritz MA, Strub K, Hoben PJ, Brenner S and Walter P (1989) Model for signal sequence recognition from amino-acid sequence of 54K subunit of signal recognition particle. Nature 340: 482-487

Blobel G and Dobberstein B (1975) Transfer of proteins across membranes. I. Presence of proteolytically process and unprocessed nascent immunoglobulin light chains on membrane-bound ribosomes of murine myeloma. J. Cell Biol. 67: 835-851

Bulleid NJ and Freedman RB (1990) Cotranslational glycosylation of proteins in systems depleted of protein disulphide isomerase. EMBO J. 9: 3527-3532

Connolly T, Rapiejko PJ and Gilmore R (1991) Requirement of GTP hydrolysis for dissociation of the signal recognition particle from its receptor. Science 252: 1171-1172

Deshaies RJ and Schekman R (1989) SEC62 encodes a putative membrane protein required for protein translocation into the yeast endoplasmic reticulum. J. Cell Biol. 109: 2653-2664

Deshaies RJ, Sanders SL, Feldheim DA and Schekman R (1991) Assembly of yeast Sec proteins involved in translocation into the endoplasmic reticulum into a membrane-bound multisubunit complex. Nature 349: 806-808

Gilmore R, Walter P and Blobel G (1982) Protein translocation across endoplasmic reticulum. II. Isolation and characterization of the signal recognition particle receptor. J. Cell Biol. 95: 470-477

Hortsch M, Crimaudo C and Meyer D (1987) Structure and function of the rough endoplasmic reticulum. In Integration and control of metabolic processes (Kon OL, ed.) pp 3-11, Cambridge University Press, Cambridge, UK

Lauffer L, Garcia PD, Harkins RN, Coussens L, Ullrich A and Walter P (1985) Topology of signal recognition particle receptor in endoplasmic reticulum membrane. Nature 318: 334-338

Meyer D, Krause E and Dobberstein B (1982) Secretory protein translocation across membranes: The role of the docking protein. Nature 297: 647-650

Nicchita CV and Blobel G (1990) Assembly of translocation-competent proteoliposomes from detergent-solubilized rough microsomes. Cell 60: 259-269

Nicchita CV, Migliaccio G and Blobel G (1991) Biochemical fractionation and assembly of the membrane components that mediate nascent chain targeting and translocation. Cell 65: 587-598

Rapoport TA (1990) Protein transport across the ER membrane. Trends in Biochem. Sci. (TIBS) 15: 355-358

Römisch K, Webb J, Herz J, Prehn S, Frank R, Vingron M and Dobberstein B (1989) Homology of 54K protein of signal-recognition particle, docking protein and two E. coli proteins with putative GTP-binding domains. Nature 340: 478-482

Sadler I, Chiang A, Kurihara T, Rothblatt J, Way J and Silver P (1989) A yeast gene important for protein assembly into the endoplasmic reticulum and the nucleus has homology to DnaJ, and Escherichia coli heat shock protein. J. Cell Bio. 109: 2665-2675

Sanz P and Meyer DI (1989) Secretion in yeast: Preprotein binding to a membrane receptor and ATP-dependent translocation are sequential and separable events in vitro. J. Cell Biol. 108: 2101-2106

Savitz AJ and Meyer DI (1990) Identification of a ribosome receptor in the rough endoplasmic reticulum. Nature 346: 540-544

Schatz G (1991) A protein translocation machine in yeast. Curr. Biol. 1: 43-44

Yu Y, Zhang Y, Sabatini DD and Kreibic G (1989) Reconstitution of translocation-competent membrane vesicles from detergent-solubilized dog pancreas rough microsomes. Proc. Natl. Acad. Sci USA 86: 9931-9935

Zimmerman DL and Walter P (1989) Reconstitution of protein translocation activity from partially solubilized microsomal vesicles. J. Biol. Chem. 265: 4048-4053

MEMBRANE PROTEIN STRUCTURE DETERMINATION BY DEUTERIUM-NMR

CASE STUDY : RETINAL IN BACTERIORHODOPSIN

Anne S. Ulrich and Anthony Watts
Department of Biochemistry
University of Oxford
South Parks Road
Oxford OX1 3QU
England

To understand the function of a protein, one needs to establish the relationship between its three-dimensional structure, its dynamics and its environment. For many soluble proteins the primary, secondary and tertiary structures have been described in considerable detail, but a crucial gap in our knowledge is still the molecular structure of membrane proteins. These cannot usually be investigated by diffraction methods, because their amphipathic nature makes crystallization inherently difficult.

Non-crystallographic approaches are being developed for protein structure determination, such as nuclear magnetic resonance (NMR) spectroscopy, primary sequence analysis and energy minimization. For small (<20 kDa) soluble proteins, the highly successful alternative to diffraction methods is multi-dimensional high-resolution NMR. To give high resolution spectra, the molecules in a sample must be tumbling rapidly to average out all anisotropic linebroadening interactions. In membranes, however, the constituent lipids and proteins undergo relatively slow and highly restricted motions on the timescale of the NMR experiment. Therefore, high-resolution ^{1}H- or ^{13}C-NMR cannot be used to investigate proteins that are associated with membranes.

Solid-state NMR, on the other hand, applied to biomembranes, provides a powerful tool to investigate their structural and dynamic properties, particularly in combination with isotope labelling. Specifically placed deuterons can be used as non-perturbing, highly localized reporters about the labelled segment on the molecule in the membrane. Complementary neutron diffraction experiments can be performed on the same deuterated sample to localize the depth of the deuterated group within the bilayer. The combined information will lead to a full three dimensional and dynamic characterization of the selected site on the protein.

We have developed a strategy to determine the structure of a prosthetic group on a membrane protein by ^{2}H-NMR, using bacteriorhodopsin (bR) as a well-characterized model system. The protein is arranged as a planar array throughout the purple membrane (PM) of the

NATO ASI Series, Vol. H 63
Dynamics of Membrane Assembly
Edited by J. A. F. Op den Kamp
© Springer-Verlag Berlin Heidelberg 1992

bacterium *Halobacterium halobium* which grows in concentrated brine. Under conditions of oxygen starvation the protein acts as a light-driven proton pump to generate an electrochemical gradient across the membrane, which is utilized in the synthesis of ATP. Upon the absorption of light the chromophore undergoes an all-*trans* to 13-*cis* isomerization, and during the photocycle a transient deprotonation of its Schiff base is involved in transferring a proton across the membrane.

Knowledge of the detailed geometry and orientation of retinal and its interactions with the surrounding amino acid residues is thus essential for the elucidation of protein function. Retinal is buried within the seven membrane-spanning α-helices of the protein, and is linked to Lys-216 on helix G via a protonated Schiff base. The three-dimensional structure of bR has recently been determined by electron microscopy to a resolution of 2.7Å in-plane (Ceska & Henderson 1990; Henderson *et al.* 1990), revealing the position of the chromophore within the protein, but the orientation and conformation of retinal is not yet discernible from the electron density map.

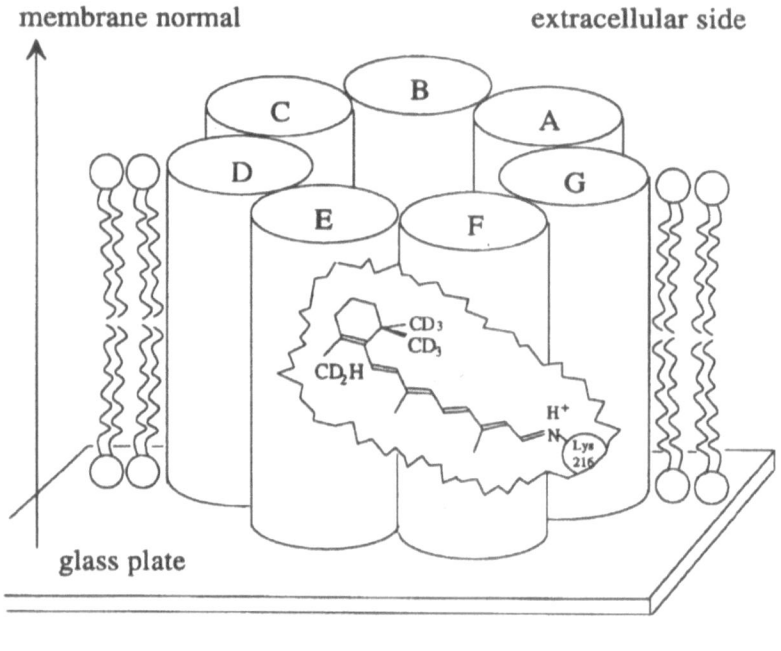

Figure1: Retinal in bR is attached via a protonated Schiff base to Lys-216, and tilted towards the extracellular side. The chromophore was selectively deuterated on the three methyl-groups of the β-ionone ring, and an oriented sample of PM prepared on small glass plates by evaporation.

Neutron scattering experiments have probed the in-plane position as well as the depth of a series of deuterium-labelled segments along the chromophore, which directly established the location of its long-axis within the protein (Heyn et al. 1988; Hauß et al. 1990). A variety of spectroscopic techniques have contributed many separate indications about the detailed structure of retinal within bR, leading to the generally accepted picture shown in Figure 1. The chromophore is tilted away from Lys-216 towards the extracellular side (Huang & Lewis 1989) by about 23° relative to the membrane surface (Heyn et al. 1977), with an approximately vertical molecular plane (Earnest et al. 1986). The proton on the Schiff base nitrogen faces the extracellular side (Lin & Mathies 1989), and the β-ionone ring was recently shown to assume a 6-S-*trans* conformation (Creuzet et al. 1991). Nevertheless, no one method has been available up to now that would allow a comprehensive characterization of all aspects of the orientation and the molecular conformation of retinal within bR, or indeed of any prosthetic group within a membrane protein.

The power of the solid state ^{2}H-NMR method described here, lies in the ability to determine the orientation of a specifically labelled molecular segment relative to the symmetry axis of a macroscopically aligned sample. A simple anisotropic ^{2}H-NMR spectrum is characterized by a pair of lines corresponding to the two transitions allowed for the deuterium nucleus with a spin of I = 1. The quadrupole splitting Δv_Q of a rapidly rotating deuteromethyl-group is directly related to the angle α between the CD_3-bond vector in the molecule and the magnetic field direction H (Seelig 1977):

$$\Delta v_Q = 80 \text{ kHz} . (3\cos^2\alpha - 1)/2$$

For our ^{2}H-NMR studies, retinal was selectively deuterated on the three methyl-groups in the β-ionone ring, and incorporated into bR using a mutant of *Halobacterium halobium* that is deficient in the synthesis of the chromophore (Seiff et al. 1986). After purification, 90 mg of PM patches (approximately 25 μmol deuterons) were deposited by controlled evaporation as oriented films on small glass plates (30mm x 8mm x 0.15mm) cut to fit longitudinally into a 10mm NMR tube. The dark-adapted sample was placed horizontally into the static solenouid and could be rotated manually.

The preparation of macroscopically oriented samples is essential for this ^{2}H-NMR approach, and can be achieved for many membrane proteins by sedimentation or other biophysical methods (Clark et al. 1980). Such a sample possesses no translational or rotational symmetry within the plane, but all CD_3-bond vectors of any one type make the same angle with the sample normal (see Figure 1). The spectrum from the deuterated bR sample at a 0 degree inclination (i.e. with its normal parallel to the spectrometer field) should therefore display three pairs of resonances, from which the three quadrupole splittings are directly measurable. Other alignments of the sample in the magnetic field give rise to more complex and wider lineshapes.

Oriented ^{2}H-NMR spectra at a series of sample inclinations are shown in Figure 2. It is apparent that at 0 degree the expected three pairs of resonances are not resolved because of spectral overlap of the broadened lines. This linebroadening is due to the intrinsic T_{2^*}-linewidth and to the mosaic spread of the PM patches in the planar film. We determined the mosaic spread to be ±10°, by analyzing the linewidths of oriented ^{31}P-NMR spectra (not shown) from the phospholipid component in the PM sample.

To reveal the values of the quadrupole splittings a lineshape simulation program was written, for which all input variables were uniquely determined from independent sources such as the ^{2}H-NMR powder spectrum (for the spectral width and the T_{2^*}-linebroadening) and ^{31}P-NMR

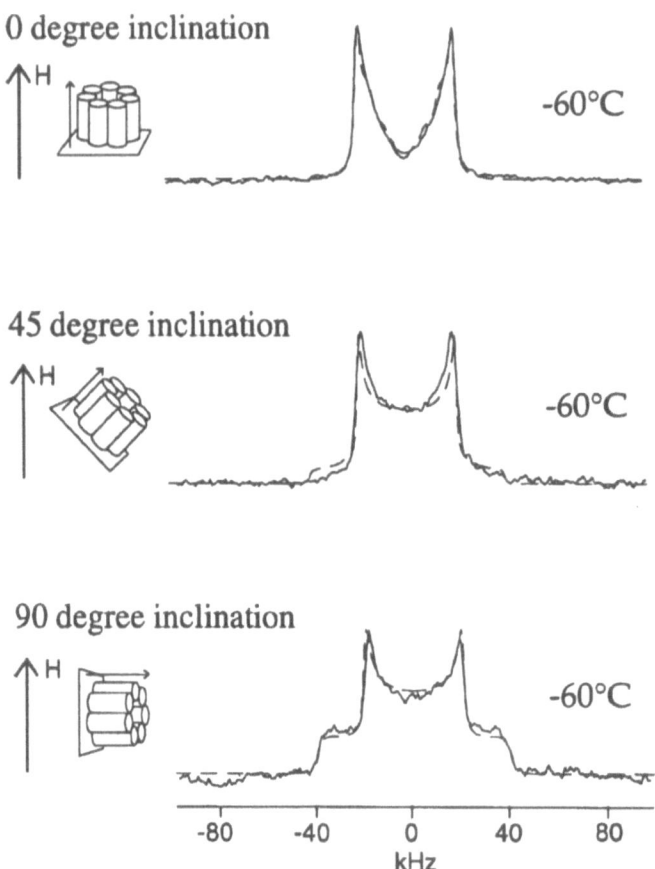

Figure 2: ^{2}H-NMR spectra of oriented PM films containing deuterated retinal, with superimposed lineshape simulations, at sample inclinations of 0, 45 and 90 degree in the solenoid. Measurements were performed at -60° C, on a 400 MSL spectrometer at a deuterium frequency of 61 MHz, using a quadrupole-echo pulse sequence with a π/2 pulse width of 4.5 μs and a repetition time of 100 ms.

spectra (for the mosaic spread). Keeping the three deuteromethyl-groups interconnected via the β-ionone ring, any conceivable orientation of this rigid set of vectors in space was probed by comparing the fit between the resulting simulation and the experimental spectrum. The optimized simulations for the whole series of spectra are shown superimposed in Figure 2, and correspond to the concluded orientation of the β-ionone ring in space. Figure 3 demonstrates how the peak positions of the underlying three pairs of resonances in the 0 degree spectrum were revealed by a simulation using zero T_2-linebroadening and zero mosaic spread together with the geometric parameters found from the best-fit. The two CD_3-groups and the CD_2H-group are readily assigned from their relative peak intensities. The resulting quadrupole splittings are directly related to the angle between the respective deuteromethyl-group and the sample director, by the equation given earlier:

$$\Delta v_Q = -31 \text{ kHz} \qquad CD_3\text{-angle (a)} = 106°$$
$$\Delta v_Q = -39 \text{ kHz} \qquad CD_3\text{-angle (b)} = 86°$$
$$\Delta v_Q = +17 \text{ kHz} \qquad CD_2H\text{-angle} = 134°$$

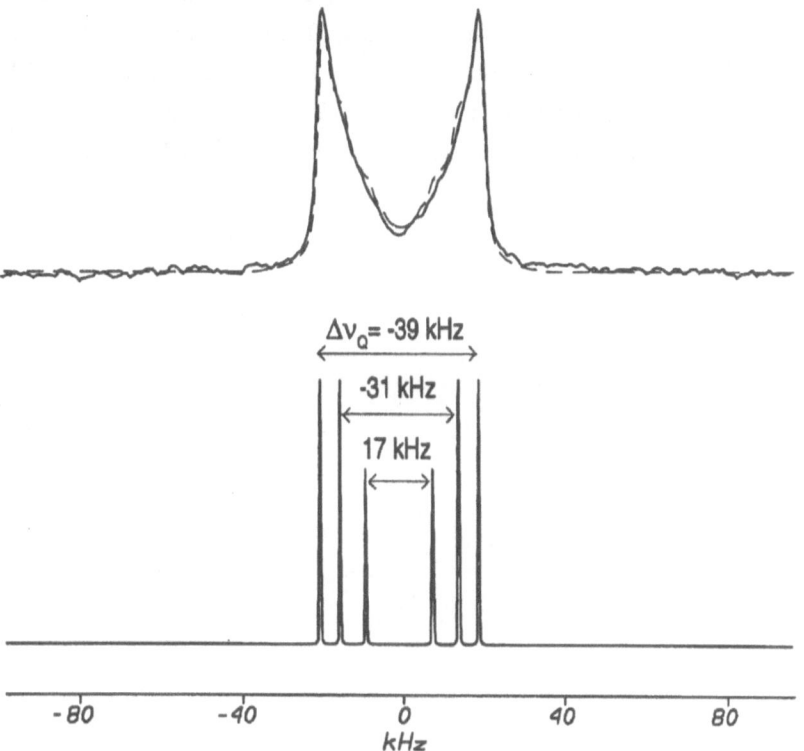

Figure 3: Experimental ^{2}H-NMR spectrum at a 0 degree sample inclination, with its lineshape simulation superimposed. Underneath, resolution of the underlying peak positions by applying a T_2-linebroadening and a mosaic spread of zero to the simulation. All input variables for the program were unique and had been determined from independent sources.

248

The above orientations for the three deuteromethyl-groups on the β-ionone ring were then used as geometrical constraints in a computer molecular modelling approach to fix the remaining part of the chromophore in space. Figure 4 shows our proposed solution, which confirms all essential features of the recognized retinal structure inside the protein as seen in Figure 1.

Our basic conclusions can be visualized by inspection of the two CD_3-groups in Figure 4, which both make an angle close to 90° relative to the sample normal. To accommodate these segments approximately horizontally in the membrane, the average plane through the β-ionone ring must sit nearly vertically in the protein, and the whole chromophore must be tilted upwards from Lys-216, with carbons C_1-C_4 pointing further up. These conclusions were derived *ab initio* by ²H-NMR, without recourse to crystalline model compounds or any other information on the system.

In a retinal molecule, the conformation of the β-ionone ring can be either 6-S-*cis* or 6-S-*trans* in different environments. A rotation of the polyene-chain by ρ = 180° (see Figure 4) would place it almost perpendicular to the membrane plane, but such a 6-S-*cis* conformation is clearly incompatible with the recognized tilt angle near 23° from spectroscopic (Heyn *et al.* 1977; Earnest *et al.* 1986; Lin & Mathies 1989) and neutron diffraction studies (Heyn *et al.* 1988; Hauß *et al.* 1990). The plausible structure for retinal within bR must therefore be close to 6-S-*trans* in our model, as has also recently been found by rotational resonance ¹³C-NMR (Creuzet *et al.* 1991).

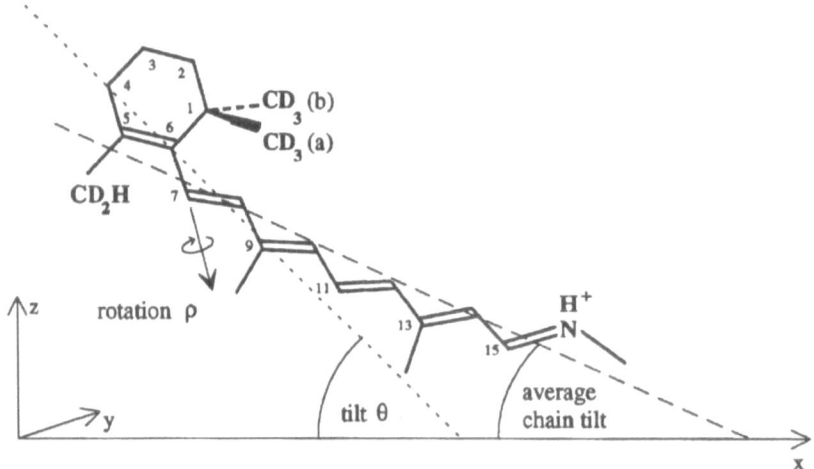

Figure 4: Retinal conformation and geometry, as determined by ²H-NMR and computer molecular modelling. The tilt angle θ of the reference axis defined by carbons C_4 and C_6 (dotted line) differs from that of the average chromophore long-axis (dashed line), due to the curvature of the polyene-chain.

We estimated by computer molecular modelling a tilt angle $\theta \approx 44°$ for the reference axis (dotted line in Figure 4), which is considerably larger than the expected value of 23° (dashed line). Since these methods only give the *average* orientation of the chromophore long-axis, we therefore conclude that the conjugated polyene-chain must be slightly curved. This in-plane bending may relieve steric crowding of the methyl-groups along the chain, as has been observed in the crystal structure of all-*trans* retinal (Stam 1972).

Independent studies have shown that retinal is tilted towards the extracellular side of bR (Huang & Lewis 1989), which thus defines the absolute sidedness of the protein in our frame of reference. Consequently, the methyl groups along the chain point downwards to the cytoplasmic side, and the proton on the Schiff base faces up towards its most likely counterion Asp-212, as reported (Lin & Mathies 1989). This is the case for both the all-*trans* and the 13-*cis* form of retinal, which make up the dark-adapted state of bR, since thermal isomerization occurs via a concerted "pedal"-mechanism in which the C_{15}=N bond also flips (Harbison 1984). Our simulations have shown that the orientation of the labelled molecular unit is sufficiently well defined to exclude the possibility of a significant change in local geometry upon thermal isomerization. During the photocycle, on the other hand, considerable structural changes occur both in the chromophore and in the protein. The possibility of trapping these intermediates and observing them by ^{2}H-NMR might provide further insight into the function of bacteriorhodopsin.

Acknowledgement:
We would like to thank Professor Heyn (Freie Universität Berlin) for kindly providing the sample.

References:

Ceska T.A. & Henderson R. (1990) *J. Mol. Biol.* **213**, 539-560
Clark N. A., Rothschild K. J., Luippold D. A. & Simon B. A. (1980) *Biophys. J.* **31**, 65-96
Creuzet, F., McDermott, A., Gebhard, R., van der Hoef, K., Spijker-Assink, M. B., Herzfeld, J., Lugtenburg, J., Levitt, M. H. & Griffin, R. G. (1991) *Science* **251**, 783-786
Earnest, T. N., Roepe, P., Braiman, M. S., Gillespie, J. & Rothschild, K. J. (1986) *Biochem.* **25**, 7793-7798
Harbison, G. S., Smith, S. O., Pardoen, J. A., Winkel, C., Lugtenburg, J., Herzfeld, J., Mathies, R. & Griffin, R. G. (1984) *Proc. Natl. Acad. Sci. USA* **81**, 1706-1709
Hauß, T., Grzesiek, S., Otto, H., Westerhausen, J. & Heyn, M. P. (1990) *Biochem.* **29**, 4904-4913
Henderson, R., Baldwin, J. M., Ceska, T. A., Zemlin, F., Beckmann E. & Downing, K.H. *J.* (1990) *Mol. Biol.* **213**, 899-929
Heyn, M. P., Cherry, R. J. & Müller (1977) *J. Mol. Biol.* **117**, 607-620
Heyn, M. P., Westerhausen, J., Wallat, I. & Seiff, F. (1988) *Proc. Natl. Sci. USA* **85**, 2146-2150
Huang, J. Y. & Lewis, A. (1989) *Biophys. J.* **55**, 835-842
Lin, S. W. & Mathies, R. A. (1989) *Biophys. J.* **56**, 653-660
Seelig, J. (1977) *Q. Rev. Biophys* **10**, 353-418
Seiff, F., Westerhausen, J., Wallat, I. & Heyn, M. P. (1986) *Proc. Natl. Sci. USA* **83**, 7746-7750
Stam, C. H. (1972) *Acta Cryst.* **B28**, 2936-2945

IMPORT OF PROTEINS FROM THE CYTOPLASM INTO MITOCHONDRIA: SIGNALS, ENERGETICS, AND CATALYSTS

Gottfried Schatz
Biocenter, University of Basel
Klingelbergstrasse 70
CH-4056 Basel
Switzerland

Introduction

Mitochondria contain hundreds of different proteins, but only a few of these (13 in humans) are encoded by mitochondrial DNA and synthesized inside the mitochondria. All the others are encoded by nuclear DNA, synthesized in the cytoplasm (usually as precursors with N-terminal extensions) and imported into one of the four major intramitochondrial compartments: outer membrane, intermembrane space, inner membrane, and matrix (Verner and Schatz 1988; Hartl and Neupert 1990; Glick and Schatz 1991). How is each of these proteins is transported to its correct intramitochondrial location?

Targeting and Sorting Signals

Proteins are targeted to mitochondria by an N-terminal stretch of amino acids that is usually (1) shorter than 30 residues (2) rich in basic and hydroxylated residues (3) devoid of acidic residues (4) capable of folding into an amphiphilic α-helix or β-sheet (5) removed in the matrix by a soluble protease (Roise and Schatz 1988). This signal transports attached proteins into the matrix. There is no recognizeable sequence homology between different "matrix-targeting signals". The "helical wheel" projection of the matrix-targeting signal of cytochrome oxidase subunit IV (COX

NATO ASI Series, Vol. H 63
Dynamics of Membrane Assembly
Edited by J. A. F. Op den Kamp
© Springer-Verlag Berlin Heidelberg 1992

IV) from yeast (Fig. 1) clearly reveals the hydrophilic and hydrophobic surfaces of the helix.

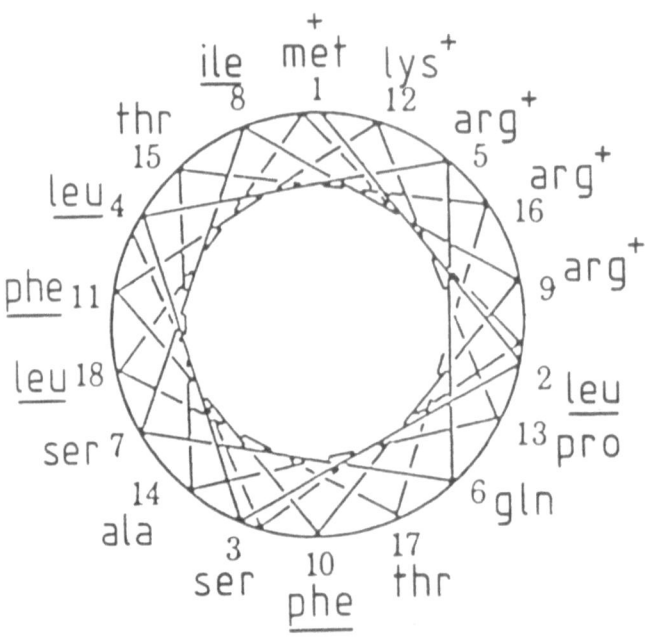

Fig.1: Helical wheel projection of the N-terminal region of the cytochrome c oxidase sububit IV precursor from *Saccharomyces cerevisiae*

Proteins destined to be transported to the intermembrane space contain an additional "sorting signal" downstream from the matrix-targeting signal (Hurt and van Loon 1986; Schatz 1987). These sorting signals have been extensively studied with cytochrome b_2 (yeast lactate dehydrogenase) and cytochrome c_1. If they are deleted, the precursor is transported to the matrix; if only the matrix-targeting signal is deleted, the protein is not transported into mitochondria. There is, thus, a clear hierarchy of signals: the sorting signal requires the matrix-targeting signal, but the reverse is not true.

In order to define the sorting signal for cytochrome b_2, we selected for mutations which inactivate this signal. For this selection, we have fused the cytochrome b_2 presequence to the mature region of COX IV and expressed the fusion protein in a yeast strain which was respiration-deficient because its nuclear COX IV gene had been deleted. Because the cytochrome b_2 presequence transports COX IV to the intermembrane space, the cells will remain respiration-deficient unless the sorting signal in the fusion protein is inactivated by a spontaneous mutation (Fig. 2).

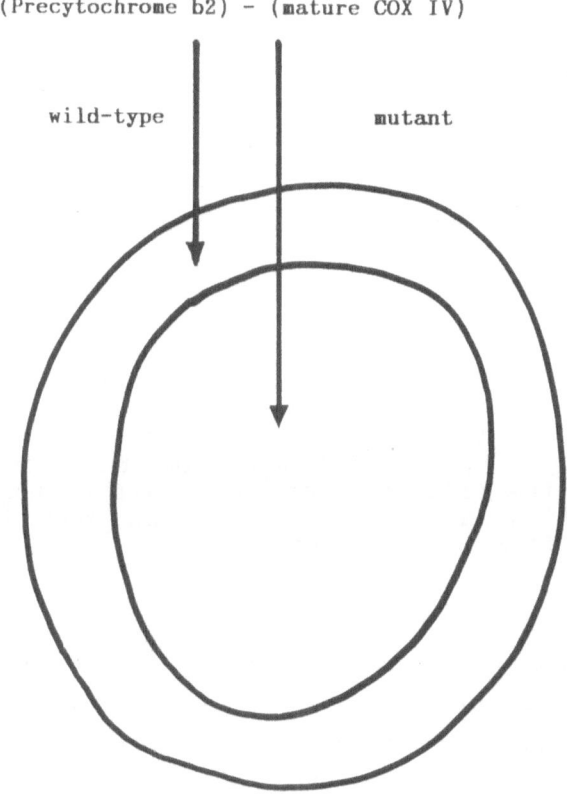

(Precytochrome b2) - (mature COX IV)

wild-type   mutant

Fig. 2: Genetic selection of mutations inactivating the sorting sequence in a cytochrome b_2-COX IV fusion protein (E. Beasley, in preparation). The two concentric circles signify the two mitochondrial membranes. For the sake of clarity, contact sites between these membranes (see below) are omitted. The wild type COX IV precursor is transported into the matrix and assembled into cytochrome oxidase from the inner side of the inner membrane.

We sequenced approximately 60 mutationally altered cytochrome b$_2$ presequences which allowed the transformants to grow on a non-fermentable carbon source. We found that the sorting signal consists of (1) a region rich in basic amino acids (3) a hydrophobic stretch of at least 15 residues (3) one or two acidic residues (Fig. 3).

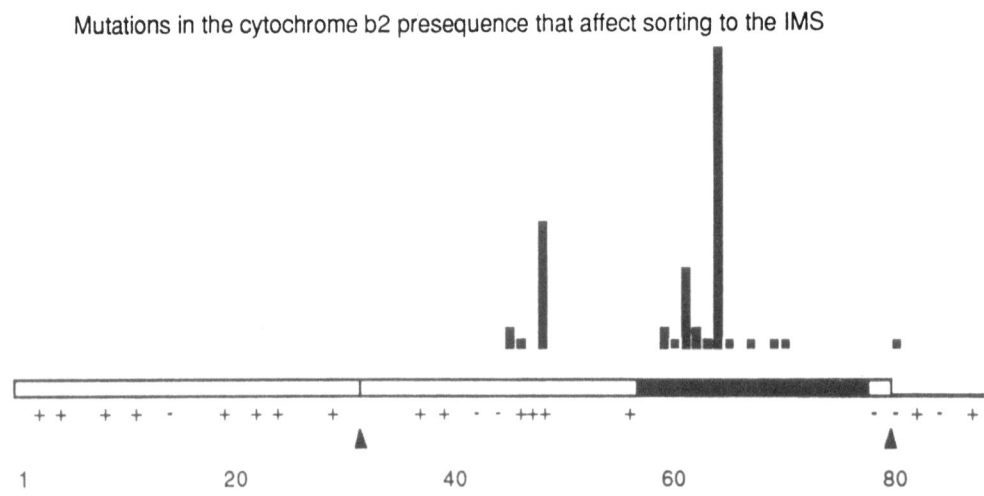

Fig. 3: Summary of mutations inactivating the sorting signal of cytochrome b$_2$. Basic and acidic residues are marked by + and - signs, the hydrophobic stretch by a black rectangle, and proteolytic cleavage sites by black triangles. The matrix-targeting sequence extends from the precursor's N-terminus up to the first cleavage site. The histogram records each inactivating mutation at a given position in the sequence by a black square on top.

Detailed analysis of the data summarized in Fig. 3 revealed that all mutations in the hydrophobic region decreased the hydrophobicity of that region. However, there was no simple correlation between the decrease in hydrophobicity and the severity of the inactivation. This suggests that the function of the sorting signal requires protein-protein interaction.

How proteins are sorted to the outer membrane is still a mystery. Studies on the outer membrane protein Mas70p (see below) suggest that this sorting process, too, is mediated by a hydrophobic sorting signal downstream from a matrix-targeting signal (Hase et al. 1984).

Intramitochondrial Sorting Pathways

When we discovered the intramitochondrial sorting signal of
cytochrome c_1, we suggested that its hydrophobic region
functions as a "stop-transfer" signal for the mitochondrial
inner membrane. According to this model (Fig. 4, right part),
the precursor is targeted to mitochondria by its N-terminal
matrix-targeting signal which penetrates into the matrix
space. Further transport is then stopped by the sorting
signal. The matrix-targeting signal is cleaved off by the
matrix protease and the transmembre intermediate is released
into the intermembrane space by a second protease on the
outer face of the inner membrane.

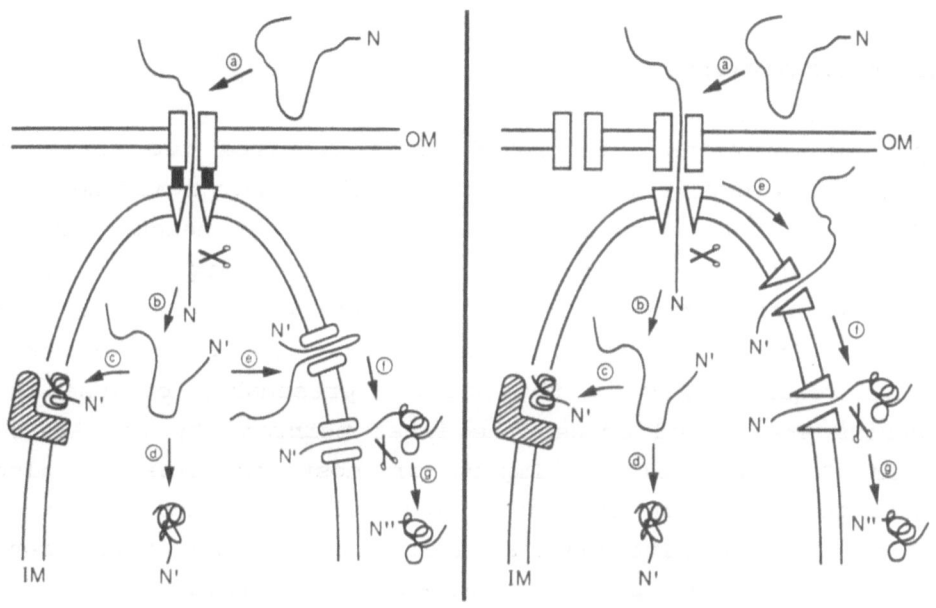

Fig. 4: The "stop-transfer" model for sorting proteins to the
intermembrane space (right) and the "conservative sorting"
model (left). OM and IM, outer and inner membrane; open
rectangles and open triangles, components of the import
channels across the outer and inner membrane; scissors,
proteolytic cleavages; cross-hatched L-shaped structure in
the inner membrane, transmembrane partner subunits with which
the protein imported into the matrix assembles; ellipsoid
symbols across the inner membrane, hypothetical proteins
proposed by the conservative sorting model to catalyze
reexport from the matrix. From Glick and Schatz (1991).

The "stop-transfer" model was challenged by Hartl et al. (review: Hartl and Neupert 1990) who proposed that the precursor is completely imported into the matrix and then reeported across the inner membrane (Fig. 4, left). Although this is an intellectually appealing model, we have been unable to confirm the experiments on which it rests (Glick and Schatz 1991). Upon extensive reeinvestigation of this problems we conclude that the precursors of cytochrome b_2 and cytochrome c_1 never completely enter the matrix and that their correct sorting to the intermembrane space does not require the matrix-localized hsp60 chaperone (see below; A. Brandt, in preparation). At least with these precursors, the sorting signal functions as a stop-transfer signal.

Energy requirements

Import of proteins into mitochondria requires two forms of energy: an electric potential across the inner membrane, and ATP (Eilers and Schatz 1988). The potential is needed for the insertion of the positively charged matrix-targeting signal into the inner membrane. The requirement for ATP is more complex. Transport of proteins across the inner membrane always requires ATP in the matrix, presumably because the proteins are pulled across the inner membrane by the ATPase of mhsp70 (see below). Import of most proteins is also stimulated (but not absolutely dependent on) hydrolysis of ATP outside the mitochondria. This external ATP requirement reflects at least partly an ATP-dependent release of newly synthesized precursors from 70 kd cytosolic heat-shock proteins which prevent premature folding of the precursors. Tightly folded proteins are translocation-incompetent (Eilers and Schatz 1986).

The insertion of proteins into the outer membrane requires neither ATP in the matrix nor a potential across the inner membrane; however, it is greatly stimulated by ATP outside the mitochondria unless the precursor is added to mitochondria in denatured form.

The Mitochondrial Protein Import Machinery

Protein import into mitochondria is mediated by many different protein catalysts in the cytosol, on the mitochondrial surface, and inside the mitochondria. After many fruitless efforts in different laboratories, these proteins are now being identified and isolated. Fig. 5 summarizes the current state of the field.

In brief, precursors are probably presented to mitochondria as complexes with antifolding proteins such as hsp70 or PBF; after release from these carriers, proteins destined for the mitochondrial interior bind to one of the mitochondrial import receptors and insert their matrix-targeting signal into the inner membrane. This process requires that the precursor's N-terminal part translocate across the outer membrane. One of the proteins mediating this translocation is ISP42 (Vestweber et al. 1989; Baker et al. 1990). The presequence then moves across a still unidentified channel in the inner membrane and binds to a mitochondrial 70 kd heat-shock protein termed mhsp70. This protein is probably a protein-dependent ATPase which couples the hydrolysis of ATP to the release of the bound precursor chain into the matrix. Release into the matrix is not directly coupled to refolding, however (Krieg et al. submitted). Upon release from mhsp70, the precursor may perhaps fold spontaneously. However, imported proteins destined to assemble into oligomeric structures first bind to another chaperone, hsp60. Release from hsp60, too, requires ATP hydrolysis (Ostermann et al. 1989). Finally, the matrix-targeting signal of the released protein is cleaved off by the matrix-localized processing protease.

Fig. 5: The mitochondrial protein import machinery. hsp70 and mhsp70, cytosolic and mitochondrial 70 kd heat-shock proteins; PBF, presequence-binding factor (Murakami and Mori 1990); R, one of several functionally redundant "import receptors" on the outer membrane; delta psi, electric potential across the inner membrane; ISP42, subunit of the import channel across the outer membrane (the *Neurospora* homolog is termed MOM38); X,Y,Z and A-F, additional hypothetical subunits of the import channels across the outer and inner membrane, respectively; hsp60, groEL-like 60 kd chaperone; Mas1p and Mas2p, the two nonidentical, but homologous subunits of the matrix processing protease (the *Neurospora* homologs are termed PEP and MMP); IM and OM, inner and outer membrane. For further details see Baker and Schatz (1991).

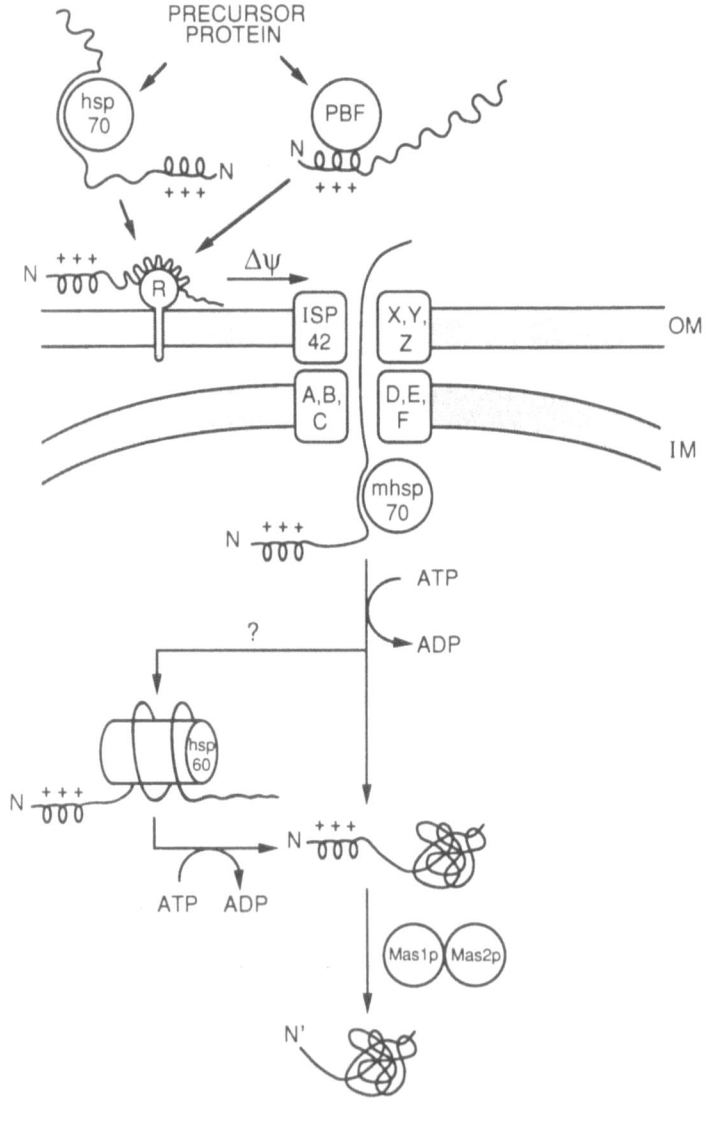

Essential and redundant import catalysts

Mitochondria appear to be essential for the life of a
eukaryotic cell even in the absence of respiratory metabolism
(Subik et al. 1978; Yaffe and Schatz 1984). It would thus be
expected that each component of the mitochondrial protein
import machinery is essential for life as well. However, this
is not so, because some components are functionally
redundant: their function can either be also performed by
other components, or can be bypassed altogether (Baker and
Schatz 1991). So far, only five mitochondrial components of
the import machinery are essential for viability of yeast
cells (Table 1). This list will undoubtedly grow as our
knowledge expands.

Component	Gene	Location	Essential?	Function
MAS1	MAS1	matrix	yes	removal of matrix-targeting signal
MAS2	MAS2	matrix	yes	removal of matrix-targeting signal
mit hsp60	MIF4	matrix	yes	refolding of pre-cursors
mit hsp70	SSC1	matrix	yes	refolding of pre-cursors?
ISP42	ISP42	outer membrane	yes	translocation across outer membrane?
MAS70	MAS70	outer membrane	no	import receptor
p32	MIR1	outer membrane	no	import receptor
heme lyase	CYC3	intermembrane space	no	import of apo-cytochrome c
inner membrane protease I	IMP1	inner membrane	no	processing of cytochrome b_2 and cytochrome oxidase subunit II

Table 1: Essential and nonessential components of the
mitochondrial protein import machinery in *Saccharomyces
cerevisiae*. Adapted from Baker and Schatz (1991).

Protein import across the two mitochondrial membranes: two import systems in tandem?

Many imported mitochondrial proteins must be pass the outer and the inner membrane. It has long been held that this is accomplished through a fixed import channel which spans both membranes at sites where the two membranes are in close contact (Hackenbrock 1968; Kellems et al. 1975; Schwaiger et al. 1987). However, our recent results suggest that the two membranes contain distinct protein transport channels which interact reversibly. The evidence for this is threefold (Glick et al. 1991). First, if protein import into mitochondria is inactivated by treating the mitochondria with protease, with antibody to outer membrane proteins, or with a chimeric precursor blocking the import sites, it can be fully reactivated by rupturing the outer membrane. Second, isolated rightside-out vesicles of the inner membrane essentially devoid of outer membrane components import precursor proteins at least as efficiently as intact mitochondria. Third, mitochondria that are depleted of ATP, but allowed to maintain an inner membrane potential accumulate partly translocated precursor proteins *between* the two membranes (Fig. 6). Current evidence suggests the the outer membrane channel is passive; it functions essentially as a gated pore for precursors whose N-terminus is being pulled into the mitochondria through the combined action of the inner membrane potential and the ATPase of mhsp70. Reversible interaction between the two import systems is essential for the stop-transfer mechanism of intramitochondrial sorting and, thus, for mitochondrial biogenesis.

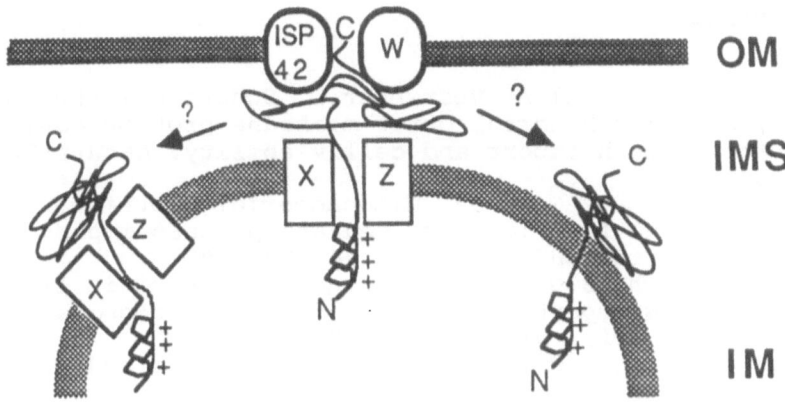

Fig. 6: Accumulation of a translocation intermediate between
the two mitochondrial membrane by mitochondria which are
depleted of ATP, but allowed to maintain an electric
potential across their inner membrane(Hwang et al.,
submitted). W, X, Z, hypothetical components of the import
channel across the outer and inner membrane, respectively;
OM, IM, IMS, outer membrane, inner membrane, and
intermembrane space. The question marks on top of the arrows
indicates that it is currently unknown whether the
intermediate stuck in the inner membrane is located directly
in the lipid bilayer or in a proteinacous pore.

Acknowledgements

This study was supported by grants from the Swiss National
Science Foundation, the US Public Health Service, the Human
Frontier Science Program Organization, the Louis Jeantet
Foundation, and postdoctoral fellowships from EMBO, the US
Public Health Service, and the Life Sciences Foundation.

References

Baker KP, Schaniel A, Vestweber D, Schatz G (1990) A yeast mitochondrial outer membrane protein essential for protein import and cell viability. Nature 348:605-609

Baker KP, Schatz G (1991) Mitochondrial proteins essential for viability mediate protein import into yeast mitochondria.Nature 349:205-208

Eilers M, Schatz G (1986) Binding a of specific ligand inhibits import of a purified precursor protein into mitochondria. Nature 322:228-232

Eilers M, Schatz G (1988) Protein unfolding and the energetics of protein translocation across biological membranes. Cell 52:481-483

Glick B, Schatz G (1991) Import of proteins into mitochondria. Annu. Rev. Genet. in press

Glick B, Wachter C and Schatz G (1991) Protein import into mitochondria: two systems acting in tandem? Trends Cell Biol. in press

Hackenbrock CR (1968) Chemical and physical fixation of isolated mitochondria in low-energy and high-energy states. Proc. Natl. Acad. Sci. USA 61:598-605

Hartl FU, Neupert W, (1990) Protein sorting to mitochondria - evolutionary conservations of folding and assembly. Science 247:930-938

Hase T, Müller U, Riezman H and Schatz G (1984) A 70-kd protein of the yeast mitochondrial outer membrane is targeted and anchored via its extreme amino terminus. EMBO J. 3:3157-3164

Hurt EC, van Loon APGM (1986) How proteins find mitochondria and intramitochondrial compartments. Trends Biochem. Sci. 11:204-206

Kellems RE, Allison VF, Butow RA (1975) Cytoplasmic type 80S ribosomes associated with yeast mitochondria. IV. Attachment of ribosomes to the outer membrane of isolated mitochondria. J. Cell Biol. 65:1-14

Murakami K, Mori M (1990) Purified presequence binding factor (PBF) forms an import-competent complex with a purified mitochondrial precursor protein. EMBO J. 9:3201-3208.

Ostermann J, Horwich A, Neupert W, Hartl FU (1989) Protein folding in mitochondria requires complex formation with hsp60 and ATP hydrolysis. Nature 341:125-130

Roise D, Schatz G (1988) Mitochondrial presequences. J. Biol. Chem. 263:4509-45011

Schatz G (1987) Signals guiding proteins to their correct locations in mitochondria. Eur. J. Biochem. 165:1-6

Schwaiger M, Herzog V, Neupert W (1987) Characterization of translocation contact sites involved in the import of mitochondrial proteins. J. Cell Biol. 105:235-246

Subik J, Takacsova G, Kovac L (1978) Intramitochondrial ATP and cell functions. I. Growing yeast cells depleted of intra-mitochondrial ARP are losing mitochondrial genes. Mol. Gen. Genet. 166:103-116

Verner K, Schatz G (1988) Protein translocation across membranes. Science 241:1307–1313

Vestweber D, Brunner J, Baker A and Schatz G (1989) A 42 kd outer membrane protein is a component of the mitochondrial protein import site. Nature 341:205–209

Yaffe MP, Schatz G (1984) The future of mitochondrial research. Trends Biochem. Sci. 9:179–181

THE USE OF HYBRID PROTEINS TO ISOLATE KINETIC EXPORT INTERMEDIATES

Joan Stader, Sheryl Justice, and Shui-Qing Wei
School of Basic Life Sciences
Division of Cell Biology & Biophysics
University of Missouri - Kansas City
Kansas City, Missouri 64110-2499
USA

INTRODUCTION

The general export pathway in E. coli is responsible for the localization of proteins to the periplasm and the outer membrane. Its operation relies on the function of multiple signals within the precursor protein to be localized (intragenic) and cellular components (extragenic) located both in the cytosol and in the inner membrane. Intragenic export signals include the N-terminal signal sequence and regions within the mature portion of the polypeptide. Extragenic components include the products of the sec and prl genes.

Given the current available technology, the only way to study the kinetics of extracytoplasmic protein localization is to employ an in vivo biochemical approach. In vivo, the synthesis to translocation reactions occur with perfect efficiency on the order of seconds; while in vitro, the process takes about 15 minutes to achieve less than 50 percent efficiency.

The general approach to kinetic studies involves pulse-labeling a growing culture with radioactive methionine, chasing the label with cold methionine, and assaying the culture at various time points following the chase. The assay is designed to detect a specific intermediate of the pathway. For example, if one wants to measure the kinetics of signal sequence processing, samples from the various time points would be immunoprecipitated and subjected to SDS-PAGE. Over time, the decrease in mobility of the protein on the gel reveals the rate of processing. This assay detects both precursor and mature species based solely on molecular weight.

The processing assay is the most straightforward, because it measures the only covalent change known to occur during export. Other intermediates that appear both prior to and

NATO ASI Series, Vol. H 63
Dynamics of Membrane Assembly
Edited by J. A. F. Op den Kamp
© Springer-Verlag Berlin Heidelberg 1992

following signal sequence removal are characterized by location, structural or folding changes, and/or associations with extragenic components of the export pathway. Developing assays for such intermediates is empirical and depends largely on luck and opportunity.

Some of the assays successfully employed include the measurement of: (1) the rate of appearance of protease sensitive protein domains on the periplasmic side of the inner membrane (Josefsson and Randall, 1981); (2) a transient cytoplasmic export intermediate of maltose binding protein (MBP) that is protease sensitive (Randall and Hardy, 1986); (3) trimerization intermediates of LamB that differ from one another in their solubility characteristics (vos Scheperkeuter and Witholt, 1984; Stader and Silhavy, 1988).

In general, a given model protein will pass through n intermediate forms between its synthesis and final localization. If a specific assay existed for each intermediate form, one could combine each assay with a pulse-chase and determine how long it takes for 50 percent of the labeled protein to pass through each one. A composite picture could then be constructed that would indicate the relative order in which each intermediate appears and which are the rate-limiting steps.

Clearly, biochemical assays for all kinetic intermediates of export are necessary so that the order in which the various intragenic and extragenic factors act can be determined. If the characteristics of these intermediates are determined, the information can also be employed in in vitro systems to study the mechanistic details of export. One of the major problems in developing biochemical assays is that intermediates in the early steps of export are extremely short-lived. Therefore it is necessary to find a way to extend the half-lives of the intermediates to create an opportunity for their biochemical characterization.

Our goal is to systematically identify export intermediates and determine the kinetics of their appearance in vivo. Below we describe an approach to the problem which utilizes hybrid proteins to jam the translocation machinery and cause the accumulation of export intermediates proximal in the pathway.

Hybrid proteins and "maltose sensitivity." Numerous studies of protein export over the past 12 years have utilized the novel properties of hybrid proteins containing the amino terminus of an exported protein and the enzymatically active carboxy terminus of ß-galactosidase. Such hybrids carry some of the export signals present in their amino terminus, causing them to initiate interaction with the cellular export apparatus. In E. coli, the attempt to export the hybrid protein fails, leading to jamming of the export machinery (Hall et al., 1982). In the case of fusions containing the amino terminus of either maltose binding protein or LamB, the induction of high levels of gene expression in the presence of the inducer maltose leads to the lethal blockage of export ("maltose sensitivity" or "overproduction lethality").

Maltose sensitivity of fusion strains has been exploited genetically numerous times, resulting in the isolation of signal sequence mutations (Emr and Silhavy, 1980). More recently, fusions have been used genetically as a means of determining the order of function of various components of the export pathway in the SDI experiments of Bieker and Silhavy (1990). In our experimental system, LamB-ß-galactosidase fusions are employed as a means of trapping localization intermediates for biochemical characterization. We have exploited the jamming phenomenon to cause the accumulation of hybrid protein export intermediates which have greater stability than the corresponding species in wild type strains. The information gained from the study of hybrid protein localization intermediates facilitates the isolation and characterization of the corresponding species in nonfusion strains when export proceeds with wild type kinetics.

Distinguishing characteristics of export intermediates. Several recent studies have established that, in order for a precursor to be competent for translocation through the lipid bilayer it must be in a denatured state (Randall and Hardy, 1986), implying an underlying requirement for flexibility. The cytoplasmic protein SecB has been shown to play a key role in anti-folding for a subset of proteins, including LamB, that utilize the general export pathway in E. coli (Kumamoto and Beckwith, 1985; Altman et al., 1990). Other proteins that are likely to interact with the denatured precursor are SecA, SecE, SecY, and signal peptidase. Thus, export intermediates of a precursor protein may be defined as a series of protein associations with cellular export components accompanied by a certain degree of conformational change. Thus, measurable differences would occur in some or all of the following categories: (i) kinetics of appearance; (ii) associations with extragenic components; (iii) precursor conformation; and (iv) localization.

Rationale. When a MBP- or LamB-ß-galactosidase hybrid protein jams the export machinery at the translocation step, it causes an irreversible block in export (Fig. 1A). If the hybrid protein is produced at high levels (full induction with maltose), SecY/PrlA becomes limiting and cell death ensues approximately 3-5 hours post induction (Bieker and Silhavy, 1989; Hall et al., 1982). For about the first thirty minutes of induction, the hybrid protein undergoes processing by signal peptidase (Rasmussen et al., 1984; Voorhout et al., 1988); but in the later stages of induction, the precursor species accumulates. We reasoned that since the export blockage is irreversible, the translocation complex should be isolable. Furthermore, if induction times are long enough, intermediates representing more proximal steps in the pathway should also appear (Fig. 1A). We predicted that the induction kinetics of these intermediates would show an increase at about the time processing decreases (approximately 30 min. post-induction, Fig. 1B and C).

Isolation of two inner membrane hybrid protein pools. The initial experiments have been limited to isolation and characterization of inner membrane intermediates. Since the rate-limiting step in hybrid protein jamming is the SecY/PrlA step, the SecY-hybrid protein intermediate should be the first to appear kinetically. Following a 2-hour induction, cells carrying a single copy of the gene fusion are harvested and the inner membranes are isolated on sucrose gradients either by flotation of lysed spheroplasts or by isopycnic centrifugation of lysed whole cells (Justice and Stader, in preparation). The majority of the hybrid protein is insoluble upon treatment with the detergent. The Triton-insoluble fraction can be further

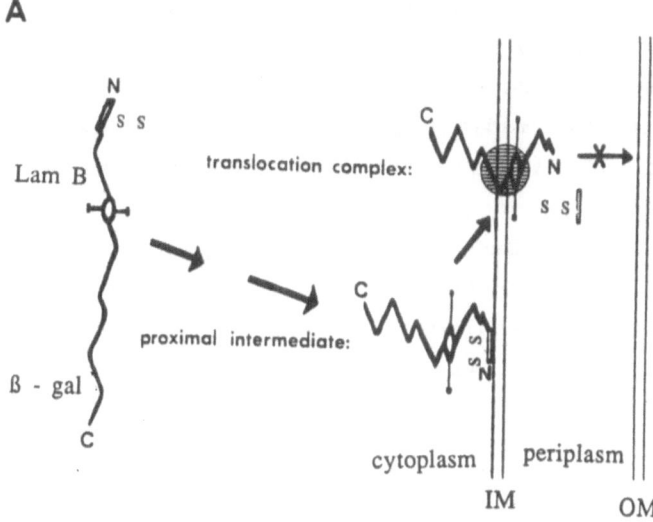

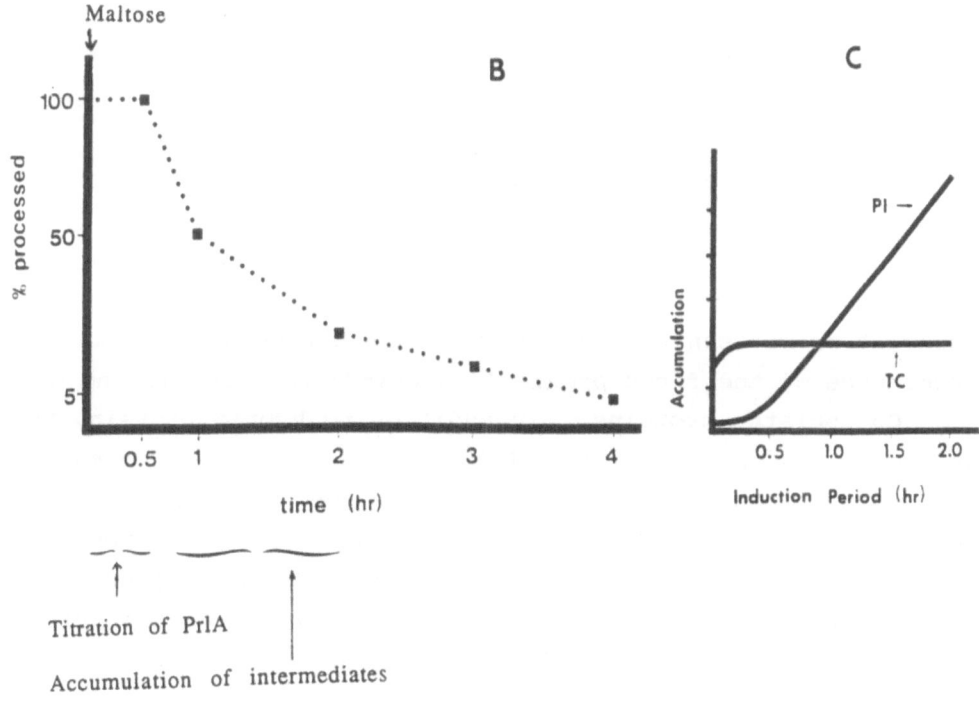

Fig. 1.   Accumulation of hybrid protein export intermediates.
See text for details.   IM = inner membrane; OM = outer membrane;
PI = proximal intermediate; TC = translocation intermediate.

separated into two pools on the basis of density. The fraction of lower density is referred to as TCP (translocation complex pool) and the fraction of higher density is referred to as PIP (proximal intermediate pool). In almost every characteristic measured, TCP and PIP are distinct, as will be described more fully below.

ß-Galactosidase activity. Enzyme activity of the hybrid protein is a convenient way of determining conformational differences. ß-galactosidase activity of the two pools was measured, and it was found that the hybrid protein in TCP possesses enzyme activity. In contrast, the hybrid protein in PIP is inactive even though it contains at least as much (and often more) hybrid protein as TCP, depending on the induction period. The absence of enzyme activity in PIP is not indicative of any particular structure, but merely indicates that it differs from that of TCP.

Time of appearance. Following the addition of the inducer, we have observed that the levels of hybrid protein associated with TCP increase slightly during the first thirty minutes, then plateau. PIP levels are not detectable during the first thirty minutes of induction, but begin to appear concomitant with the plateau of TCP. PIP levels rise continuously throughout the remainder of the induction period. There is a remarkable correlation between this observation and our prediction for the appearance of the first proximal intermediate (Fig. 1B and C).

Co-isolated proteins. In addition to hybrid protein, the only other major protein in TCP that is detectable by coomassie blue staining is OmpA. Immunoblotting has revealed that low levels of LamB, OmpF, and MBP are also present in this pool (S. Justice, unpublished). We are currently testing the prediction that SecY/PrlA is present in TCP as well. The fact that mature species are isolated in this fraction indicates that TCP is a late intermediate.

PIP contains hybrid protein, pOmpF, and a 17K polypeptide which we have not yet identified. Merodiploids that carry a wild type copy of _lamB_ in addition to the gene fusion also accumulate pLamB in this pool (S.-Q. Wei, unpublished data). The levels of all of these species increase linearly beginning 30 minutes post-induction. The presence of precursor species in this fraction supports the conclusion that PIP is a more proximal export intermediate than TCP.

Signal sequence requirement. TCP can be isolated from fusion and non-fusion strains. In a fusion strain, if a mutation is present in the signal sequence of the hybrid protein, the hybrid protein is still able to associate with TCP. We did not predict that this would be the case. On the other hand, association of wild type LamB with this pool is absolutely dependent on signal sequence function. We believe the discrepancy is due to the leakiness of signal sequence mutations and the long induction period in the fusion strain. Since signal sequences are not the only intragenic export signals, there is still measurable interaction between the mutant precursor and the export machinery after a two-hour induction. The residual activity of the mutant signal sequence is undetectable in nonfusion strains because the export machinery is clear and operating efficiently. Under these conditions the mutant precursor is unable to compete with other proteins for export.

When a signal sequence mutation is present on the hybrid protein, the cell does not accumulate PIP. This situation may be a reflection of the inefficiency of hybrid protein association with the export machinery resulting in its inability to compete with other precursors for SecY/PrlA. Indeed, the cells are not sensitive to maltose and do not titrate SecY/PrlA. This situation does not allow us to test whether signal sequence activity is required for hybrid protein association with PIP when the mutation is on the hybrid protein. To properly test for signal sequence dependence, we have used a merodiploid containing a wild type fusion and a copy of lamB that carries a signal sequence mutation. In such a construct, the mutant precursor fails to accumulate in PIP (S.-Q. Wei, unpublished).

In conclusion, with nonfusion precursors, association with both TCP and PIP requires a functional signal sequence.

Requirement for mature sequences. Hybrid proteins containing varying lengths of LamB moiety at the N-terminus and a constant length of ß-galactosidase at the C-terminus (Benson et al. 1984) were employed to test the requirement for mature sequences. The data are summarized in Table 1. Incorporation of hybrid protein into TCP requires the region between 15 and 28 amino acids of mature LamB, while incorporation into PIP requires the regions between 28 and 39. Since the formation of PIP requires jamming (maltose sensitivity) it is not possible to conclude that incorporation of hybrid protein into PIP requires the region between amino acids 15 to 28. We have not yet tested the requirement for mature regions in nonfusion strains. It is possible that length of induction and low export efficiency in the fusions are affecting the results in a manner similar to the signal sequence mutations described above.

Length of Mature Region[a,b]	Maltose Phenotype[b]	Incorporation into TCP[c]	Incorporation into PIP[d]
15 a.a.	resistant	NO	NO
28 a.a.	resistant	YES	NO
39 a.a.	sensitive	YES	YES
49 a.a.	sensitive	YES	YES
181 a.a.	sensitive	YES	YES

Table 1. Mature regions of LamB required for hybrid protein association into TCP and PIP.

[a]All hybrid proteins carry a wild type signal sequence.
[b]S. A. Benson et al. 1984.
[c]S. Justice and J. Stader, in preparation.
[d]S. Justice, unpublished.

Correlation between intermediates found in fusion and nonfusion strains. To test the usefulness of our approach, it is necessary to see if export intermediates similar to the ones found in the fusion strains can be found in nonfusion strains. TCP can be isolated from nonfusion strains. Indeed, when fractionation is combined with pulse-chase analysis, we find a transient association of LamB with this fraction (S. Justice and J. Stader, in preparation). We have found that the kinetics of appearance of LamB in TCP and its solubility characteristics correlate with mt-LamB, the metastable trimer form (Stader and

Silhavy, 1988).

PIP has not yet been found in wild type nonfusion strains.

Characteristic	TCP[a]	PIP[b]
β-Galactosidase enzyme activity	ACTIVE	INACTIVE
Time of appearance	IMMEDIATE	30′ POST-INDUCTION
Co-isolated proteins	OmpA	pOmpF 17K Protein
Signal sequence requirement	NOT REQUIRED FOR HYBRID PROTEINS; REQUIRED FOR LamB	REQUIRED
Mature sequence requirement	REQUIRES AMINO ACIDS BETWEEN 15 AND 28	REQUIRES AMINO ACIDS BETWEEN 28 AND 39
Correlate found in non-fusion wild type strains	YES	NO

Table 2. Comparison of the characteristics of the two Triton X-100 insoluble inner membrane fractions derived from E. coli LamB-LacZ hybrid protein producing strains. See text for further details.

[a]S. Justice and J. Stader, manuscript in preparation
[b]S. Justice, unpublished data

DISCUSSION

We have described the isolation of two inner membrane hybrid protein pools that differ in protein associations, conformation, and kinetics of induction (Table 2). The appearance of both pools depends on active intragenic export signals. Although the hybrid proteins are associated with other polypeptides in the two pools, it is not yet clear whether their association is fortuitous or related to an export mechanism. Further work is currently being done to examine this question.

Translocation Complex Pool (TCP). This pool appears with the expected kinetics of the translocation complex (Fig. 1). The plateau seems to correspond with the titration of PrlA/SecY (Bieker and Silhavy, 1989) and the decrease in processing previously reported for hybrid proteins (Rasmussen et al. 1984).

We are currently looking for PrlA/SecY in this fraction.   TCP most likely represents the rate-limiting step in LamB localization because it is the step that is blocked when the hybrid protein jams the export machinery and it correlates, in nonfusion strains, with the formation of mt-LamB, a process that has a $t_{1/2}$ of formation of about 2 minutes. All of the proximal steps in export occur on the order of seconds.

Proximal Intermediate Pool (PIP). In order to observe this pool, TCP must be jammed. It is therefore important to consider when working with this fraction that factors affecting PIP may be indirect.   PIP appearance during maltose induction agrees with what we expect for a more proximal intermediate in the export pathway (Fig. 1).   If it is not the rate-limiting step, we would expect it to have a very short half-life in strains that are localizing proteins normally.

PIP seems to represent an aggregate of porin precursors that are blocked from proceeding further down the pathway owing to the titration of the SecY/PrlA. The major cellular protein found in this fraction is pOmpF; however, pOmpC and pLamB are also present, while periplasmic precursor proteins are not found in this fraction. One possibility is that all precursors are in equilibrium with this fraction but only the outer membrane protein get stuck because they are more compatible with the bilayer. Accordingly, the periplasmic proteins would partition back into the cytoplasm.

Preliminary data seem to indicate that the lethality associated with maltose induction of fusion strains is a direct result of the accumulation of PIP. The kinetics of cell death corresponds more closely to the appearance of PIP than to the jamming of SecY/PrlA (4 to 5 hours). Also, we have found that decreasing the porin content of the cells by using an ompF null strain decreases the sensitivity to maltose (J. Stader, unpublished) and that increasing the porins by using a lamB merodiploid increases maltose sensitivity (S.-Q. Wei, unpublished). We are currently trying to find a way to induce the accumulation of a related intermediate in nonfusion strains.

The future direction of this research will involve the characterization of export intermediates in other cellular fractions (the Triton X-100 soluble inner membrane fraction, and

the cytoplasmic fraction) and more detailed characterization of
the intermediates described above.

LITERATURE REFERENCES

Altman E, Emr SD, and Kumamoto CA (1990) The presence of both
    the signal sequence and a region of mature LamB protein is
    required for the interaction of LamB with the export factor
    SecB. J Biol Chem 265:18154-18160.
Benson SA, Bremer E, and Silhavy TJ (1984) Intragenic regions
    required for LamB export. Proc Natl Acad Sci (USA) 81:3830-
    3834.
Bieker KL and Silhavy TJ (1989) PrlA is important for the
    translocation of exported proteins across the cytoplasmic
    membrane of Escherichia coli. Proc Natl Acad Sci (USA)
    86:968-972.
Bieker KL and Silhavy TJ (1990) PrlA (SecY) and PrlG (SecE)
    interact directly and function sequentially during protein
    translocation in E. coli. Cell 61:833-842.
Emr SD and Silhavy TJ (1980) Mutations affecting localization of
    an Escherichia coli outer membrane protein, the
    bacteriophage λ receptor. J Mol Biol 141:63-90.
Hall MN, Schwartz M, and Silhavy TJ (1982) Sequence information
    within the lamB gene is required for proper routing of the
    bacteriophage λ receptor protein to the outer membrane of
    Escherichia coli K-12. J Mol Biol 156:93-112.
Josefsson L-G and Randall LL (1981) Different exported proteins
    in E. coli show differences in the temporal mode of
    processing in vivo. Cell 25:151-157.
Kumamoto CA and Beckwith J (1985) Evidence for specificity at an
    early step in protein export in Escherichia coli. J
    Bacteriol 163:267-274.
Randall LL and Hardy SJS (1986) Correlation of competence for
    export with lack of tertiary structure of the mature
    species: A study in vivo of maltose-binding protein in E.
    coli. Cell 46:921-928.
Rasmussen BA, Bankaitis VA, and Bassford PJ Jr. (1984) Export
    and processing of MalE-LacZ hybrid proteins in Escherichia
    coli. J Bacteriol 160:612-617.
Stader J and Silhavy TJ (1988) A progenitor of the outer
    membrane LamB trimer J Bacteriol 170:1973-1974.
Voorhout W, De Kroon T, Leunissen-Bijvelt J, Verkleij A, and
    Tommassen J (1988) Accumulation of LamB-LacZ hybrid
    proteins in intracytoplasmic membrane-like structures in
    Escherichia coli K12. J Gen Microbiol 134:599-604.
vos Scheperkeuter GH and Witholt BJ (1984) Assembly pathway of
    newly-synthesized LamB protein an outer membrane protein of
    Escherichia coli K12. J Mol Biol 175:511-528.

THE GENETIC DISSECTION OF PROTEIN TRANSLOCATION ACROSS THE YEAST ER MEMBRANE

C.J. Stirling[*] and R. Schekman[#]
[*]Department of Biochemistry & Molecular Biology,
Medical School, University of Manchester,
Oxford Rd., Manchester.
England M13 9PT

Introduction

The first stage in the eukaryotic secretory pathway is the translocation of polypeptides across the endoplasmic reticulum (ER) membrane. This process has been extensively studied in mammalian systems using an *in vitro* assay which faithfully reproduces the cotranslational translocation of specific precursor proteins into the ER lumen (Blobel and Dobberstein, 1975). Biochemical analysis of the components required to support translocation, has revealed requirements for both cytosolic, and membrane associated factors. Of these, the best characterised to date is signal recognition particle (SRP), a cytosolic ribonucleoprotein complex comprising six polypeptide sub-units (9, 14, 19, 54, 68, 72kDa), plus one molecule of 7SL RNA (Walter & Blobel, 1980; Walter & Blobel, 1982). Evidence suggests that SRP binds to the signal sequence of a nascent secretory protein as it emerges from the ribosome. The binding of SRP to the nascent chain/ribosome complex significantly reduces the rate of polypeptide chain elongation (elongation arrest, Walter and Blobel, 1981). The arrested complex is then targeted to the ER membrane *via* the interaction of SRP with an integral membrane G-protein known as SRP receptor, or "docking protein", (Gilmore *et al.,* 1982; Meyer *et al.,* 1982). The SRP-receptor mediates the GTP-dependent displacement of SRP from the signal sequence/ribosome complex (Connolly & Gilmore, 1989). The displacement of SRP relieves elongation arrest, thus permitting the targeted precursor to be co-translationally translocated across the ER membrane.

[#] Division of Biochemistry and Molecular Biology,
University of California, Berkeley,
California 94730 USA

NATO ASI Series, Vol. H 63
Dynamics of Membrane Assembly
Edited by J. A. F. Op den Kamp
© Springer-Verlag Berlin Heidelberg 1992

Whilst the SRP-dependent targeting cycle is relatively well characterised, the precise mechanism by which the translocating polypeptide chain actually penetrates the lipid bilayer remains unclear. According to one school of thought, the translocating polypeptide may insert directly into the lipid bilayer. Indeed various synthetic peptides have been shown to insert spontaneously into phospholipid bilayers in the absence of participating membrane proteins. Furthermore, the relative affinities of these peptides for the lipid phase correlates directly with their ability to function as signal sequences (Briggs *et al.,* 1986). Nonetheless, the subsequent translocation of hydrophilic protein sequences through the hydrocarbon core of the lipid bilayer must surely face a significant thermodynamic obstacle. It has been proposed that specific rearrangements of phospholipids within the membrane may create a transient hydrophilic environment around a translocating protein; for example the formation of a hexagonal lipid phase (Nesmeyanova, 1982; de Vrije *et al.,* 1990). In support of this view, at least some proteins are capable of translocating into protein-free liposomes (Wickner *et al,* 1978).

On the other hand, there exists a wealth of data implicating membrane proteins in the translocation process. The putative translocation complex, or translocon (Walter & Lingappa, 1986), might form an entirely proteinaceous channel through the lipid bilayer, or indeed, might function *via* the induction of local changes in the lipid phase required to facilitate translocation. The role for at least some protein component to the translocation channel is supported by the observation that aqueous perturbants, such as urea, are capable of removing nascent chains from the membrane (Gilmore & Blobel, 1985). Cross-linking studies using mammalian-derived components have identified an ER membrane protein, termed signal sequence receptor (SSR), which interacts directly with the signal peptide of nascent proteins (Weidmann *et al.,*1987). More recent studies suggest that SSR exists as an oligomeric complex comprising at least two integral membrane glycoproteins, namely SSRα (34kDa), and SSRβ (22kDa), (Rapoport, 1990). Furthermore, SSRα remains in close proximity to the mature portion of a translocating preprotein, and may therefore represent a constituent of the translocon *per se* (Weidmann *et al.,*1989; Prehn *et al.,*1990; Rapoport, 1990).

Yeast Mutants Defective in Protein Translocation

The genetic analysis of the yeast secretory pathway has clearly demonstrated its overall similarity to that of mammalian cells (Novick *et al.,* 1980; Novick *et al.,* 1981). More recently this similarity has been confirmed at the molecular level by the identification of functionally interchangeable homologues involved in vesicular transport (Dunphy *et al.,* 1986); amongst these SEC18p represents the yeast homologue of mammalian NSF, whereas *sec17* mutants are defective in the functional equivalent of SNAP (Eakle *et al.,* 1988; Wilson *et al.,* 1989; Clary *et al.,* 1990). Yeast therefore offers a potentially powerful model system for the genetic dissection of the translocation machinery. The isolation of a large number of translocation-defective mutants might lead to the identification of many, if not all, of the essential components of the yeast translocation apparatus. Of the 23 complementation groups of *sec* mutants originally defined by Novick *et al* (1980), none exhibit translocation defects. Therefore, genetic selections have been developed which enable the direct selection of translocation mutants *via* their mis-localisation of a selectable enzyme activity. Deshaies & Schekman (1987), demonstrated that the normally cytosolic product of the *HIS4* gene, His4p, can be efficiently translocated into the lumen of the ER when fused to the N-terminal signal sequence of pre-pro-α-factor. Crucially, they observed that upon sequestration within the ER lumen, the histidinol dehydrogenase activity of HIS4p becomes phenotypically inert. Consequently, wild type cells expressing such a fusion protein are unable to utilise histidinol. On the contrary, host mutants defective in protein translocation retain a proportion of the HIS4-fusion protein within the cytoplasmic compartment; such mutants possess histidinol dehydrogenase activity and can be selected by their growth on histidinol medium. A defect in protein translocation would be expected to result in pleiotropically lethal effects, therefore conditional lethal mutants were selected on histidinol-medium at an arbitrarily chosen "semi-permissive" temperature (30°C), then screened for conditional lethality at 37°C. The procedure therefore focuses upon conditional alleles which produce a minor translocation defect at 30°C, but which exert a lethal effect at 37°C.

Using this selection, three complementation groups of temperature sensitive translocation mutants have been defined, namely *sec61, sec62,* and *sec63* (Deshaies & Schekman, 1987; Rothblatt *et al.,* 1989). Mutants in all three classes accumulate untranslocated precursor forms of various secretory proteins including pre-pro-α-factor, CPY, acid phosphatase, and to a lesser extent invertase. The accumulated precursors are unglycosylated, lack signal peptide processing, and are exclusively located within the cytosolic compartment (Deshaies & Schekman, 1987; Rothblatt *et al.,* 1989).

A wider survey of the types of precursors whose translocation is blocked in these mutants revealed a striking dichotomy between soluble *versus* membrane proteins. Whilst all soluble precursors were affected to varying degrees, no significant defects were observed for any of a number of integral membrane glycoproteins examined;

including dipeptidylaminopeptidaseB (DPAPB), and KEX2 (Stirling *et al.*, in preparation). This prompted a second mutant selection specifically designed to identify mutants defective in membrane protein assembly. This latter selection is identical in principle to that described above, except that it employs a fusion protein in which His4p is directed to the lumen of the ER *via* the correct insertion of the integral membrane domain of hydroxy-methyl-glutaryl-CoA reductase (HMG1; Stirling *et al.*, in preparation). More specifically, HIS4 was fused to a normally lumenal domain of HMG1 between transmembrane domains six and seven (Fig. 1). Wild type cells expressing this fusion (from the multicopy plasmid pCS5), insert it into the ER membrane such that the HIS4 domain is presented to the ER lumen; as demonstrated by its extensive elaboration with N-linked core oligosaccharides. As predicted, pCS5 containing cells are unable to grow on histidinol. On the contrary, similar cells carrying a control plasmid, pCS4, in which HIS4 is fused to a cytoplasmic domain of HMG1, express an unglycosylated fusion protein and grow well on histidinol (see Fig 1). Significantly, pCS5 fails to confer a histidinol+ phenotype upon any of the original Deshaies & Schekman (1987) mutants, a finding consistent with their phenotypes with respect to the insertion of other integral membrane proteins.

Mutants which fail to insert, or which misassemble, the pCS5 fusion protein were then selected by their ability to grow on histidinol. This HMG1-based selection has to date identified two further conditional lethal translocation mutants. Of these, one represents a new, more restrictive, allele of *sec61* (*sec61-3*), whereas the other defines an entirely novel complementation group designated *sec65*. The *sec61-3* and *sec65-1* mutants both exhibit defects in the processing of various integral membrane proteins, including DPAPB. However, they also exhibit severe defects in the processing of all soluble precursors tested (including pre-pro-α-factor, invertase etc), thereby suggesting that the *SEC61* and *SEC65* gene products are required for both secretory protein translocation, and membrane protein insertion (Stirling *et al.*, in preparation).

In addition to the four *SEC* genes described above, two classes of topologically distinct HSP70s have also been implicated in protein translocation in *Saccharomyces cerevisiae*, namely the ER lumenal KAR2p, and the cytosolic products of the *SSA* gene family. In the case of *KAR2*, mutant alleles have been isolated which exhibit severe defects in the translocation of secretory proteins (Vogel *et al.*, 1990). *KAR2* encodes an HSP70 of the ER lumen, and represents the yeast homologue of the so-called mammalian immunoglobulin heavy chain binding protein (Bip; Normington *et al.*, 1989; Rose *et al.*, 1989). The *SSA1* and *SSA2* genes encode functionally interchangeable cytosolic

HSP70s, at least one of which is required for the proper targeting of pre-pro-α-factor, both *in vivo* and *in vitro* (Deshaies *et al.,* 1988; Chirico *et al.,* 1988).

(A)

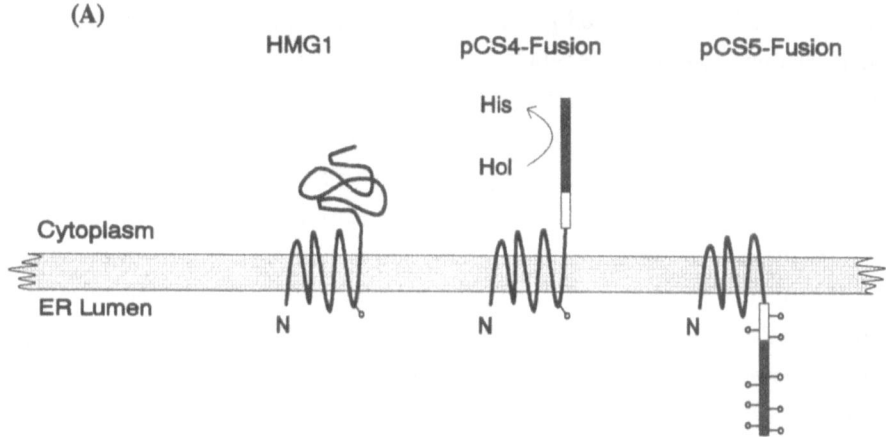

(B)

	Growth on histidinol conferred by plasmid			
Strain	YEp352	pGD2	pCS4	pCS5
SEC+	-	-	+	-
sec61-2	-	+	+	-
sec62-1	-	+	+	-
sec63-1	-	+	+	-

Figure 1 (A) Diagrammatic representation of the topologies of various HMG1 derivatives. HMG1 sequences are indicated by a black line. In fusion protein derivatives, the open bar indicates SUC2 sequences (included to provide an immunological determinant); the darkly shaded bar indicates the catalytic HIS4 domain. N-linked core sugar additions are indicated (↑) (B). The histidinol phenotypes conferred by various plasmids upon previously defined translocation mutants. The various strains listed were each transformed with either vector alone, YEp352, or with pCS4, pCS5, or pGD2. The latter plasmid encodes a fusion protein which directs HIS4 to the ER lumen *via* the signal sequence of pre-pro-α-factor. The fusion encoded by pGD2 is identical to that employed in the selection of *sec61-2, sec62-1,* and *sec63-1,* (Deshaies and Schekman, 1987; Rothblatt *et al,* 1990).

Membrane Components of the Yeast Translocation Machinery

The *SEC61, SEC62,* and *SEC63* genes have all been cloned and sequenced, and their products characterised in some detail. All three are essential genes, and all encode ER-specific integral membrane proteins (Stirling *et al.,* in preparation; Deshaies & Schekman, 1989; Deshaies & Schekman, 1990; Sadler *et al.,* 1989; Feldheim *et al.,* in preparation). *SEC61* encodes a 53kDa polypeptide (SEC61p), whose sequence predicts the presence of 4-8 transmembrane domains. The extremely hydrophobic nature of this protein results in a severe gel mobility aberration in SDS-PAGE; the observed relative molecular weight varying from 38 to 48kDa depending upon the precise gel conditions employed (Stirling *et al.,* in preparation). The membrane topology of SEC61p remains unknown, however those of SEC62p (32kDa) and SEC63p (73kDa) have been empirically determined confirming the existence of two and three transmembrane domains respectively (Fig. 2; Deshaies and Schekman, 1990; Feldheim *et al.,* in preparation).

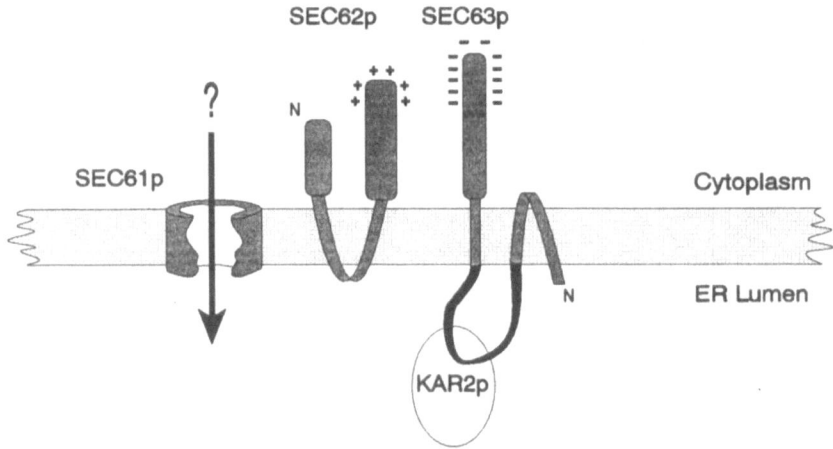

Figure 2. ER proteins required for protein translocation in *Saccharomyces cerevisiae*. The topologies of SEC62p and SEC63p are indicated with the amino-termini denoted (N). Both SEC62p and SEC63p possess highly charged C-terminal, cytoplasmic domains which may mediate the observed interaction between these two proteins (see text). The DnaJ-like domain within SEC63p is indicated by a darker shading; this domain may mediate an interaction with KAR2p (see text). SEC61p is predicted to be a polytopic transmembrane protein whose precise topology has not yet been determined. Crosslinking studies suggest that SEC61p is in contact with a translocating precursor molecule, thereby suggesting that SEC61p represents a component of the translocon.

A multisubunit complex in the ER membrane

A number of genetic observations led to predictions that the *SEC61, 62,* and *63* gene products might interact directly with one another. Firstly, the various pairwise combinations of mutants exhibit exaggerated phenotypes; ie *sec62sec63* double mutants exhibit poor viability at 24°C, whilst *sec61sec62* are inviable above 17°C. The *sec61sec63* double mutant is inviable at any temperature (Rothblatt *et al.,* 1989). This phenomenon, termed synthetic lethality, has been observed between genes known to encode interacting proteins, eg α and β tubulins (Huffaker *et al.,* 1987). For further discussion of the phenomenon of synthetic lethality see Kaiser and Schekman, (1990). Furthermore, certain mutant alleles of *sec63* can be suppressed by the overexpression of SEC62p (Deshaies and Schekman, 1990). The biochemical confirmation of such interactions has come from recent cross-linking studies which have identified a membrane-associated multi-subunit complex comprising SEC61p, SEC62p, SEC63p, together with a 31.5 kDa glycoprotein (gp31.5), and a 23kDa protein (p23). The same complex, minus SEC61p, can be immunoprecipitated from detergent solubilised membranes under non-denaturing conditions, suggesting that the association with SEC61p is labile (Deshaies *et al.,* 1991). Whilst the identities of gp31.5 and p23 are unknown, the authors note the coincidence between the molecular weights of these components and those reported for the sub-units of the dog pancreas signal sequence receptor, SSRα (34kDa), and SSRβ (23kDa) (Weidmann *et al.,* 1989; Rapoport, 1990). Both sub-units of SSR are N-glycosylated transmembrane proteins (Rapoport, 1990). The observed N-glycosylation of gp31.5 suggests that it is either a transmembrane, or ER lumenal protein.

The sequence of SEC63p suggests a possible interaction with KAR2p (Sadler *et al.,*1989; Deshaies *et al.,* 1989). The argument is based upon the identification of a domain within SEC63p which is highly similar to the N-terminal portion of *E.coli* DnaJ (43% identity over 71 residues; Sadler *et al.,*1989). In *E.coli,* DnaJ interacts with DnaK in a number of processes including DNA replication of phage-λ (Dodson *et al.,* 1986). Like KAR2p, DnaK is a member of the highly conserved HSP70 superfamily. Therefore, *if* the N-terminal 70 residues of DnaJ mediate its interaction with DnaK, then the equivalent domain in SEC63p may interact with an HSP70 within the ER lumen, eg KAR2p. Site-specific mutagenesis of several conserved residues within the DnaJ-like domain has confirmed that they are essential for SEC63p function (Feldheim *et al.,* in preparation). Once again genetic data provides indirect evidence for this proposed interaction; firstly, certain *sec63 kar2* double mutants exhibit allele-specific synthetic lethality (M. Rose, personal communication), secondly, certain mutant alleles of *kar2* are able to suppress the growth defect of *sec63-1* cells (M. Rose, personal communication). Such allele-specific suppression is generally considered to be evidence of a direct interaction between the mutated gene products (Huffaker *et al.*, 1987).

Two groups have independently developed techniques with which to trap a translocating precursor protein within the ER membrane *in vitro*. When these putative translocation intermediates are exposed to cross-linking agents, the trapped precursor becomes crosslinked to SEC61p. Furthermore, the cross-linking of precursor to SEC61p is completely dependent upon the presence of ATP (Sanders & Schekman, personal communication; Tom Rapoport, personal communication). ATP has previously been shown to be essential for the post-translational translocation of pre-pro-α-factor *in vitro* (see below). These cross-linking studies indicate that SEC61p is in close proximity to a translocating chain, and thus suggest that SEC61p may form at least part of the translocon.

Yeast Cytosolic Factors Involved in ER Targeting and Translocation

Several groups have reconstituted the yeast translocation reaction *in vitro* (Hansen *et al.*, 1986; Rothblatt and Meyer, 1986a; Waters and Blobel, 1986). Whilst the co-translational translocation of invertase and pre-pro-α-factor has been demonstrated, the latter is unique in its ability to efficiently translocate post-translationally. This uncoupling of translocation from translation has enabled the identification of an ATP requirement for translocation (Hansen *et al.*, 1986; Waters and Blobel, 1986; Rothblatt and Meyer, 1986b). The post-translational translocation of pre-pro-α-factor has been interpreted as evidence for the absence of an SRP-equivalent, leading to suggestions that the translocation mechanism in yeast is significantly different to that in other eukaryotes. However, it has become increasingly evident that the significance of translational arrest has been exaggerated. For example, Lipp *et al.* (1987), have shown that in wheat germ lysates, SRP binding does not arrest translation of secretory protein mRNA, but instead acts to reduce the rate of chain elongation to varying degrees depending upon the mRNA species. Perhaps more significant is the finding that full-length pre-pro-α-factor can be translocated into mammalian microsomes in an SRP-dependent fashion, provided it remains tethered to the ribosome as a peptidyl-tRNA (Garcia and Walter, 1988).

When translated in wheat germ lysates, the post-translational translocation competence of pre-pro-α-factor is not an innate property of the polypeptide, but is dependent upon the activity of two functionally interchangeable cytosolic HSP70s, (encoded by the *SSA1* and *SSA2* genes), plus an unidentified NEM-sensitive factor (Chirico *et al.,* 1988; Deshaies *et al.,* 1988). It has been proposed that the role of HSP70 is that of an ATP-dependent protein "unfoldase"; a view that is supported by the observation that the translocation of urea-denatured prepro-α-factor is independent of HSP70 (Chirico *et al.,* 1988). Genetic evidence suggests that the SSA family of HSP70s are also involved in mitochondrial protein import in yeast (Deshaies *et al.,* 1988), therefore suggesting mechanistic similarities between mitochondrial protein import and ER translocation.

Components of an SRP-like particle in yeast?

The amino-acid sequence of the 54kDa subunit of SRP (SRP54), has recently been deduced from cDNA clones, and has revealed striking sequence similarities to a number of other proteins namely, the α-subunit of mammalian SRP-receptor (SRα; Lauffer *et al.,* 1985; Hortsch *et al.,* 1988), and the putative products of two *E.coli* genes of unknown function, *ftsY* and the previously unnamed *ffh* (*fifty four homologue*) (Romisch *et al.,* 1989; Bernstein *et al.,* 1989). Using oligonucleotide probes designed from the most conserved regions between SRP54 and Ffh, Hann *et al,* (1989) have cloned *SRP54* homologues from both *Schizosaccharomyces pombe,* and *Saccharomyces cerevisiae.* The *S. cerevisiae* gene (SRP54$_{sc}$) predicts a protein which exhibits 47% identity to mammalian SRP54. We have recently cloned and sequenced the yeast *SEC65* gene and found it to encode a 30kDa cytosolic protein with significant sequence similarity to the 19kDa subunit of human SRP (SRP19; Lingelbach *et al.,* 1988). The fact that a *sec65* mutant exhibits a severe translocation defect, might indicate that this sequence similarity is biologically significant. More convincing evidence has come from the observation that over expression of SRP54$_{sc}$ can suppress the *sec65-1* mutant defect (B. Hann, & CJ Stirling, unpublished data). SRP54$_{sc}$ null mutants exhibit a translocation-defective phenotype, confirming a role for this molecule in protein translocation in yeast (Byron Hann, personal communication; Amaya & Nakano, 1991). In mammalian signal recognition particle, SRP19 and SRP54 interact directly with one another in the assembly of the ribonucleoprotein complex. The suppression of *sec65-1* by *SRP54sc* provides potent genetic evidence for a similar interaction in yeast. A functional interaction between SRP54p and SEC65p would appear to indicate the existence of some SRP-related factor in yeast. However, whether or not such a factor comprises any additional sub-units is as yet unknown. A candidate for an SRP-like RNA has been proposed in the form of *SCR1,* (Felici *et al.,* 1989). SCR1 represents the major *small cytoplasmic RNA* in *Saccharomyces cerevisiae,* just as 7SL RNA is the most abundant small cytoplasmic RNA in mammalian cells. However, it displays only limited sequence similarity to other 7SL RNAs, and does not share the same features of secondary structure that are so highly conserved amongst 7SL homologues from a wide range of species (Zwieb, 1989). Therefore, if SCR1 represents the scaffold around which yeast "SRP" is assembled, then one would predict significant structural differences between yeast and mammalian SRPs. In this regard it is interesting to note that SEC65p (30kDa) is considerably larger than its 19kDa homologue. Whether or not the putative SRP-like factor corresponds the translocation-promoting NEM-sensitive factor detected in yeast extracts by Chirico *et al,*(1988), remains to be determined.

Table 1. **S. cerevisiae genes involved in ER translocation**

	Gene product	Function
ER components		
SEC61	53K integral membrane protein	Multi-subunit Complex:
SEC62	31K " " "	Components of which can be
SEC63	73K " " "	crosslinked to precursor
--	gp31.5	chains in translocation
--	p23	intermediates.
KAR2	ER Lumenal Hsp70	
SEC11	19K	Signal peptidase[*]
Cytosolic components		
SSA1-4	Cytosolic Hsp70s	Molecular chaperone?
SRP54sc	60K homologue of mammalian SRP54	
SEC65	31K homologue of mammalian SRP19	SRP-like particle?
SCR1	519 nucleotide cytosolic RNA	

[*]*SEC11* mutants are not defective in translocation *per se,* but rather fail to process precursor molecules appropriately.

Gene disruption experiments have shown that *SRP54, SEC65,* and *SCR1* are non-essential genes, but that the null mutants grow very slowly (Byron Hann, personal communication and contrary to Hann *et al,* 1989; Stirling, in preparation, Felici *et al.,* 1989). These observations imply that any yeast SRP-like moiety is not essential for protein translocation *in vivo.* This does not appear unreasonable if the principle role of SRP is to "chaperone" a nascent chain to some membrane receptor in a translocation-competent state. It may be that sufficient nascent precursors reach their membrane receptor stochastically, such that a slow rate of growth can be sustained. SRP would then be necessary only for the efficient targeting required to maintain a normal growth rate. However, the 7SL RNA gene of *Schizosaccharomyces pombe* is clearly essential for cell viability (Brennwald *et al.,* 1988; Ribes *et al.,* 1988). It is not clear whether or not this points to a genuine mechanistic difference between these yeast, or merely varying tolerance for inefficient targeting. It is perhaps notable that the *sec65* gene disruption is clearly essential in some strains of *Saccharomyces cerevisiae,* the cause of these strain-background affects is obscure (Stirling, unpublished data). Alternatively, the differences observed between *S. cerevisiae* and *Ss. pombe* may reflect the existence of an SRP-independent bypass pathway in *Saccharomyces cerevisiae* capable of sustaining a viable level of translocation.

Conclusions

The translocation-defective mutants described above offer potentially powerful tools for the analysis of the translocation machinery. Firstly, the mutants define genes whose products are required for translocation, and in turn provide a means with which to clone the wild type *SEC* genes by complementation. The molecular analysis of the cloned genes has identified the nature of the SEC proteins themselves, which in turn has produced a number of important insights into the structure and function of the yeast translocation apparatus. In addition, the detailed analysis of mutant phenotypes, both *in vivo* and *in vitro,* may ultimately reveal the precise nature of the biochemical role(s) played by each component in the translocation process.

References

Amaya Y, Nakano A (1991) *SRH1* protein, the yeast homologue of the 54kDa subunit of signal recognition particle, is involved in ER translocation of secretory proteins. Febs Letts 283:325-328.

Bernstein HD et al. (1989) Model for signal sequence recognition from amino-acid sequence of 54K subunit of signal recognition particle. Nature 340:482-486

Brennwald P et al. (1988) Identification of an essential Schizosaccharomyces pombe RNA homologous to the 7SL component of signal recognition particle. Mol Cell Biol 8 (4):1580-1590

Briggs MA, Cornell DG, Dluhy RA, Gierasch LM (1986) Conformation of signal peptides induced by lipids suggest initial steps in protein export. Science 233:206-208

Chen W-J, Douglas MG (1987) Phosphodiester bond cleavage outside mitochondria is required for completion of protein import into the mitochondrial matrix. Cell 49:651-658

Chirico WJ, Waters MG, Blobel G (1988) 70K heat shock related proteins stimulate protein translocation into microsomes. Nature 332:805-810

Clary DO, Griff IC, Rothman JE (1990) SNAPs, a family of NSF attachment proteins involved in intracellular membrane fusion in animals and yeast. Cell 61:709-721

Connolly T, Gilmore R (1989) The signal recognition particle receptor mediates the GTP-dependent displacement of SRP from the signal sequence of the nascent polypeptide. Cell 57:599-610

Deshaies RJ et al. (1988) A subfamily of stress proteins facilitates translocation of secretory and mitochondrial precursor polypeptides. Nature 332:800-805

Deshaies RJ, Sanders SL, Feldheim DA, Schekman R (1991) Assembly of yeast Sec proteins involved in translocation into the endoplasmic reticulum into a membrane-bound multisubunit complex. Nature 349:806-808

Deshaies RJ, Schekman R (1987) A yeast mutant defective at an early stage in import of secretory protein precursors into the endoplasmic reticulum. J Cell Biol 105 (2):633-645

Deshaies RJ, Schekman R (1989) SEC62 encodes a putative membrane protein required for protein translocation into the yeast endoplasmic reticulum. J Cell Biol 109:2653-2664

Deshaies RJ, Schekman R (1990) Structural and functional dissection of Sec62p, a membrane-bound component of the yeast endoplasmic reticulum protein import machinery. Mol Cell Biol 10 (11):6024-6035

Dodson M et al. (1986) Specialized nucleoprotein structures at the origin of replication of bacteriophage lambda:localized unwinding of duplex DNA by a six-protein reaction. Proc Natl Acad Sci USA 83:7638-7642

Dunphy WG et al. (1986) Yeast and mammals utilize similar cytosolic components to drive protein transport through the Golgi complex. Proc Natl Acad Sci USA 83:1622-1626

Eakle KA, Bernstein M, Emr SD (1988) Characterisation of a component of the yeast secretion machinery:identification of the SEC18 gene product. Mol Cell Biol 8:4098-4109

Eilers M, Oppliger W, Schatz G (1987) Both ATP and an energized inner membrane are required to import a purified precursor protein into mitochondria. EMBO J 6 (4):1073-1077

Eilers M, Schatz G (1986) Binding of a specific ligand inhibits import of a purified precursor protein into mitochondria. Nature 322:228-232

Felici F, Cesareni G, Hughes JM (1989) The most abundant small cytoplasmic RNA of Saccharomyces cerevisiae has an important function required for normal cell growth. Mol Cell Biol 9 (8):3260-3268

Gilmore R, Blobel G (1985) Translocation of secretory proteins across the microsomal membrane occurs through an environment accessible to aqueous perturbants. Cell 42:497-505

Gilmore R, Walter P, Blobel G (1982) Protein translocation across the endoplasmic reticulum II. Isolation and characterisation of the signal recognition particle receptor. J Cell Biol 95:470-477

Hann BC, Poritz MA, Walter P (1989) Saccharomyces cerevisiae and Schizosaccharomyces pombe contain a homologue to the 54-kD subunit of the signal recognition particle that in S. cerevisiae is essential for growth. J Cell Biol 109:3223-3230

Hansen W, Garcia PD, Walter P (1986) In vitro protein translocation across the yeast endoplasmic reticulum:ATP-dependent post-translational translocation of the prepro-alpha-factor. Cell 45:397-406

Hortsch M, Labeit S, Meyer DI (1988) Complete cDNA sequence coding for human docking protein. Nucleic Acids Res 16:361-361

Huffaker TC, Hoyt MA, Botstein D (1987) Genetic analysis of the yeast cytoskeleton. Ann.Rev Genet 21:259-284

Kaiser CA, Schekman R (1990) Distinct sets of SEC genes govern transport vesicle formation and fusion early in the secretory pathway. Cell 61:723-733

Lauffer L et al. (1985) Topology of signal recognition particle receptor in endoplasmic reticulum membrane. Nature 318:334-338

Lingelbach K et al. (1988) Isolation and characterization of a cDNA clone encoding the 19 kDa protein of signal recognition particle (SRP):expression and binding to 7SL RNA. Nucleic Acids Res 16 (20):9431-9442

Meyer DI, Krause E, Dobberstein B (1982) Secretory protein translocation across membranes- the role of the "docking protein". Nature 297:503-508

Murakami H, Pain D, Blobel G (1988) 70 kDa heat shock-related protein is one of at least two distinct cytosolic factors stimulating protein import into mitochondria. J. Cell Biol 107:2051-2057

Nesemeyanova MA (1982) On the possible participation of acid phospholipids in the translocation of secreted proteins through the bacterial cytoplasmic membrane. Febs letts 142:189-193

Normington K et al. (1989) S. cerevisiae encodes an essential protein homologous in sequence and function to mammalian Bip. Cell 57:1223-1236

Novick P, Ferro S, Schekman R (1981) Order of events in the yeast secretory pathway. Cell 25:461-469

Novick P, Field C, Schekman R (1980) Identification of 23 complementation groups required for post-translational events in the secretory pathway. Cell 21:205-215

Pfanner N, Tropschug M, Neupert W (1987) Mitochondrial protein import:nucleoside triphosphates are involved in conferring import-competence to precursors. Cell 49 (6):815-823

Prehn S et al. (1990) Structure and biosynthesis of the signal sequence receptor. Eur J Biochem 188:439-445

Rapoport TA (1990) Protein transport across the ER membrane. Trends Biochem Sci 15 (9):355-358

Ribes V, Dehoux P, Tollervey D (1988) 7SL RNA from Schizosaccharomyces pombe is encoded by a single copy essential gene. EMBO J 7:231-237

Romisch K et al. (1989) Homology of 54K protein of signal-recognition particle, docking protein and two E. coli proteins with putative GTP-binding domains. Nature 340:478-482

Rose MD, Misra LM, Vogel JP (1989) KAR2, a karyogamy gene, is the yeast homolog of the mammalian Bip/GRP78 gene. Cell 57:1211-1221

Rothblatt JA et al. (1989) Multiple genes are required for proper insertion of secretory proteins into the endoplasmic reticulum in yeast. J Cell Biol 109:2641-2652

Rothblatt JA, Meyer DI (1986) Secretion in yeast:translocation and glycosylation of prepro-alpha-factor in vitro can occur via an ATP-dependent post-translational mechanism. EMBO J 5:1031-1036

Sadler I et al. (1989) A yeast gene important for protein assembly into the endoplasmic reticulum and the nucleus has homology to DnaJ, and Escherichia coli heat shock protein. J Cell Biol 109:2665-2675

Toyn J et al. (1988) In vivo and in vitro analysis of ptlI, a yeast ts mutant with a membrane-associated defect in protein translocation. EMBO J 7:4347-4353

Vogel JP, Misra L, Rose MD (1990) Loss of Bip/GRP78 function blocks translocation of secretory proteins in yeast. J Cell Biol 110:1885-1895

Vrije de et al. (1990) In: Op den Kamp JAF (ed) Dynamics and biogenesis of membranes, NATO ASI series Vol H40, Plenum press, p247-258

Walter P, Blobel G (1982a) Signal recognition particle contains a 7S RNA essential for protein translocation across the endoplasmic reticulum. Nature 99:691-698

Walter P, Blobel G (1980) Purification of a membrane-associated protein complex required for protein translocation across the endoplasmic reticulum. Proc Natl Acad Sci USA 77:7112-7116

Walter P, Blobel G (1981) Translocation of proteins across the endoplasmic reticulum III. Signal recognition particle (SRP) causes signal sequence-dependent and site-specific arrest of chain elongation that is released by microsomal membranes. J Cell Biol 91:557-561

Walter P, Blobel G (1982a) Mechanism of protein translocation across the endoplasmic reticulum. Biochem Soc Symp 47:183-191

Walter P, Blobel G (1982b) Signal recognition particle contains a 7S RNA essential for protein translocation across the endoplasmic reticulum. Nature 99:691-698

Walter P, Lingappa VR (1986) Mechanism of protein translocation across the endoplasmic reticulum membrane. Annu Rev Cell Biol 2:499-516

Waters MG, Blobel G (1986) Secretory protein translocation in a yeast cell-free system can occur post translationally and requires ATP hydrolysis. J Cell Biol 102:1543-1550

Weidmann M, Gorlich D, Hartmann E, Kurzchalia TV, Rapoport T (1989) Photocrosslinking demonstrates proximity of a 34kDa membrane protein to different portions of preprolactin during translocation through the endoplasmic reticulum. Febs Letts 257:263-268

Weidmann M, Kurzchalia TV, Hartmann E, Rapoport TA (1987) A signal sequence receptor in the endoplasmic reticulum membrane. Nature 328:830-833

Wickner W (1979) The assembly of proteins into biological membranes:The membrane trigger hypothesis. Ann Rev Biochem 48:23-45

Wilson DW et al. (1989) A fusion protein required for vesicle-mediated transport in both mammalian cells and yeast. Nature 339:355-359

Zwieb C (1989) Structure and function of signal recognition particle RNA. Progress in nucleic acid research and molecular biology 37:207-235.

β-GALACTOSIDE TRANSPORT IN *ESCHERICHIA COLI*: THE INS AND OUTS OF LACTOSE PERMEASE

H. Ronald Kaback
Howard Hughes Medical Institute
Molecular Biology Institute and
Departments of Physiology and Microbiology & Molecular Genetics
University of California Los Angeles
Los Angeles, California 90024-1570
U.S.A.

Accumulation of β-galactosides against a concentration gradient in *Escherichia coli* is carried out by the lactose (*lac*) permease, a hydrophobic polytopic cytoplasmic membrane protein that catalyzes the coupled translocation of β-galactosides and H$^+$ with a stoichiometry of unity (i.e. β-galactoside/H$^+$ symport or cotransport) (cf. Kaback, 1983, 1986, 1990 for reviews). Under physiological conditions, where the H$^+$ electrochemical gradient across the cytoplasmic membrane $(\Delta\bar{\mu}_H^+)$[1] is interior negative and/or alkaline, *lac* permease utilizes free energy released from downhill translocation of H$^+$ to drive accumulation of β-galactosides against a concentration gradient. In the absence of $\Delta\bar{\mu}_H^+$, the permease catalyzes the converse reaction, utilizing free energy from downhill translocation of β-galactosides to drive uphill translocation of H$^+$ and generating $\Delta\bar{\mu}_H^+$, the polarity of which depends upon the direction of the substrate concentration gradient. As such, *lac* permease is a paradigm for a wide variety of biological machines in both prokaryotic and eukaryotic membranes that transduce the free energy of an electrochemical ion gradient into work or into other forms of chemical energy (i.e., ATP).

Lac permease is encoded by the *lacY* gene, the second structural gene in the *lac* operon, and it has been cloned into a recombinant plasmid and sequenced. By combining overexpression of *lacY* with the use of a highly specific photoaffinity probe for the permease and reconstitution of transport activity in artificial phospholipid vesicles (i.e., proteoliposomes), the permease has been solubilized from the membrane, purified to homogeneity and shown to catalyze all the translocation reactions typical of the β-galactoside transport system with comparable turnover numbers. Therefore, a single gene product--the product of *lacY*--is solely responsible for all of the

---

[1]Abbreviations: $\Delta\bar{\mu}_H^+$, the proton electrochemical gradient; NPG, *p*-nitrophenyl-α,D-galactopyranoside.

NATO ASI Series, Vol. H 63
Dynamics of Membrane Assembly
Edited by J. A. F. Op den Kamp
© Springer-Verlag Berlin Heidelberg 1992

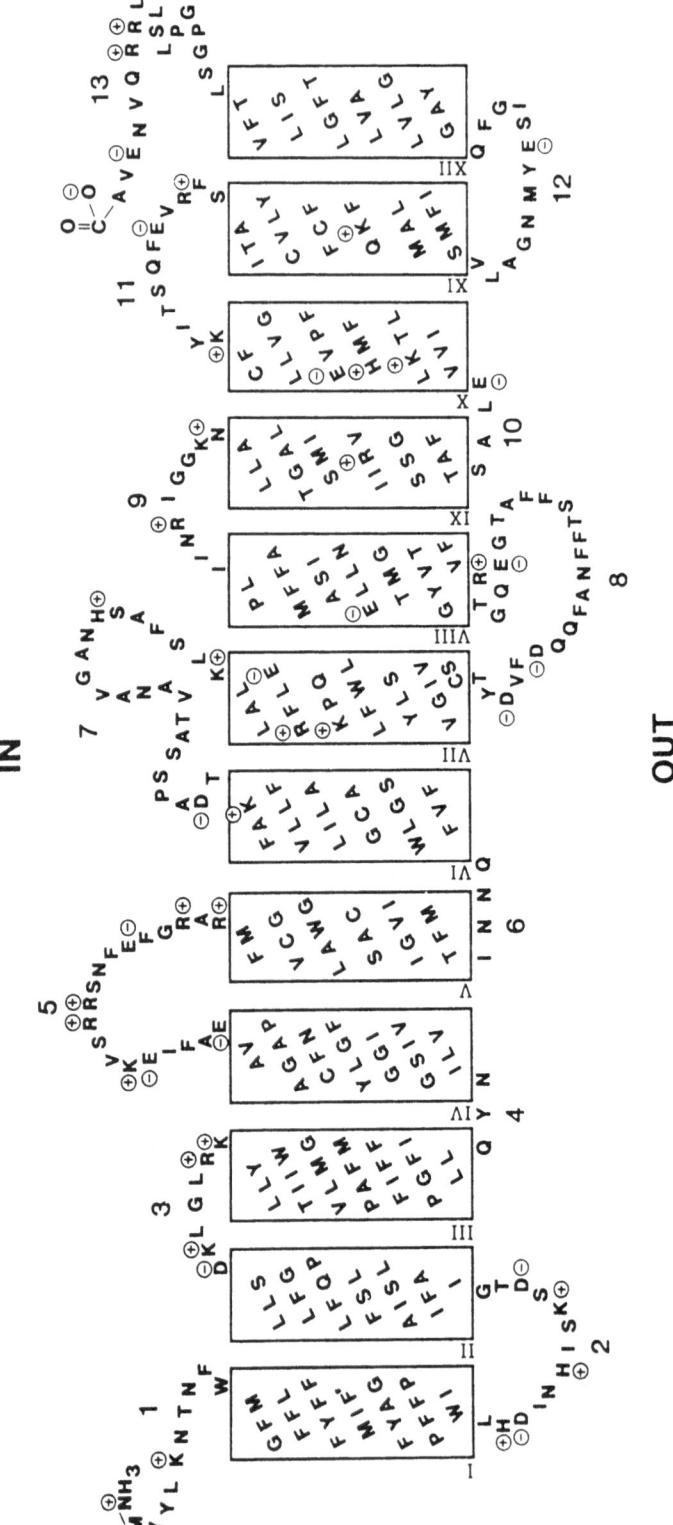

translocation reactions catalyzed by the β-galactoside transport system in intact cells and right-side-out membrane vesicles.

This article is a discussion of selected observations with a specific membrane transport protein at the molecular level, and it should be emphasized at the outset that structural information at the atomic level is particularly difficult to obtain with hydrophobic membrane proteins (Deisenhofer & Michel, 1989). Since the great majority of membrane proteins, *lac* permease in particular, have yet to be crystallized and high-resolution structure is a prerequisite for mechanistic considerations, conclusions derived from some of the studies to be described must be regarded as speculative.

Structure of Lac Permease. Circular dichroic measurements on purified *lac* permease indicate that the protein is over 80% helical in conformation, an estimate consistent with the hydropathy profile of the permease which suggests that approximately 70% of its 417 amino acid residues are found in hydrophobic domains with a mean length of 24 ± 4 residues (Foster, et al 1983). Thus, a secondary structure was proposed in which the permease is composed of a hydrophilic N-terminus followed by 12 hydrophobic segments in α-helical conformation that traverse the membrane in zig-zag fashion connected by hydrophilic domains (loops) with a 17-residue C-terminal hydrophilic tail (Fig. 1). Support for the general features of the model and evidence that both the N and C termini of the permease are exposed to the cytoplasmic face of the membrane was obtained subsequently from laser Raman (Vogel, et al 1985) and Fourier transform infrared (P. D. Roepe, H. R. Kaback & K. J. Rothschild, unpublished information) spectroscopy, immunological studies (Carrasco, et al 1982, 1984a,b; Seckler, et al 1983, Seckler & Wright 1983, Seckler, et al 1986; Herzlinger, et al 1984), limited proteolysis (Goldkorn, et al 1983; Stochaj, et al 1986) and chemical modification (Page & Rosenbusch, 1988). However, none of these approaches is able to differentiate between the 12-helix structure and other models containing 10 or 14 putative transmembrane helices.

Calamia & Manoil (1990) have provided elegant support for the topological predictions of the 12-helix model by analyzing an extensive series of *lac* permease-alkaline phosphatase (*lacY-phoA*) chimeras. Alkaline phosphatase is synthesized as an inactive precursor in the cytoplasm of *E. coli* with a short signal sequence that directs its secretion into the periplasmic space where it dimerizes to form active enzyme. If the signal sequence is deleted, the enzyme remains in the cytoplasm in an inactive form. When alkaline phosphatase devoid of the signal sequence is fused

Figure 1. Secondary-structure model of *lac* permease based on the hydropathy profile of the protein. The single amino-acid code is used, and hydrophobic segments are shown in boxes as transmembrane α-helical domains connected by hydrophilic loops.

to the C-termini of fragments of a cytoplasmic membrane protein, enzyme activity reflects the ability of the fragments to translocate alkaline phosphatase to the outer surface of the membrane. Alkaline phosphatase activity in cells independently expressing each of 36 *lacY-phoA* fusions exclusively favors the model containing 12 transmembrane domains. In addition, it was demonstrated (Calamia & Manoil, 1990) that approximately half of a transmembrane domain is needed to translocate alkaline phosphatase to the external surface of the membrane. Thus, the alkaline phosphatase activity of fusions engineered at every third amino-acid residue in putative helices III and V increases as a step function as the fusion junction proceeds from the 8th to the 11th residue of each of these transmembrane domains. Furthermore, when fusions are constructed at each amino-acid residue in putative helix X, a sharp discontinuity in alkaline phosphatase activity is observed at His322-Met323, thereby implying that these residues are located in the middle of the membrane (M. L. Ujwal, E. Bibi, C. Manoil & H. R. Kaback, unpublished information).

Lac permease exhibits a notch or cleft under high magnification (Costello, et al 1987), an observation independently documented by Li & Tooth (1987) using completely different techniques. The presence of a solvent-filled notch in the permease may have important implications with regard to the molecular mechanism of β-galactoside/H⁺ symport, as the barrier within the permease may be thinner than the full thickness of the membrane. Therefore, the number of amino-acid residues in the protein directly involved in translocation may be fewer than generally thought to be the case.

Insertion and Stability of Lac Permease. Stochaj & Ehring (1987) and Stochaj, et al (1988) demonstrated that sequences within the first 170 amino-acid residues of *lac* permease may be important for insertion. Moreover, a truncated permease containing only the N-terminal 50 amino-acid residues is inserted into the membrane, and it was proposed that this region contains an internal "start transfer" sequence that results in the insertion of the N-terminus as a "helical hairpin" (von Heijne & Blomberg, 1979; Engleman & Steitz, 1981). With respect to the C-terminus of *lac* permease, the 17 amino-acid C-terminal hydrophilic tail is not involved in insertion of the protein into the membrane, its stability, or its ability to catalyze transport. On the other hand, a 3 amino-acid sequence at the end of the last putative transmembrane helix (...VFT...; cf. Fig. 1) is critical for stability and hence activity once the protein is inserted into the membrane (Roepe, et al 1989; McKenna, et al 1991). When stop codons (TAA) are placed sequentially at amino-acid codons 396-401, permease truncated at residue 396 or 397 is completely defective with respect to lactose transport, while molecules truncated at residues 398, 399, 400 and 401, respectively, exhibit 15-

25%, 30-40%, 40-45% and 70-100% of wild-type activity. As judged by pulse-chase experiments with [^{35}S]methionine, wild-type permease or permease truncated at residue 401 is stable, while permease molecules truncated at residue 400, 399, 398, 397 or 396 are degraded at increasingly rapid rates. Finally, recent experiments (E. McKenna, D. Hardy & H. R. Kaback, unpublished information) demonstrate that replacement of residues 397-399 with three Leu residues yields a stable, fully functional permease, while replacement with GlyProGly yields an unstable molecule with minimal transport activity. The results indicate that the last turn of putative helix XII is important for proper folding and protection against proteolytic degradation.

The overall topology of polytopic membrane proteins like *lac* permease is thought to result from either the oriented insertion of the amino-terminal α-helical domain followed by passive, serpentine insertion of subsequent helices (Wickner & Lodish, 1985; Rapoport, 1986; Singer, et al 1987) or from the function of independent topogenic determinants dispersed throughout the molecules (Blobel, 1980; Popot & Engelman, 1990; Popot & de Vitry, 1990). In order to test these alternatives, even or odd numbers of putative transmembrane domains in *lac* permease were deleted, and the effect of the deletions on insertion, stability and the orientation of the C-terminus with respect to the plane of the membrane were examined (Bibi, et al 1991). The strategy is that deletion of odd numbers of transmembrane domains might be expected to alter the position of the C terminus relative to the plane of the membrane, while deletion of even numbers of transmembrane domains would not be expected to do so. So long as the first N-terminal and the last four C-terminal putative α-helical domains are retained in the constructs, stable polypeptides are inserted into the membrane, even when an odd number of helical domains is deleted. Moreover, even when an odd number of helices is deleted, the C-terminus remains on the cytoplasmic surface of the membrane, as judged by *lacY-phoA* fusion analyses. Interestingly, although none of the deletions catalyzes active lactose accumulation, permease molecules devoid of even or odd numbers of putative transmembrane helices retain a specific pathway for downhill lactose translocation. One construct, in particular, which is devoid of putative helices II-V exhibits about 80% of the downhill transport activity of intact permease. Thus, relatively short C-terminal domains of the permease contain topological information sufficient for insertion in the native orientation regardless of the orientation of the N-terminus and that the pathway for lactose translocation may be largely contained within the last six transmembrane domains.

Oligomeric State of Lac Permease. Although early evidence (cf. Kaback, 1989) suggested that oligomerization might be important for *lac* permease activity, the permease is probably fully functional as a monomer, as demonstrated by rotational diffusion measurements with eosinylmaleimide-labeled permease (Dornmair, et al 1985) and by direct observation (Costello, et al 1987). Regarding the latter approach, purified *lac* permease and/or purified cytochrome *o* was reconstituted into proteoliposomes under conditions in which both proteins are fully functional, and the preparations were examined by freeze-fracture electron microscopy. In nonenergized proteoliposomes, both proteins appear to reconstitute as a monomers based on: (i) the variation of intramembrane particle density with protein concentration; (ii) the ratio of particles corresponding to each protein in proteoliposomes reconstituted with a known ratio of permease to oxidase; and (iii) the dimensions of the particles observed in tantalum replicas. None of these parameters are altered in the presence of $\Delta\bar{\mu}_H{}^+$. Importantly, moreover, the initial rate of $\Delta\bar{\mu}_H{}^+$-driven lactose transport in proteoliposomes varies linearly with the ratio of *lac* permease to phospholipid, particularly over the range at which there is statistically between 0 and 3 molecules of permease/proteoliposome.

In Vivo Expression of the LacY Gene in Two Fragments Leads to Functional Lac Permease. Bibi & Kaback (1990) restricted the *lacY* gene into two approximately equal-size fragments which were subcloned individually or together under separate *lac* operator/promoters. Under these conditions, *lac* permease is expressed in two portions: (i) the N-terminus, the first 6 putative transmembrane helices and most of putative loop 7; and (ii) the last 6 putative transmembrane helices and the C-terminus. Cells expressing both fragments transport lactose at about 30% the rate of cells expressing intact permease to a comparable steady-state level of accumulation. In contrast, cells expressing either half of the permease independently do not transport lactose. [^{35}S]Methionine labeling and immunoblotting experiments demonstrate that intact permease is completely absent from the membrane of cells expressing *lacY* fragments either individually or together. Thus, transport activity must result from an association between independently synthesized pieces of *lac* permease. When the gene fragments are expressed individually, the N-terminal portion of the permease is observed sporadically and the C-terminal portion is not observed. When the gene fragments are expressed together, polypeptides identified as the N- and C-terminal moieties of the permease are found in the membrane. The results are consistent with the conclusion that the N- or C-terminal halves of *lac* permease are proteolyzed when synthesized independently and that association between the two complementing polypeptides leads to a more stable, catalytically-active complex. More recent experiments demonstrate that co-expression of independently cloned fragments of the *lacY* gene encoding N_2 and C_8 (Wrubel, et al 1990), N_1 and C_{11} or $N_{6.5}$ and $C_{5.5}$ (E. McKenna, D. Hardy, E. Bibi and H. R. Kaback, unpublished information) also form stable molecules in the membrane which interact to form functional permease, while expression of the fragments by themselves yields polypeptides that are relatively unstable and exhibit no transport activity.

Functional Complementation of Deletion Mutants. As discussed above, convincing evidence has been presented indicating that *lac* permease is functional as a monomer. Nonetheless, recent experiments (Bibi & Kaback 1990) demonstrate that certain paired in-frame deletion mutants expressed from separate but compatible plasmids exhibit functional reconstitution when transformed into an appropriate host cell. The nomenclature of the constructs (N_xC_y) describes the number of putative transmembrane helices retained in the N-terminal (N_x) and C-terminal (C_y) portions of the permease before and after the deletion (cf., Fig. 1). Although cells expressing each deletion individually are unable to catalyze active lactose accumulation, cells simultaneously expressing N_2C_8/N_8C_2, catalyze lactose transport 60% as well as cells expressing wild-type permease, while N_2C_6/N_8C_2 or N_4C_6/N_8C_2 exhibit diminished but significant transport activity. On the other hand, N_4C_6/N_6C_4 or N_2C_6/N_6C_2 exhibit only marginal activity, and the combinations N_4C_4/N_8C_2, N_2C_4/N_8C_2 or N_6C_4/N_8C_2 exhibit no activity whatsoever. Moreover, the following pairs of missense mutations or single amino-acid deletions also exhibit no activity: P28S/H322K, E325C/H322K, E325C/K319L, E325C/R302L or ΔW38/ΔH322. Importantly, it has been shown that complementation between N_2C_8/N_8C_2 does not occur at the DNA level, but probably at the protein level. Therefore, the ability to complement functionally is apparently a specific property of pairs of permease molecules containing relatively large deletions separated by at least two transmembrane hydrophobic domains, and it is not observed with pairs of missense mutations or point deletions.

One possible interpretation of the results is that there are specific interactions between transmembrane helices in wild-type permease and that disruption of these interactions by deletion leaves a "potential gap" in the structure that can be filled by interaction with another molecule containing the deleted segment. For instance, perhaps putative helix VIII has a high affinity for helix IX (cf., Fig. 1; note that the loop between putative transmembrane helices VIII and IX is relatively short) and poor affinity for helix XI. By this means, a permease molecule deleted of helices IX and X (eg., N_8C_2) might "accept" these helices from a "donor" molecule deleted of helices III and IV (eg., N_2C_8) and/or *vice versa*. However, *E. coli* T184 transformed with plasmids encoding P28S and N_8C_2 or H322K and N_2C_8 as potential donor/acceptor pairs do not complement functionally. Similarly, ΔH322 does not exhibit functional complementation with

N_2C_8 nor does $\Delta W78$ functionally complement with N_8C_2. Thus, the simplistic explanation does not appear to be the case.

Although nothing is known about the 3-dimensional structure of *lac* permease, as discussed above, certain independently cloned fragments of the *lacY* yield functionally active complexes when expressed together. Since these observations may be related to the functional complementation phenomenon, it is suggested that permease mutants containing missense mutations or point deletions, like the intact wild-type molecule, form relatively compact structures that are unable to form intermolecular complexes. On the other hand, molecules containing deletions in certain hydrophobic transmembrane domains (eg., N_2C_8 and N_8C_2) may be in a more "relaxed" state and therefore able to interact to form functional dimers. In any case, the ability of a given set of deletion mutants to complement functionally is clearly dependent upon the specific nature of the deleted domains (i.e., N_2C_8/N_8C_2, N_2C_6/N_8C_2, and N_4C_6/N_8C_2 exhibit significant transport activity, N_4C_6/N_6C_4 and N_2C_6/N_6C_2 exhibit marginal activity and N_4C_4/N_8C_2, N_2C_4/N_8C_2 and N_6C_4/N_8C_2 exhibit no activity).

Site-Directed Mutagenesis. Notwithstanding the importance of high-resolution structure , it has become apparent that oligonucleotide-directed site-specific mutagenesis can be used to delineate amino-acid residues that play an important role in active transport (cf. refs. Kaback, 1989; Roepe & Kaback, 1990a for reviews) Out of the more than 150 site-directed mutants in *lac* permease, approximately 90% exhibit significant activity. Therefore, it is unlikely that individual amino-acid replacements indiscriminately cause conformational changes in the protein. Given the limited scope of this discussion, the reader is referred to other more extensive reviews (Kaback, 1989; Roepe & Kaback, 1990a; Roepe, et al 1990b) for descriptions of site-directed mutants in Cys, Tyr, Pro and Trp residues of the permease.

The most provocative findings from site-directed mutagenesis studies on *lac* permease began when the 4 His residues in the protein were mutagenized (Padan, et al 1985; Püttner, et al 1986; Püttner, et al 1989). Replacement of His35 and His39 (putative loop 2; Fig. 1) with Arg or replacement of His205 (putative loop 7) with Arg[2], Asn or Gln has no effect on active lactose transport, whereas replacement of His322 (putative helix X) with Arg, Asn, Gln or Lys causes dramatic loss of activity. Conversely, a permease mutant with a single His at position 322 exhibits properties identical to wild-type permease (Püttner & Kaback, 1988). Strikingly, H322R

[2]Although initial experiments (Padan, et al 1985) indicated that permease with Arg205 is defective in transport, the mutant construct was found to have two additional mutations in the 5' end of the gene. When the secondary mutations are removed, Arg205 permease exhibits normal activity (Püttner, et al 1986).

permease catalyzes downhill lactose influx at high substrate concentrations without concomitant H+ translocation.

Efflux, exchange and counterflow are useful for studying permease turnover because specific steps in the overall kinetic cycle can be delineated. Permease with Arg, Asn, Gln or Lys in place of His322 is markedly defective in all translocation reactions presumed to involve protonation or deprotonation (Fig. 2). Furthermore, the primary kinetic effect of $\Delta\tilde{\mu}_H{}^+$ (i.e. a decrease in apparent K_m for lactose) is not observed. Interestingly, permeases with Asn, Gln or Lys in place of His322 catalyze downhill efflux, as well as influx, but both processes appear to occur without concomitant H+ translocation (Püttner, et al 1989).

Since His322 may be directly involved in lactose-coupled H+ translocation and this residue is located in putative helix X (Fig. 1), attention focused on Glu325, which should be on the same face of helix X as His322 and may be ion-paired with this residue. In addition, structure/function studies on chymotrypsin (Blow, et al 1969) and other serine proteases have led to the notion that acidic amino-acid residues may function with His as components of a charge-relay system, a type of mechanism that might conceivably be related to H+ translocation. For these reasons, Glu325 was subjected to site-directed mutagenesis (Carrasco, et al 1986, 1989). Permease with Ala, Gln, Val, His, Cys or Trp in place of Glu325 catalyzes downhill influx of lactose without H+ translocation, but does not catalyze either active transport or efflux. Remarkably, the rate of equilibrium exchange with the altered permeases is at least a great as that observed with wild-type permease. Moreover, permease mutated at position 325 catalyzes counterflow at the same rate and to the same extent as wild-type permease, but the internal concentration of [^{14}C]lactose is maintained for a prolonged period due to the defect in efflux. It is also noteworthy that permease mutated at position 325 catalyzes counterflow 3-to 4-times better than wild-type permease when the external lactose concentration is below the apparent K_m.

The results can be rationalized by the simple kinetic scheme shown in Fig. 2. Efflux down a concentration gradient is thought to consist of a minimum of 5 steps: (1) binding of substrate and H+ on the inner surface of the membrane (order unspecified); (2) translocation of the ternary complex to the outer surface; (3) release of substrate; (4) release of H+; (5) return of the unloaded permease to the inner surface. Alternatively, exchange and counterflow with external lactose at saturating concentrations involve steps 1-3 only. Furthermore, release of H+ appears to be rate-limiting for the overall cycle (cf. ref. Kaback 1989).

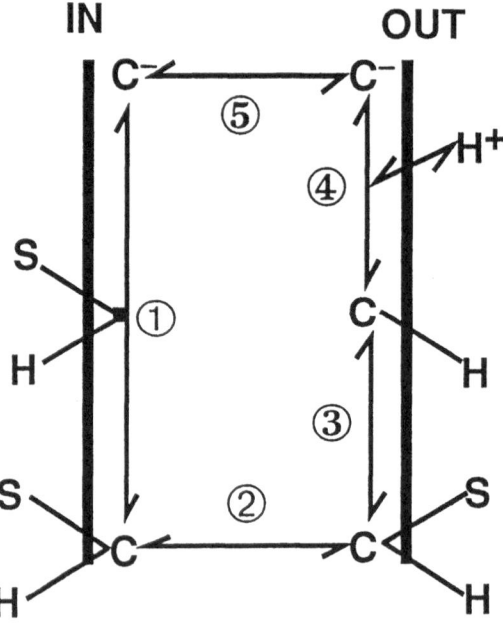

Figure 2. Schematic representation of reactions involved in lactose efflux, exchange and counterflow. C represents *lac* permease; S is substrate (lactose). The order of substrate and H^+ binding at the inner surface of the membrane is not implied.

All steps in the mechanism presumed to involve protonation or deprotonation appear to be blocked in the His322 mutants. Therefore, it seems reasonable to suggest that protonation of His322 may be involved in step 1. In contrast, replacement of Glu325 results in a permease that is defective in all steps involving net H^+ translocation but catalyses exchange and counterflow normally. Clearly therefore, permease mutated at position 325 is probably blocked in step 4 (i.e. it is unable to lose H^+).

Experiments in which Glu325 was replaced with Asp have yielded unexpected results (P. D. Roepe, D. Mechling, L. Patel & H. R. Kaback, unpublished information). Permease with Asp325 is partly uncoupled and catalyzes symport about 30% as well as wild-type permease. The observation is not surprising, as the side-chain containing the carboxylate is about 1.5 Å shorter in Asp relative to Glu. However, E325D permease catalyzes equilibrium exchange normally below pH 7.7, but as ambient pH is increased, exchange activity is progressively and reversibly inhibited with a mid-point at about pH 8.5, while equilibrium exchange with wild-type or E325A permease is unaffected by bulk-phase pH over a wide range. The findings indicate that translocation of the fully loaded permease does not tolerate a negative charge at position 325 and suggest that the carboxylate at 325 may undergo protonation and deprotonation during lactose/H^+ symport. The observation that equilibrium exchange with wild-type permease is insensitive to pH over the same range, is consistent with the notion that Glu325 is strongly H-bonded to His322.

Replacement of Arg302 with Leu, His or Lys (putative helix IX; figure 1) yields permease with properties similar to those of H322R permease (Menick, et al 1987). In marked contrast, replacement of Ser300 with Ala (helix IX), Ser306 with Ala (helix IX) or Cys333 with Ser (helix X) has no significant effect on active lactose transport, thereby highlighting the specificity of Arg302, His322 and Glu325 and providing further support for the contention that single amino-acid changes do not indiscriminately cause conformational alterations. Furthermore, by molecular modeling of putative helices IX and X, it can be shown that the guanidino group in Arg302 may be sufficiently close to His322 to participate in H-bonding with the imidazole ring that, in turn, may be H-bonded to the carboxylate of Glu325 (Menick, et al 1987). Minimally, therefore, the putative charge-relay in the permease would involve interactions between Arg302, His 322 and Glu325.

As evidenced by binding studies with the high-affinity ligand p-nitrophenyl-α,D-galactopyranoside (NPG), permease mutated at position 325 binds with a K_d approximating that of wild-type permease (Carrasco, et al 1989). The finding is consistent with the observation that counterflow, a process that exhibits an apparent K_m similar to that observed for active transport, is intact in the mutants, but is in marked contrast to findings with permeases mutated at Arg302 or

His322 which exhibit markedly decreased affinities (Püttner, et al 1989; Menick, et al 1987). Therefore, it is tempting to speculate that the pathways for H^+ and lactose may overlap [(i.e. that Arg302 and His322 may also be components of the substrate binding site in addition to being involved in H^+ translocation (cf. refs. Collins, et al 1989; Franco, et al 1989 in addition)] and that protonation of His322 may be required for high-affinity binding.

If Arg302, His322 and Glu325 are sufficiently close to H-bond and function as components of a charge-relay, the polarity, distance and orientation between the residues should be critical (Lee, et al 1989). The importance of polarity between His322 and Glu325 was studied by interchanging the residues, and the modified permease is inactive in all modes of translocation. The effect of distance and/or orientation between His322 and Glu325 was investigated by interchanging Glu325 with Val326, thereby moving the carboxylate one residue around putative helix X. The resulting permease molecule is also completely inactive, and control mutations indicate that a Glu residue at position 326 inactivates the permease. The wild-type orientation between His and Glu was then restored by further mutation to introduce a His residue into position 323 or by interchanging Met323 with His 322. The resulting permease molecules contain the wild-type His/Glu orientation, but the putative His/Glu ion-pair is rotated about the helical axis by 100° relative to Arg302 in putative helix IX. Both mutants are inactive with respect to all modes of translocation. The results provide support for the contention that the polarity between His322 and Glu325 and the geometric relationships between Arg302, His322 and Glu325 are critical for permease activity. In addition, the results suggest that perturbation of the putative His322/Glu325 ion-pair alone is insufficient to account for inactivation (i.e. Glu322/His325 should remain ion-paired) and are consistent with the possible role of His322 and Glu325 as components of a H^+ relay.

Previous studies (Menick, et al 1987) suggested that replacement of Lys319, which should be on the same face of putative helix X as His322 and Glu325, with Leu has no effect on permease activity; however, this conclusion has been found to be incorrect (B. Persson, P. D. Roepe, L. Patel, J. Lee & H. R. Kaback, unpublished information). Rather, K319L permease is defective in lactose/H^+ symport. The mutant is unable to catalyze lactose accumulation or efflux, but exhibits downhill lactose influx and catalyzes equilibrium exchange at about half the rate of wild-type permease. Unlike the Glu325 mutants, which catalyze exchange at least as well as wild-type permease, K319L permease binds NPG poorly (cf. ref. Collins, et al 1989, in addition) and is defective in counterflow activity. The results suggest that Lys319 is important for substrate recognition, as well as lactose-coupled H$^+$ translocation. In addition, it is noteworthy that the double mutant K319L/E325A catalyzes equilibrium exchange.

Although the notion that a type of charge-relay mechanism may be involved in lactose/H$^+$ symport is consistent with the findings discussed above, recent experiments question the notion that His322 is obligatory for lactose-coupled H$^+$ translocation. From studies on permease mutants with Tyr or Phe in place of His322, King & Wilson (King & Wilson, 1989a, b; 1990) have concluded that sugar-dependent H$^+$ transport is observed, albeit with low efficiency, that melibiose efflux, in particular, remains coupled to H$^+$ translocation in the mutants and that reactions involving exchange are limiting for lactose but not melibiose efflux suggesting that slow exchange is substrate-specific in the His322 mutants. In addition, Brooker (1990) has shown that permease with Val in place of Ala177 *and* Asn in place of His322 catalyzes lactose-dependent H$^+$ influx with a stoichiometry close to unity. Taken at face value, the results are difficult to reconcile with the contention that His322 is obligatory for lactose-driven H$^+$ translocation. However, it should be emphasized that *all* of the His322 mutants isolated thus far are *grossly defective* with regard to sugar accumulation. Thus, whatever its precise role in the mechanism, a His residue at position 322 appears to be very important for β-galactoside accumulation against a concentration gradient.

On a broader level, H$^+$- and Na$^+$-coupled symport are conceptually and thermodynamically analogous, but it is unclear whether the two types of transport occur by the same general mechanism. Since melibiose (*mel*) permease uses H$^+$, Na$^+$ or Li$^+$ as the coupling cation depending on the sugar transported, and like *lac* permease, *mel* permease is inactivated by diethylpyrocarbonate, Pourcher *et al.* (Pourcher, et al 1990) replaced each of the 7 His residues in *mel* permease with Arg and demonstrated that His94 alone is important for transport and binding. Thus, *mel* permease and *lac* permease appear to require a single specific His residue for activity, although the residues are located at opposite ends of the respective molecules. Importantly, however, *mel* permease with H94R does not exhibit an uncoupled phenotype, and the defect observed may be due specifically to a loss in ability to bind substrate. As outlined above, certain data with *lac* permease are consistent with the idea that a type of H$^+$ relay may be operative. On the other hand, it is not obvious how the same mechanism can be directly involved in Na$^+$ symport unless N or O atoms that can coordinate with Na$^+$ are present in the pathway in addition to the minimal structures necessary for a charge-relay or a "H$^+$-wire" of the type suggested by Onsager (1969). Despite certain marked kinetic differences between H$^+$- and Na$^+$-coupled sugar translocation, a few lines of evidence suggest that substrate-coupled Na^+ and H^+ translocation may occur by similar mechanisms in *mel* permease: (i) NPG binding studies suggest that H^+ and Na^+ compete for a common binding site; (ii) Na^+ is a competitive inhibitor of melibiose-coupled H^+ translocation; and (iii) certain mutations in *melB* have been isolated and characterized that alter cation specificity. Boyer (1988) has suggested that H_3O^+, rather than H^+, may be the symported species. Appropriately placed N or O atoms in symporters like *lac* or *mel* permease could provide cation-binding domains akin to those in the crown ethers or cryptates, both of which form coordination complexes with Na^+ and H_3O^+. In this context, however, some of the translocation reactions catalyzed by *lac* permease exhibit a significant D_2O effect (cf. Kaback, 1989) which is not expected if coordination with H_3O^+ is the rate-limiting step in translocation. In summary, therefore, although the contrast between H^+- and Na^+-coupled symport is of singular importance, the mechanistic relationship between the two is presently unclear.

REFERENCES

Bibi, E, Kaback, HR (1990) Proc Natl Acad Sci USA 87:4325-4329

Bibi, E, Kaback, HR (1991) Science, submitted for publication

Bibi, E, Verner, G, Chang, C-Y, Kaback, HR (1991) Proc Natl Acad Sci, USA, in press

Blobel, G (1980) Proc Natl Acad Sci USA 77:1496-1500

Blow, DM, Birktoft, JJ, Hartley, BS (1969) Nature, Lond 221:337-340

Boyer, PD (1988) Trends in Biochem Sci 13:5-7

Brooker, RJ (1990) J Biol Chem 265:4155-4160

Calamia, J, Manoil, C (1990) Proc Natl Acad Sci USA 87:4937-4941

Carrasco, N, Tahara, SM, Patel, L, Goldkorn, T, Kaback, HR (1982) Proc Natl Acad Sci USA 79:6894-6898

Carrasco, N, Viitanen, P, Herzlinger, D, Kaback, HR (1984a) Biochemistry 23:3681-3687

Carrasco, N, Herzlinger, D, Mitchell, R, DeChiara, S, Danho, W, Gabriel, TF, Kaback, HR (1984b) Proc Natl Acad Sci USA 81:4672-4676

Carrasco, N, Antes, LM, Poonian, MS, Kaback, HR (1986) Biochemistry 25:4486-4488

Carrasco, N, Püttner, IB, Antes, LM, Lee, JA, Larigan, JD, Lolkema, JS, Kaback, HR (1989) Biochemistry 28:2533-2539

Collins, JC, Permuth, SF, Brooker, RJ (1989) J Biol Chem 264:14698-14703

Costello, MJ, Viitanen, P, Carrasco, N, Foster, DL, Kaback, HR (1984) J Biol Chem 259:15570-15586

Costello, MJ, Escaig, J, Matsushita, K, Viitanen, PV, Menick, DR, Kaback, HR (1987) J Biol Chem 262:17072-17082

Deisenhofer, J, Michel, H (1989) Science 245:1463-1473

Dornmair, K, Corin, AS, Wright, JK, Jähnig, F (1985) EMBO J 4:3633-3638

Engleman, DM, Steitz, TA (1981) Cell 23:411-422

Foster, DL, Boublik, M, Kaback, HR (1983) J Biol Chem 258:31-34

Franco, PJ, Eelkema, JA, Brooker, RJ (1989) J Biol Chem 264:15988-15992

Goldkorn T, Rimon, G, Kaback, HR (1983) Proc Natl Acad Sci USA 80:3322-3326

Herzlinger, D, Viitanen, P, Carrasco, N, Kaback, HR (1984) Biochemistry 23:3688-3693

Kaback, HR (1983) J Membr Biol 76:95-112

Kaback, HR (1986) Physiology of Membrane Disorders, Andreoli, TE, Hoffman, JF, Fanestil, DD, Schultz, SG, (eds) Plenum Press, New York, NY p 387

Kaback, HR (1989) Harvey Lectures 83:77-103

Kaback, HR (1990) The Bacteria XII, Krulwich, TA, (ed) Academic Press, Inc., New York, NY p 151

King, SC, Wilson, TH (1989a) J Biol Chem 264:7390-7394

King, SC, Wilson, TH (1989b) Biochim Biophys Acta 982:253-264

King, SC, Wilson, TH (1990) J Biol Chem 265:3153-3160

Lee, JA, Püttner, IB, Carrasco, N, Antes, LM, Kaback, HR (1989) Biochemistry 28:2540-2544

Li, J, Tooth, P (1987) Biochemistry 26:4816-4823

McKenna, E, Hardy, D, Pastore, JC, Kaback, HR (1991) Proc Natl Acad Sci USA 88:2969-2973

Menick, DR, Carrasco, N, Antes, L, Patel, L, Kaback, HR (1987) Biochemistry 26:6638-6644

Onsager, L (1969) Science 166:1359-1364

Padan, E, Sarkar, HK, Viitanen, PV, Poonian, MS, Kaback, HR (1985) Proc Natl Acad Sci USA 82:6765-6768

Page, MGP, Rosenbusch, JP (1988) J Biol Chem 263:15906-15914

Popot, JL, Engelman, DM (1990) In: Protein Form and Function RA Bradshaw, M Purton (eds), Elsevier Publishing Comp, Cambridge, p 147

Popot, JL, de Vitry, C (1990) Ann Rev Biophys Biophys Chem 19:369-403

Pourcher, T, Sarkar, HK, Bassilana, M, Kaback, HR, Leblanc, G (1990) Proc Natl Acad Sci USA 87:468-472

Püttner, IB, Sarkar, HK, Poonian, MS, Kaback, HR (1986) Biochemistry 25:4483-4485

Püttner, IB, Kaback, HR (1988) Proc Natl Acad Sci USA 85:1467-1471

Püttner, IB, Sarkar, HK, Padan, E, Lolkema, JS, Kaback, HR (1989) Biochemistry 28:2525-2533

Rapoport, TA (1986) CRC Crit Rev Biochem 20:73- 137

Roepe, PD, Zbar, R, Sarkar, HK, Kaback, HR (1989) Proc Natl Acad Sci USA 86:3992-3996

Roepe, PD, Kaback, HR (1990a) CRC Critical Rev Biochem, Page, MGP, Henderson, PJF, (eds), in press

Roepe, PD, Consler, TG, Menezes, ME, Kaback, HR (1990b) Res in Microbiol 141:290-308

Seckler, R, Wright, JK, Overath, P (1983) J Biol Chem 258:10817-10820

Seckler, R, & Wright, JK (1984) Eur J Biochem 142:269-279

Seckler, R, Möröy, T, Wright, JK & Overath, P (1986) Biochemistry 25:2403-2409

Singer, SJ, Maher, PA, Yaffe, MP (1987) Proc Natl Acad Sci USA 84:1960-1964

Stochaj, V, Bieseler, B, Ehring, R (1986) Eur J Biochem 158:423-428

Stochaj, U, Ehring, R (1987) Eur J Biochem 163:653-658

Stochaj, U, Fritz, H-J, Heibach, C, Markgraf, M, Schaewen, AV, Sonnewald, U, Ehring, R (1988) J Bacteriol 170:2639-2645

Vogel, H, Wright, JK, Jähnig, F (1985) EMBO J 4:3625-3631

Von Heijne, G, Blomberg, C (1979) Eur J Biochem 97:175-181

Wickner, WT, Lodish, HF (1985) Science 230:400-407

Wrubel, W, Stochaj, U, Sonnewald, U, Theres, C, Ehring, R (1990) J Bacteriol 172:5374-5381

RECONSTITUTION OF *E. COLI* PREPROTEIN TRANSLOCATION USING PURIFIED COMPONENTS

Lorna Brundage and William Wickner
Department of Biological Chemistry and
the Molecular Biology Institute
University of California
Los Angeles, CA 90024 USA

Virtually all proteins found in membrane bound compartments are synthesized as precursor proteins in the cytoplasm. In route to their final destination, these protein must be directed to and pass through at least one membrane. How are precursor proteins directed to the correct membranes? How do they then translocate through the membrane, a barrier which is otherwise impervious to such large charged molecules? To approach these issues, we are biochemically dissecting the process of targeting and translocation of proteins across the *E. coli* inner membrane. *E. coli* is well suited to biochemical analysis because it is genetically malleable, easily fractionated and most (or perhaps all) of the genes involved in targeting and translocation have been identified, cloned and sequenced (Schatz and Beckwith, 1990; Bieker *et al.*, 1990).

The requirements of bacterial preprotein translocation are well documented (reviewed in Wickner *et al.*, 1991). Precursor proteins must contain both correct linear sequences, including a leader sequence and appropriate sequences from the mature region, and also they must be in the correct, translocation competent conformation. Two forms of metabolic energy, ATP and a proton-motive force, drive translocation. Ongoing protein synthesis is not required, and proteins can translocate entirely post-translationally. Translocation of most proteins (called *sec* dependent proteins) is catalyzed by a specific set of enzymes. Subunits of these enzymes are encoded by the genes *sec*A (*prl*D), *sec*B, *sec*D,

NATO ASI Series, Vol. H 63
Dynamics of Membrane Assembly
Edited by J. A. F. Op den Kamp
© Springer-Verlag Berlin Heidelberg 1992

*sec*E (*prl*G), *sec*F, *sec*Y (*prl*A), and perhaps additional genes. Acidic phospholipids are essential for the membrane associated steps in this catalysis.

Although the general parameters are known, the molecular mechanism of the translocation reaction itself has remained elusive. In part this is due to the chemical complexity of the membranes and cellular extracts which have been required to study the translocation process *in vitro*. In view of these limitations, we began a directed effort a few years ago to identify and purify a set of *E. coli* derived components with which we could reconstitute the translocation of a typical, *sec* dependent precursor protein, proOmpA. We have recently found that only three macromolecular components, all localized to the membrane, are required. Using these three components- purified SecA protein, SecY/E protein and a bilayer containing acidic phospholipids such as diolylphosphotidal glycerol- we have reconstituted ATP and proton-motive force driven translocation of purified proOmpA. SecB and Leader peptidase, two secretion proteins we have also isolated in an active form, are not required for this reconstituted reaction (Brundage *et al.,* 1990; Hendrick *et al.,* submitted).

ProOmpA translocation requires proteins which are both peripheral and integral to the membrane. When isolated inner membrane vesicles are treated with urea, which removes peripheral membrane proteins, they are no longer able to translocate proOmpA. If they are supplemented with a cytosolic high speed supernatant, their translocation activity is completely restored. By using biochemical complementation of urea treated membranes as an assay, we purified the peripheral component and ultimately found only one polypeptide was required, the product of the SecA gene (Cunningham *et al.,* 1989; Cabelli *et al.,* 1988; Schmidt *et al.,* 1988).

SecA plays a central role in protein translocation. In all, SecA specifically associates with at least four other components during the translocation reaction: the precursor protein- through both the leader region and the mature region (Cunningham and Wickner, 1990; Lill *et al.,* 1990); SecY/E (Hartl *et al.,* 1990); SecB (Hartl *et al.,* 1990); and acidic phospholipids (Lill *et al.,* 1990; Hendrick and Wickner, submitted). In addition, SecA is the subunit of the translocase which hydrolyzes ATP (Lill *et al.,* 1989). How does SecA transduce the energy of ATP to produce protein movement? We believe that upon binding ATP, SecA drives translocation of short segments of the precursor through the membrane. Hydrolysis of the bound ATP allows SecA to be released and subsequently reassociate "downstream" on the precursor when ATP is rebound (Schiebel *et al.,* 1991).

In contrast to the peripheral domain of the translocase, identification of the integral domain of the translocase was more challenging. Two advances made isolation of this domain ultimately possible. The first was a method to extract translocation activity from inner membranes and functionally reconstitute it into liposomes (Driessen and Wickner, 1990). The second was a rapid assay for a sub-reaction of translocation, the 'translocation ATPase' (Lill *et al.,* 1989). We fractionated detergent extracted total *E. coli* membrane proteins using their ability to support the translocation ATPase as an assay. After three chromatographic steps, only four polypeptides remained in our preparation. N-terminal sequence analysis of the silver stainable bands revealed that they originated in the membrane as three polypeptides-the *sec*E gene product, the SecY gene product (which was clipped into two peptides during isolation), and the third peptide, 'band 1', derived from a gene not yet identified (Brundage et al., 1990; Brundage et al., submitted). Cross-precipitation studies have shown that these polypeptides are tightly associated and form subunits of a single protein (Brundage et al., submitted) which we have termed the SecY/E protein.

Proteoliposomes prepared from a mixture of SecY/E and SecA efficiently catalyze translocation of proOmpA (Brundage *et al.,* 1990), as detected by the protection of proOmpA to externally added protease. This translocation is authentic by several criteria. First, it requires ATP and is accelerated by an electrochemical potential across the bilayer. Second, it requires the gene products of *sec*A, *sec*E and *sec*Y and can be made dependent upon the product of *sec*B. Third, it requires acidic phospholipids in the bilayer. Fourth, it is about as efficient, per SecY polypeptide, as translocation into unfractionated membrane vesicles (Brundage *et al.,* 1990; Hendrick and Wickner, submitted).

We developed this reconstituted system to enable us to address fundamental questions about how presecretory proteins translocate across the bacterial inner membrane. For example, how does the translocase discriminate between presecretory and cytoplasmic proteins? Do proteins translocate directly through the lipid, or through a proteinacious pore, presumably formed by the translocase? What is the conformation of a translocating protein as it passes through the membrane? Experiments to address these and other questions are currently underway. The answers will undoubtably shed light not only for our understanding of translocation across the bacterial inner membrane, but also protein translocation in general.

REFERENCES

Bieker, K.L., Philips, G.J. and Silhavy, T. (1990) The *sec* and *prl* genes of *E. coli*. J. Bioenerg. and Biomemb. 22:291-310.

Brundage, L., Hendrick, J.H., Schiebel, E., Driessen, A., and Wickner, W. (1990) The purified *E. coli* integral membrane protein SecY/E is sufficient for reconstitution of SecA-dependent precursor protein translocation. Cell 62:649-657.

Cabelli, R.J., Chen, L., Tai, P.C., and Oliver, D.B. (1988) SecA protein is required for secretory protein translocation into *E. coli* membrane vesicles. Cell 55: 683-692.

Cunningham, K., and Wickner, W. (1990) Specific recognition of the leader region of precursor proteins is required for the activation of translocation ATPase of *Escherichia coli*. Proc. Natl. Acad. Sci., USA 86: 8630-8634.

Cunningham, K., Lill, R., Crooke, E., Rice, M., Moore, K., Wickner, W., and Oliver, D. (1989) SecA protein, a peripheral protein of the *Escherichia coli* membrane, is essential fro the functional binding and translocation of proOmpA. EMBO J. 8:955-959.

Hartl, F.-U., Lecker, S., Schiebel, E., Hendrick, J.P. and Wickner, W. (1990) The binding cascade of SecB to SecA to SecY/E mediates preprotein targeting to the *E. coli* plasma membrane. Cell 63: 269-279.

Lill, R., Dowhan, W., and Wickner, W. (1990) The ATPase activity of SecA is regulated by acidic phospholipids, SecY, and the leader and mature domains of precursor proteins. Cell 60: 271-280.

Lill, R., Cunningham, K., Brundage, L. A., Ito, K., Oliver, D., and Wickner, W. (1989) SecA protein hydrolyzes ATP and is an essential component of the protein translocation ATPase of *Escherichia coli.* EMBO. J. 8: 961-966.

Schatz, P.J. and Beckwith, J. (1990) Genetic analysis of protein export in *Escherichia coli.* Ann. Rev. Genet. 24:215-248.

Schiebel, E., Driessen, A.J.M. , Hartl, F.-U. and Wickner, W. (1990) ΔμH+ and ATP function at different steps of the catalytic cycle of preprotein translocase. Cell 64:927-939.

Schmidt, M.G., Rollo, E.E., Grodberg, J., and Oliver, D.B. (1988) Nucleotide sequence of the *sec*A gene and *sec*A(Ts) mutations preventing protein export in *Escherichia coli.* J. Bacteriol. 170: 3404-3414.

Wickner, W., Driessen, A.J.M., and Hartl, F.-U. (1991) The enzymology of protein translocation across the *Escherichia coli* plasma membrane. Ann. Rev. Biochem., 60:101-124.

IMPORT OF COLICINS INTO *ESCHERICHIA COLI*

Hélène BENEDETTI, Lucienne LETELLIER+, Roland LLOUBES, Vincent GELI, Daniel BATY, Jean-Marie PAGES and Claude LAZDUNSKI

Centre de Biochimie et de Biologie Moléculaire du C.N.R.S.
  31 Chemin Joseph Aiguier, B.P. 71, 13402 Marseille Cedex 9, France.
+ Laboratoire des Biomembranes, U.A. 1116, C.N.R.S., Université PARIS-SUD, 91405 Orsay Cedex, France.

The permeability barrier of the outer membrane

All bacterial organisms have an envelope which has the primary role to constitute a physical barrier between the cytoplasm and the extracellular medium thereby protecting cells from harmful compounds from this medium. However, exchanges with the latter are required since cells must take up nutrients useful for growth.

Structure of the cell envelope

The cell envelope of *E. coli* consists of two membranes : the cytoplasmic or inner membrane and the outer membrane. The latter is an asymmetric bilayer that serves as a permeability barrier to allow passage of small (< 700 Da in *E. coli*) hydrophilic molecules into the periplasmic space. The outer layer of this membrane is constituted of lipopolysaccharide (LPS) while its inner layer contains phospholipids. Between the two membranes, the periplasmic space is an aqueous compartment containing the bacterial cytoskeleton, the peptidoglycan, conferring the shape and osmotic resistance to the cell envelope. In addition, this compartment contains numerous detoxifying enzymes, scavenging enzymes, and binding proteins that facilitate active transport [For a review, see (Lugtenberg and van Alphen, 1983)].

NATO ASI Series, Vol. H 63
Dynamics of Membrane Assembly
Edited by J. A. F. Op den Kamp
© Springer-Verlag Berlin Heidelberg 1992

The cytoplasmic membrane is a permeability barrier containing proteins required for generating and maintaining an electrochemical potential as well as proteins that use the electrochemical potential for active transport of nutrients into the cell.

Transport of compounds across the outer membrane

The outer membrane of *E. coli* fulfils two functions : i) increased resistance to a number of toxic agents (antibiotics, host-defence proteins, digestive enzymes, hydrophobic dyes and bile salts, complement, etc...) and ii) limit the variety of attainable nutrients to those that can traverse the proteinaceous pores of the outer membrane in an energy-independent fashion.

The two major porins are OmpF and OmpC which are 60% homologous however another porin, PhoE, is induced when cells are grown under phosphate limitation. Facilitated diffusion with a certain degree of transport specificity is observed in the uptake of maltodextrins, phosphates and nucleosides. Selective uptake comes from the stereospecific recognition of the substrates by the transport proteins. Interestingly, these proteins are subject to regulation by the substrates, delivered in the growth medium. However, these proteins are dispensable for growth on these substrates provided the concentration of the substrates is sufficiently high and their molecular mass does not exceed 600-700 Da. Under these conditions, these substrates diffuse through the porin channels with sufficiently high rates to support growth (Nikaido and Vaara, 1985).

Iron is one of the most abundant element on the earth's surface. However, all of the iron in the aerobic environment is present in insoluble ferric hydroxide complexes. To cope with the insolubility of iron, many microorganisms synthesize and secrete into the environment molecules called siderophores (from the Greek "iron bearer"). Members of the structurally diverse group of known siderophores share the property of an exceedingly high affinity for iron, sufficient to wrest the environmental iron from its insoluble complexes or from host proteins that sequester iron. Iron-siderophores complexes are then transported back into the microorganisms.

The situation for vitamin B12 is somewhat similar to that of iron. Although it is not required for the growth of *E. coli*, under aerobic conditions, vitamin B12 greatly enhances its growth. Its transport across the outer membrane is mediated by a specific protein, the vit B12 receptor, BtuB.

In contrast to maltodextrins, nucleosides or phosphates of low molecular mass, no uptake of iron (III)-siderophores (and impaired uptake of vitamin B12) is observed in cells devoid of outer membrane receptor proteins.

Their transport across the outer membrane is energy-coupled and requires recognition of the iron (III)-siderophores (and of vit B12 by BtuB) by the related receptor proteins. Binding of the substrates to the receptors occurs also in energy-deprived cells, so that the vectorial release from the receptors into the periplasmic space seems to be the energy-consuming step. However, no energy-providing system, nor energy-rich metabolites are known to exist in the outer membrane or in the periplasmic space. Energy is generated in the cytoplasm and the cytoplasmic membrane, and charged molecules such as ATP and phosphoenol pyruvate are not exported. By coupling cytoplasmic membrane energy to active transport of iron-bearing siderophores and vitamin B12 across the outer membrane, the TonB protein allows Gram-negative bacteria to enjoy the benefits of an outer membrane, as well as cope with the problem it poses (as a permeability barrier to iron and vit B12) (Braun *et al.*, 1991).

The three domains of colicin polypeptides

Colicins are toxic proteins produced by an active against *E. coli* and closely related bacteria. Current evidence suggests that their mode of action can be divided into three steps : (i) binding to a specific cell surface receptor on the outer membrane of the target cell; (ii) translocation across the membrane(s); (iii) interaction with its target in the cell. The lethal action can involve various processes : formation of voltage-dependent channels in the cytoplasmic membrane; degradation of cellular DNA; inhibition of protein synthesis through 16sRNA cleavage and autolysis resulting from inhibition of peptidoglycan synthesis (Konisky, 1982).

There is a linear organization of three distinct domains along the polypeptide chain of colicins (Ohno-Iwashita and Imahori, 1980; Ohno-Iwashita and Imahori, 1982; Dankert *et al.*, 1982; Martinez *et al.*, 1983; Brunden *et al.*, 1984; de Graaf and Oudega, 1986; Mankovich *et al.*, 1986; Pugsley, 1987; Baty *et al.*, 1988). A specific function has been assigned to each of these domains. In general, the central domain of colicin appears to be involved in receptor binding, the N-terminal domain seems to be required for translocation across the outer membrane, and the C-terminal domain carries the lethal activity. This domain is also involved in interaction of colicins and their respective immunity proteins (de Graaf *et al.*, 1978; Ohno-Iwashita and Imahori, 1980).

The two different pathways of colicin uptake

Different systems have evolved for the uptake of macromolecules such as colicins or the genomes of certain bacteriophages. In general, transport of these molecules to their target requires interaction with high affinity receptors in the outer membrane. Most of the outer membrane porins involved in nutrient uptake are used as colicin receptors (Konisky, 1982; Datta *et al.*, 1977; Braun and Hantke 1977; Kadner *et al.*, 1979). Thus the polypeptide that serves as the receptor for colicin E1, E2 and E3 functions in uptake of vitamin B12, whereas the colicin K receptor (Tsx) serves as a specific diffusion pathway for nucleosides (Hantke, 1976).

Several colicin receptors are involved in iron uptake, serving as siderophore-binding proteins. For example, FhuA (formerly called TonA) is the receptor for colicin M and ferrichrome (Hantke and Braun, 1975a; Wayne and Neilands, 1975; Braun *et al.*, 1980) whereas enterochelin and colicins B and D utilize FepA for adsorption (Hantke and Braun, 1975b; Hollifield and Neilands, 1978; Pugsley and Reeves, 1977). The colicin Ia, Ib receptor (Cir) is also involved in iron accumulation (Konisky, 1982).

Bacteria carrying specific mutations in the genes coding for these outer membrane receptors do not bind their respective colicins and have been termed colicin resistant bacteria. Mutations in other loci have been isolated which allow the colicins to adsorb to their receptors but are insensitive or refractory to their effect. Many of these mutant bacteria have been termed colicin tolerant (*tol*). Presumably, mutations causing such tolerance are in genes which encode functions necessary for translocating the colicin to its target after it has bound to its outer membrane receptor.

The colicins have been assigned to one of two groups, A or B, based on the activity of each colicin to various mutant bacteria (Davies and Reeves, 1975a, 1975b). For example, the group B colicins (B, D, G, H, Ia, Ib, M, Q, V) are inactive against bacteria containing mutations in *tonB*. The TonB protein, as mentioned above, is part of a family of high affinity transport systems for vitamin B12, iron-siderophore complexes as well as for many group B colicins. In contrast, the group A colicins (A, E1, E2, E3, K, L, N, S4) are inactive against strains containing lesions in the *tolA* gene. TolA is part of another system, termed the *tol* import system, composed of several proteins which are involved in the import of group A colicins as well as the DNA of the filamentous bacteriophage. In general, bacteria containing mutations in the *tol* import system are only tolerant to one or more of the group A colicins while those in the *tonB* system are only tolerant to any colicin in group B.

The TonB-dependent pathway

The *tonB* gene locus was originally related to phage T1 (hence the mnemonic "ton" for "T-one") infection in that *tonB* mutants were resistant to the phage. T1 reversibly adsorbed to *tonB* mutants without release of DNA from the phage head and uptake into the cytoplasm. Irreversible adsorption accompanied by infection required in addition cellular energy, which could be generated by the membrane-bound electron transport chain or by ATP hydrolysis (Hancock and Braun, 1976).

Studies with this system (Hantke and Braun, 1978) indicated that the adsorption rather than the uptake of phage DNA required the TonB protein (and cellular energy).

The second line of evidence about the role of TonB came from studies concerned with the uptake of colicin M. This colicin binds to the same FhuA receptor as phage T1. It binds and stays bound to the receptor in *tonB* mutant cells or in unenergized cells (Braun, 1989). Mutants in *tonB* became colicin M-sensitive only upon osmotic shock treatment which renders the outer membrane temporarily permeable. This result suggested that colicin M transport across the outer membrane depends on the TonB protein.

The *tonB* gene was cloned and sequenced (Postle and Good, 1983; Hannavy *et al.*, 1990). It encodes a 26 kDa protein comprising a high percentage of prolyl residues (17%). There is a particularly proline-rich region in the amino-terminal third of TonB (a 33-residue peptide segment) which appears to adopt an extended constrained conformation that presumably spans the periplasmic space and has recently been shown to interact specifically with the FhuA protein (Brewer *et al.*, 1990), the outer membrane receptor for ferrichrome-iron and colicin M. As TonB is anchored to the cytoplasmic membrane (Postle, 1990), these data together with studies reported above suggest a model by which TonB serves to transduce conformational information over extended distances, from the cytoplasmic membrane to the outer membrane (Hannavy *et al.*, 1990).

For the interaction between the TonB protein and various receptors, a common structure in the receptor is expected which is recognized by the TonB protein. Indeed, such a structure, comprising a pentapeptide close to the N-terminus, designated the "TonB box", has been found in all receptors whose activity depends on TonB. For example, the FhuA TonB box is comprised of Asp-Thr-Ile-Thr-Val, the FhuE sequence reads Glu-Thr-Met-Val-, the BtuB sequence Asp-Thr-Leu-Val-Val (Braun *et al.*, 1991).

It was somewhat suprising to find the same TonB box in colicins which are taken up by a TonB-dependent mechanism (Ross *et al.*, 1989). This could mean that colicins are released from the receptors by the same mechanism as the other substrates but further transport across the outer membrane requires an additional TonB-dependent step. With the exception of the TonB box, the amino acid sequences of the receptors and colicins display little, if any similarity.

Auxilliary proteins in TonB-dependent outer membrane transport

Mutations in unlinked genes exist which appear to modify the action of TonB. Among these, the best characterized is the *exbB* gene. Originally the term *exb* defined mutations which conferred insensitivity to colicin B through a compound (enterochelin/enterobactin) which was excreted into the culture medium and competed for FepA with colicin B. Mutations in *exbB* have a "leaky" TonB phenotype, with reduced but not entirely absent levels of vitamin B12 and siderophore-mediated iron transport (Hantke and Zimmerman, 1981). The *exbB* mutants are insensitive to B-group colicins, but to a lesser extent than *tonB* mutants. The function of ExbB protein appears to be to stabilize TonB protein (Fischer *et al.*, 1989).

ExbB is a 26 kDa, it has been localized to the cytoplasmic membrane. It shares 26% aminoacid identity with TolQ, a protein involved in the uptake of group A colicins. The *tolQ, exbB* strains are completely insensitive to group B colicins, φ 80, and T1 (Braun, 1989). The stabilization of TonB by ExbB/TolQ proteins can be taken as evidence for a mechanical interaction between the TonB protein and ExbB/TolQ.

Additional genes whose products probably modify or enhance TonB activity are *exbC* and *exbD*. Other than the observation that *exbC* mutants hyperexcrete enterochelin, little is known about the locus. The *exbD* gene was discovered upon sequence analysis of an *exbB* clone (Eick-Helmerich and Braun, 1989). It is located downstream the *exbB* gene and encodes a 15.5 kDa protein localized in the cytoplasmic membrane. ExbD shares 25% amino acid identity with TolR. The structural similarity between ExbBD proteins and TolQR proteins suggests a common ancestor. Unlike ExbB, ExbD does not stabilize TonB (Fisher *et al.*, 1989).

The hypothesis of a functional connection between Exb and Tol proteins gained support by the finding that the TonB-dependent sensitivity to colicins, which is strongly reduced but not fully abolished by mutations in the *exbBD* locus, was completely lost by additional mutation in the *tolQ* gene (Braun, 1989). Double mutants in *exbBD tolQ* were completely insensitive to colicins B, D, M and to phages T1 and φ 80. Insensitivity concerns colicins with a TonB-dependent and a TonB-independent uptake mechanism. Although *tolQ* and *tolR* single mutants have been listed as being resistant to colicins E1, E2 and K, one can frequently observe turbid zones of growth inhibition with undiluted solutions which never appear in *tolQ exbB* and *tolR exbB* double mutants indicating partial substitution of the missing *tol* gene function by the *exbB* gene function.

The mechanism of TonB-dependent uptake of colicins

There is both genetic and biochemical evidence that TonB directly interacts with the outer membrane receptors and group B colicins which both contain the so-called "TonB box" (Brewer *et al.*, 1990).

To date, the mechanism of TonB-dependent uptake of colicins is still poorly understood. It is likely that the first step, the binding of colicins to the receptors causes, like with the group A colicins (Benedetti *et al.*, 1991b), a conformational change with partial unfolding making the TonB box available for interaction with TonB in the second step. The mechanism of translocation of the polypeptide chain across the outer membrane is still unknown. The transition of the translocation complex from the binding conformation to the translocating conformation would be induced by energization of the cell (potential across the inner membrane). The conformational change in the receptor would be triggered by a conformational change of TonB transmitted via the X-Pro region from the inner to the outer membrane (Hannavy *et al.*, 1990).

The role of auxilliary proteins such as ExbBD and TolQR is not known at present. It is likely that these proteins form a complex with TonB and modify its functional half-life. But nothing is known beyond this.

The Tol-dependent pathway

This translocation pathway is also used by filamentous phages M13, fd, f1. Tol proteins (TolA, B, Q, R and C) are involved in uptake of phage DNA and colicins, but each of them is not required for the entry of each colicin. For example, while TolA and TolQ are required for the entry of colicins A, E1, E2, E3, K and N; TolB and TolR are only required for the entry of colicins A, E2, E3 and K, and TolC is only required for colicin E1 uptake (Nagel del Zwaig and Luria, 1967; Davies and Reeves, 1975a, 1975b; Sun and Webster, 1986, 1987).

Expression, characterization and location of the Tol proteins

The tolQ, R, A and B genes are constituting a cluster localized at 16.8 min on the chromosomal map of *E. coli* (Sun and Webster, 1986). The complete sequence of this cluster has been determined (Sun and Webster, 1987; Levengood and Webster, 1989). Each of the genes can be independently transcribed but the eventual existence of an operon has not yet been ruled out (Sun and Webster, 1987). The gene products are weakly expressed but they could be characterized either after cloning the gene under the control of a strong promoter or using the "maxicell" or "minicell" systems (Sun and Webster, 1987; Levengood and Webster, 1989). Tol Q, R, A and B are proteins of 230, 142, 421 and 431 amino acid residues, respectively.

Sun and Webster (1986) reported that TolQ remained associated to the membranes after fractionation and that this protein had 3 putative transmembrane hydrophobic segments. The protein was fused to the 30 N-terminal residues of colicin A, which constitute an epitope, and was overproduced under the control the colicin A promoter (Bourdineaud *et al.*, 1989). It appeared to be preferentially localized in adhesion zones between inner and outer membranes. TolR appears to span the inner membrane only once through a single hydrophobic region (Sun and Webster, 1987). As previously mentioned, TolQ and TolR are homologous to ExbB and ExbD.

After fractionation, TolA remains associated to the inner membrane where it is likely anchored through a N-terminal 21 amino acid hydrophobic region. The rest of the protein is periplasmic and degraded when trypsin gains access to this compartment. TolA also harbors a region of 223 residues very rich in alanine, lysine, glutamate and aspartate (45%, 20% and 16% respectively). This latter is constituted of ten repeating units : ED(K)1-2 (A)2-4 which contributes to strongly stabilize a α-helical structure (Levengood and Webster, 1989). This region of the protein would then form an uninterrupted α-helix of approximately 34 nm allowing the protein to span the periplasmic space (Webster, 1990).

The case of TolB is more complex. In "minicells", two different products of *tolB* of 43 and 47,5 kDa were characterized but no degradation kinetics of the 47.5 kDa into 43 kDa could be evidenced. Two different initiations in the same reading frame could thus account for the existence of the two forms of TolB.

The TolC protein is important for the translocation of colicin E1. Its gene, tolC, is located at 66 min, far from the locus of other tol genes. It is a minor outer membrane protein produced under precursor form. The mature hydrophilic form contains 467 amino acid residues (Hackett and Reeves, 1983). Recently, from sequencing data, the central region of the protein (i.e. 90 amino acid residues) has been corrected (Niki *et al.*, 1990). Besides its role in colicin E1 translocation, TolC is also involved in the secretion of hemolysin, a toxin devoid of signal sequence but nevertheless released to the extracellular medium through a two-components secretion machinery (Wandersman and Delepaire, 1990). TolC is involved in the transcriptional regulation of *micF* and as a consequence has a strong effect on OmpF synthesis (Misra and Reeves, 1987).

Other mutations leading to a Tol phenotype

Dominant mutations at the locus *cet*, confering to cells specific tolerance to colicin E2, have been described (Hill and Holland, 1967; Buxton and Holland, 1973). These mutations occur in the promoter sequence of *cet*, the products of which is then overexpressed and accumulates in the inner membrane. Then, in contrast to Tol proteins, Cet is not required for colicin E2 uptake but overproduction and accumulation prevent colicin E2 from reaching its target in the cytoplasm.

Other *tol* mutants have been described. They display a tolerant phenotype only towards group A colicins but are phenotypically and genetically different from the other *tol* mutants described above (Burman and Nordström, 1971; Eriksson-Grennberg and Nordström, 1973; Foulds and Barret, 1973). Indeed, their loci are located between the 20th and 22nd min on the chromosomal map of *E. coli* and *tolD* and *tolE* mutants are tolerant to colicins E while *tolF* mutants are tolerant to colicins A and K. The *tolE* mutants appear to be deficient in the regulation of catobilism of galactose and LPS biosynthesis which may alter the assembly of porins in the outer membrane (Eriksson-Grennberg and Nordström, 1973). Mutations in tolZ are a bit unusual, in that they confer tolerance to certain colicins from both groups A and B. This may be related to the possible function of TolZ in the generation of an electrochemical proton gradient (Matsuzawa *et al.*, 1984).

Possible consensus sequences in group A colicins

Group A colicins of known sequence, carry an N-terminal region rich in glycine residues (Pugsley, 1987). A similar situation is encountered with the filamentous phage protein G3p, responsible for the binding of these phages to their receptor (F pilus) (Boeke *et al.*, 1982). G3p is involved in the tol-dependent translocation of phage DNA across the cell envelope, and especially the N-terminal glycine-rich sequences appear to play a role in this process (Stengele *et al.*, 1990).

Physiological role of Tol proteins

Proteins of the TonB-system are clearly involved in the uptake of siderophores and cobalamines, too bulky to enter through the porins. This system was parasitized for the entry of phages and group B colicins. It is logical to think along the same line for Tol proteins. However, rather surprisingly, so far the true physiological role of Tol proteins has not been elucidated, beyond the fact that they are required for the structural integrity of the outer membrane. Mutations in TolA, B, Q or R cause a pleiotropic phenotype of altered permeability of the outer membrane which is not observed in TonB mutants. The permeability and sensitivity of Tol mutants to a number of chemical products such as EDTA, detergents, drugs and dyes is increased. Moreover, these mutants allow periplasmic proteins to be released to the extracellular environment (Lazzaroni and Portalier, 1981).

COMPONENTS AND DYNAMICS OF TRANSPORT THROUGH THE TOL SYSTEM

Transport of colicin A through the Tol system

The import of colicin A, a pore-forming colicin, has been extensively investigated over recent years. The receptor is composed of two proteins, BtuB (vit. B12 receptor) and OmpF (major porin) (Cavard and Lazdunski, 1981). However, these two proteins do not play the same role. While BtuB is only used as a receptor, OmpF is used both as a receptor and for translocation across the outer membrane (Benedetti *et al.*, 1989).

The kinetics of K^+ efflux caused by colicin A in *E. coli* sensitive cells have been investigated by using a K^+ selective electrode. The dependence of K^+ efflux upon multiplicity, pH, temperature and membrane potential ($\Delta\psi$) was determined. The translocation of colicin A from the outer membrane receptor to the inner membrane and insertion into the inner membrane required a fluid membrane, but once inserted, the channel properties showed little dependence upon the state of the lipids. At a given multiplicity, the lag time before the onset of K^+ efflux was found to reflect the time required for translocation and/or insertion of colicin into the cytoplasmic membrane. Opening of the channel only occured above a threshold value of $\Delta\psi$ of 85 mV at pH 6.8. Conditions were designed for closing and reopening of the channels *in vivo*. These conditions allowed us to test separately the $\Delta\psi$ requirements for translocation and channel opening : translocation and/or insertion did not appear to require $\Delta\psi$ (Bourdineaud *et al.*, 1990). Therefore it seems that import of group A colicins across the outer membrane, in contrast to that of group B colicins, does not require energy.

The N-terminal domains of colicins confer specificity in colicin uptake

About 15 different hybrid colicins have been constructed by recombining various domains of group A and group B colicins through genetic ingeneering. These hybrid colicins were purified and their properties were studied (Frenette *et al.*, 1991; Benedetti *et al.*, 1991a). The results unambigously demonstrated that the information specifying the uptake pathway beyond the receptor (Tol-dependent or TonB-dependent pathway) was contained in the N-terminal domains of colicins. A given colicin could be shifted from the "passive diffusion" pathway that depends upon porins and Tol proteins to the energy-requiring pathway that depends upon a high affinity receptor (for siderophore or vit. B12) and TonB (Benedetti, Geli and Lazdunski, manuscript in preparation).

Colicin A unfolds during its translocation in *E. coli* cells and spans the whole cell envelope when its pore has formed

As mentioned above, the addition of colicin A to *E. coli* cells results in an efflux of cytoplasmic potassium. This efflux is preceeded by a lag time which corresponds to the time needed for the translocation of the toxin through the envelope. Denaturing the colicin A with urea before addingt it to the cells did not affect the properties of the pore but decreased the lag time. After renaturation, the lag time was similar to that of the native colicin. This suggests that the unfolding of colicin A accelerates its translocation (Benedetti *et al.*, 1991b). The addition of trypsin which has access neither to the periplasmic space nor to the cytoplasmic membrane, resulted in an immediate arrest of the potassium efflux induced by colicins A and B. The possibility that trypsin may act on a bacterial component required for colicin reception and/or translocation was ruled out. It is thus likely that the arrest of the efflux corresponds to a closing of the pores and that the colicin polypeptide chain remains at the translocation sites when the pore has formed in the inner membrane (Benedetti *et al.*, 1991b).

Colicins A and E1 interact with a component of their translocation system

Since the N-terminal domain of colicins specify their interactions with components of the import machinery it was likely that a direct interaction between this domain and one or several of these components may exist. To test this hypothesis, a system to obtain specific labeling and overexpression of Tol proteins was designed. The T7 RNA polymerase system was used (Tabor and Richardson, 1985). Plasmids were constructed which led to an overproduction of the Tol proteins involved in the import of group A colicins. *In vitro* binding of overexpressed Tol proteins to either Tol-dependent (group A) or TonB-dependent (group B) colicins was analyzed. The Tol-dependent colicins A and E1 were able to interact with TolA but the tonB-dependent colicin B was not. The C-terminal region of TolA, which is necessary for colicin uptake, was also found to be necessary for colicin A and E1 binding to occur. Furthermore, only the isolated N-terminal domain of colicin A, which is involved in the translocation step, was found to bind to TolA (Benedetti *et al.*, 1991c). These results demonstrate the existence of a correlation between the ability of group A colicins to translocate and their *in vitro* binding to TolA protein, suggesting that these interactions might be part of the colicin import process.

A hypothetical model for the translocation of colicin A

A hypothetical model for the import of colicin A (a Tol-dependent colicin), based upon the results reported above can be presented (Fig. 1).

In the first step, the central domain (R) of colicin A could bind to BtuB, the outer membrane receptor and OmpF. This binding would cause partial unfolding thereby allowing the N-terminal domain of colicin A(T domain) to interact with the OmpF region involved in translocation (Fourel *et al.*, 1990). The translocation of the polypeptide chain would then be initiated through the binding of the T domain to the C-terminal region of TolA (Benedetti *et al.*, 1991c). After transport across the outer membrane the pore-forming domain would undergo the first step in the insertion mechanism in the inner membrane. This step is the electrostatic interaction between the ring of positive charges located on the face of the protein defined by the loop connecting the two hydrophobic helices, and negatively charged phospholipid headgroups (Parker *et al.*, 1990).

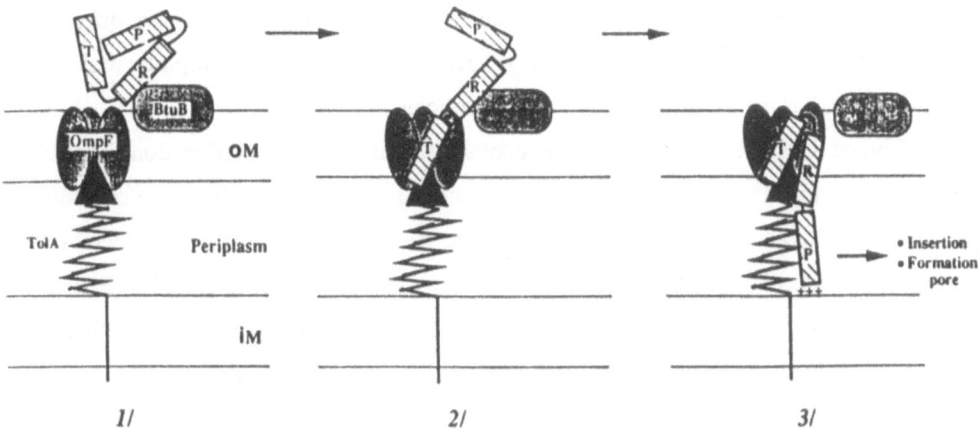

Fig. 1. Hypothetical model of translocation of colicin A.

OM and IM designate the outer and inner membranes of bacteria, respectively. Colicin A is depicted with its 3 domains : **T** is the N-terminal domain; **R** is the receptor-binding domain and **P** is the pore-forming domain. The 3 domains of TolA are represented by : a straight line (N-terminal anchor), a zig-zag line (putative α-helix) and a black triangle (C-terminal domain), respectively. The localization of the latter in the OmpF lumen is hypothetical. The other Tol proteins (B, Q, R) are not represented since the possible interactions with TolA, with colicin A and between themselves have not yet been evidenced.

We do not yet know the molecular mechanisms of transport across the outer membrane. Whether the OmpF pore itself is used or additional proteins are required is not clear but it is very likely that this transport occurs in an aqueous environment. The TolA and TolQ proteins seem to play a central role in the translocation of group A colicins since they are required for each of them. In contrast, others (TolB, TolR and TolC) are not required for some colicins.

The TolC protein, in the case of colicin E1, may constitute a surrogate of OmpF for translocation across the outer membrane. Indeed, TolC has been shown to be used for the secretion of hemolysin (Wandersman and Delepelaire, 1990). The Tol proteins may form an aqueous pore complex connecting the inner and outer membranes through their interactions with OmpF. Transport through this aqueous pore would require a minimum activation energy, in agreement with our results (Bourdineaud *et al.*, 1990). This "passive diffusion" pathway is in fact the porin diffusion pathway which is used for small hydrophilic molecules of nutrients and has been parasitized by group A colicins.

Several lines of evidence indicate that colicin A remains at the translocation sites while the C-terminal domain has already been inserted and has formed a pore. First, the addition of trypsin, which has access neither to the periplasmic space nor to the cytoplasmic membrane, resulted in an immediate arrest of the potassium efflux induced by colicins A and B (Benedetti *et al.*, 1991b). Second, competition experiments between colicins A and N indicated that after the penetration of colicin A in bacteria (immune to colicin A), colicin N using the same translocation pathway could no longer reach its target (H. Benedetti, Ph.D thesis, 1991a).

Conclusion and perspectives

The two pathways of nutrients transport across the outer membrane have been parasitized by colicins. The passive diffusion pathway used for low Mr (< 700 Mr) nutrients is the Tol-dependent pathway used by group A colicins. The energy-requiring pathway is the TonB-dependent pathway used for nutrients (siderophores and vit. B12) that exceed the diffusion limit of the outer membrane. It has been parasitized bg group B colicins. Whether a colicin is channelled through one pathway or another is entirely specified by the nature of its N-terminal domain. This domain contains specific sequences (glycine-rich sequence and TonB-box) allowing interaction with either the TolA or TonB components of translocation machineries. A physical interaction between outer membrane receptors and TonB has been evidenced through genetic studies (Heller *et al.*, 1988; Schöffler and Braun, 1989). Since TonB is anchored to the inner membrane (Postle and Skare, 1988), these results suggest the existence of translocation contact sites between the inner and outer membranes. As well, it has been demonstrated that TolQ is also preferentially located at contact sites (Bourdineaud *et al.*, 1989). Recent evidence suggest that other Tol proteins, in the presence of colicin A also have the same location (H. Benedetti, P. Boulanger, C. Lazdunski and L. Letellier, manuscript in preparation).

Thus, the components of the translocation machineries of colicins have now been identified and localised in the cell envelope. Targeting signals have been identified. The field is moving steadily ahead with emphasis clearly shifting from signals to molecular mechanisms and *in vitro* reconstituted systems.

Acknowledgements

This work was supported by the CNRS (GDR n° G-0964), the INSERM (CRE n°893011), the Fondation pour la Recherche Médicale, and the EEC program Science (contract n° SC1-0334-C). We are grateful to Monique Payan for preparing the manuscript.

REFERENCES

Baty D, Frenette M, Lloubes R, Geli V, Howard SP, Pattus F, Lazdunski C (1988) Functional domains of colicin A. Mol Microbiol 2:807-811

Benedetti H (1991a) Importation des colicines à travers l'enveloppe d'*Escherichia coli*. PhD Thesis, Université d'Aix-Marseille I

Benedetti H, Frenette M, Baty D, Lloubès R, Geli V, Lazdunski C (1989) Comparison of uptake systems for the entry of various BtuB group colicins into *Escherichia coli*. J Gen Microbiol 135:3413-3420

Benedetti H, Frenette M, Baty D, Knibiehler M Pattus F, Lazdunski C (1991a) Individual domains of colicins confer specificity in colicin uptake, in pore-properties and in immunity requirements. J Mol Biol 217:429-439

Benedetti H, Lazdunski C, Letellier L (1991b) Colicin A unfolds during its translocation in *E. coli* cells and spans the whole cell envelope when its pore has formed. Submitted for publication

Benedetti H, Lazdunski C, Lloubès R (1991c) Protein import into *Escherichia coli* : colicins A and E1 interact with a component of their translocation system. EMBO J 10 (in press)

Boeke J, Model P, Zinder N (1982) Effects of bacteriophage f1 gene III protein on the host cell membrane. Mol Gen Genet 186:185-192

Bourdineaud JP, Howard SP, Lazdunski C (1989) Localization and assembly into the *Escherichia coli* envelope of a protein required for entry of colicin A. J Bacteriol 171:2458-2465

Bourdineaud JP, Boulanger P, Lazdunski C, Letellier L (1990) *In vivo* properties of colicin A : channel activity is voltage dependent but translocation may be voltage independent. Proc Natl Acad Sci USA 87:1037-1041

Braun V, Hantke K (1977) In "Microbial Interactions", JL Reissing (ed), Chapman and Hal : London, p 101-137

Braun V (1989) The structurally related *exbB* and *tolQ* genes are interchangeable in conferring *tonB*-dependent colicin, bacteriophage, and albomycin sensitivity. J Bacteriol 171:6387-6390

Braun V, Frenz S, Hantke K, Schaller K (1980) Penetration of colicin M into cells of *Escherichia coli*. J Bacteriol 142:162-168

Braun V, Günter K, Hantke K (1991) Transport of iron across the outer membrane. Bio Metals 4:14-22

Brewer S, Tolley M, Trayer I, Barr G, Dorman C, Hannary K, Higgins C, Evans J, Levine B, Wormald M (1990) Structure and function of X-Pro dipeptide repeats in the TonB proteins of *Salmonella typhimurium* and *Escherichia coli*. J Mol Biol 216:883-895

Brunden KR, Cramer WA, Cohen FS (1984) Purification of a small receptor binding peptide from the central region of the colicin E1 molecule. J Biol Chem 259:190-196

Burman L, Nordström K (1971) Colicin tolerance induced by ampicillin or mutation to ampicillin resistance in a strain of *E. coli* K-12. J Bacteriol 106:1-13

Buxton R, Holland I (1973) Genetic studies of tolerance to colicin E2 in *Escherichia coli* K-12 : re-location and dominance relationships of *cet* mutations. Mol Gen Genet 127:69-88

Cavard D, Lazdunski C (1981) Involvment of BtuB and OmpF proteins in binding and uptake of colicin A. FEMS Microbiol Lett 12:311-316

Dankert J, Uratani Y, Graban C, Cramer W, Hermodson M (1982) On a domain structure of colicin E1. A COOH-terminal peptide fragment active in membrane depolarization. J Biol Chem 257:3857-3863

Datta DB, Arden B, Henning U (1977) Major proteins of the *Escherichia coli* outer cell envelope membrane as bacteriophage receptors. J Bacteriol 131:821-829

Davies JK, Reeves P (1975a) Genetics of resistance to colicins in *Escherichia coli* K12 : cross-resistance among colicins of group B. J Bacteriol 123:96-101

Davies JK, Reeves P (1975b) Genetics of resistance to colicins in *Escherichia coli* K12 cross-resistance among colicins of group A. J Bacteriol 123:102-117

De Graaf FK, Stukart MJ, Boogerd FC, Metselaar K (1978) Limited proteolysis of cloacin DF13 and characterization of the cleavage products. Biochemistry 17:1137-1142

De Graaf FK, Oudega B (1986) Production and release of cloacin DF13 and related colicins. Curr Top Microbiol Immunol 125:183-205

Eick-Helmerich K, Braun V (1989) Import of biopolymers into *Escherichia coli* : nucleotide sequences of the *exbB* and *exbD* genes are homologous to those of the *tolQ* and *tolR* genes, respectively. J Bacteriol 171:5117-5123

Eriksson-Grennberg K, Nordström K (1973) Genetics and physiology of a *tolE* mutant of *E. coli* K-12 and phenotypic suppression of its phenotype by galactose. J Bacteriol 115:1219-1222

Fischer E, Günter K, Braun V (1989) Involvment of ExbB and TonB in transport across the outer membrane of *E. coli* : phenotype complementation of *exbB* mutants by overexpressed *tonB* and physical stabilization of TonB by ExbB. J Bacteriol 171:5127-5134

Foulds J, Barret C (1973) Characterization of *Escherichia coli* mutants tolerant to bacteriocin JF246 : two new classes of tolerant mutants. J Bacteriol 116:885-892

Fourel D, Hikita C, Bolla JM, Mizushima S, Pagès JM (1990) Characterization of OmpF domains involved in *Escherichia coli* K-12 sensitivity to colicins A and N. J Bacteriol 172:3675-3680

Frenette M, Benedetti H, Bernadac A, Baty D, Lazdunski C (1991) Construction, expression and release of hybrid colicins. J Mol Biol 217:421-428

Hackett J, Reeves P (1983) Primary structure of the *tolC* gene that codes for an outer membrane protein of *Escherichia coli* K-12. Nucl Acids Res 11:6487-6495

Hancock R, Braun V (1976) Nature of the energy requirement for the irreversible adsoprtion of bacteriophages T1 and φ 80 to *Escherichia coli*. J Bacteriol 125:309-315

Hancock R, Braun V (1978) Functional interaction of the tonA/tonB receptor system in *Escherichia coli*. J Bacteriol 135:190-197

Hannavy K, Barr G, Dorman C, Adamson J, Mazengera L, Gallagher M, Evans J, Levine Bn Trayer I, Higgins C (1990) TonB protein of *Salmonella typhimurium* : A model for signal transduction between membranes. J Mol Biol 216:897-910

Hantke K (1976) Phage T6-colicin K receptor and nucleotide transport in *Escherichia coli*. FEBS Lett 70:109-112

Hantke K, Braun V (1975a) Membrane receptor dependent iron transport in *Escherichia coli*. FEBS Lett 49:301-305

Hantke K, Braun V (1975b) A function common to iron-enterochelin transport and action of colicins B, I, V in *Escherichia coli*. FEBS Lett 59:277-281

Hantke K, Zimmerman L (1981) The importance of *exbB* gene for vitamin B12 and ferric iron transport. FEMS Microbiol Lett 12:31-35

Heller KJ, Kadner RJ, Günther K (1988) Suppression of the *btuB*451 : mutations in the *ton*B gene suggests a direct interaction between TonB and TonB-dependent receptor proteins in the outer membrane of *Escherichia coli*. Gene 64:147-153

Hill C, Holland IB (1967) Isolation and properties of colicin refractory mutants and the preliminary mapping of their mutations. J. Bacteriol 94:677-686

Hollifield WC, Neilands JB (1978) Ferric enterobactin transport system in *Escherichia coli* K12. Extraction, assay and specificity of outer membrane receptor. Biochemistry 17:1922-192

Kadner RJ, Bassford PJ, Pugsley AP (1979) Colicin receptors and mechanism of colicin uptake. Zentralbl Bakteriol Parasitenkd Infektionskr Hy Abt 244:90-104

Konisky J (1982) Colicins and other bacteriocins with established modes of action. Ann Rev Microbiol 36:125-144

Lazzaroni JC, Portalier RC (1981) Genetic and biochemical characterization of periplasmic-leaky mutants of *Escherichia coli* K12. J Bacteriol 145:1351-1358

Levengood S, Webster R (1989) Nucleotide sequences of the *tolA* and *tolB* genes and localization of their products, components of a multistep translocation system in *Escherichia coli*. J Bacteriol 171:6600-6609

Lugtenberg B, van Alphen L (1983) Molecular architecture and functioning of the outer membrane of *E. coli* and other Gram-negative bacteria. Biochim Biophys Acta 737:51-115

Mankovich JA, Hsu CH, Konisky J (1986) DNA and aminoacid sequence analysis of structural and immunity genes of colicins Ia and Ib. J Bacteriol 168:228-236

Martinez MC, Lazdunski C, Pattus F (1983) Isolation of molecular and functional properties of the C-terminal domain of colicin A. EMBO J 2:1501-1507

Matsuzawa H, Ushiyama S, Koyama Y, Ohta T (1984) *Escherichia coli* K-12 *tolZ* mutants tolerant to colicins E2, E3, D, Ia, Ib defect in generation of electrochemical proton gradient. J. Bacteriol 160:733-739

Misra R, Reeves P (1987) Role of *micF* in the *tolC* mediated regulation of OmpF a major outer membrane protein of *Escherichia coli* K-12. J Bacteriol 169:4722-4730

Nagel del Zwaig R, Luria JE (1967) Genetics and physiology of colicin-tolerant mutants of *Escherichia coli*. J Bacteriol 94:1112-1123

Nikaido H, Vaara M (1985) Molecular basis of bacterial outer membrane permeability. Microbiol Rev 49:1-32

Niki H, Imamura R, Ogura T, Hiraga S (1990) Nucleotide sequence of the *tolC* gene of *Escherichia coli*. Nucl Acids Res 18:5547

Ohno-Iwashita Y, Imahori K (1980) Assignment of the functional loci in colicin E1 and E3 molecules by the characterization of their proteolytic fragments. Biochemistry 19: 652-659

Ohno-Iwashita Y, Imahori K (1982) Assignment of the functional loci of the colicin E1 molecule by characterization of their proteolytic fragments. J Biol Chem 257:6446-6451

Parker M, Tucker A, Tsernoglou D, Pattus F (1990) Insights into membrane insertion based on studies of colicin. TIBS 15:126-129

Postle K (1990) TonB and the gram-negative dilemna. Mol Microbiol 4:2019-2025

Postle K, Good R (1983) DNA sequence of the *Escherichia coli* tonB gene. Proc Natl Acad Sci USA 80:5235-5339

Postle K, Skare J (1988) *Escherichia coli* TonB protein is exported from the cytoplasm without proteolytic cleavag'e of its amino terminus. J Biol Chem 263:11000-11007

Pugsley AP (1987) Nucleotide sequencing of the structural gene for colicin N reveals homology between the catalytic C-terminal domains of colicins A and N. Mol Microbiol 1:317-325

Pugsley A, Reeves P (1977) The role of colicin receptors in the uptake of ferrienterochelin by *Escherichia coli* K12. Biochem Biophys Res Commun 74:903-911

Ross U, Harkness R, Braun V (1989) Assembly of colicin genes from a few DNA fragments. Nucleotide sequence of colicin D. Mol Microbiol 3:891-902

Schöffler A, Braun V (1989) Transport across the outer membrane of *Escherichia coli* via the Fhu A receptor is regulated by the TonB protein of the cytoplasmic membrane. Mol Gen Genet 217:378-383

Stengele I, Bross P, Garces X, Giray J, Rasched I (1990) Dissection of functional domains in phage fd adsorption protein : discrimination between attachment and penetration sites. J Mol Biol 212:143-149

Sun TP, Webster RE (1986) *fii* (*tolQ*) a bacterial locus required for filamentous phage infection and its relation to colicin-tolerant *tol*A *tol*B. J Bacteriol 165:107-115

Sun TP, Webster RE (1987) Nucleotide sequence of a gene cluster involved in the entry of the E colicins and the single stranded DNA of infecting filamentous phage into *Escherichia coli*. J Bacteriol 169:2667-2674

Tabor S, Richardson C (1985) A bacteriophage T7 DNA polymerase/promoter system for controlled exclusive expression of specific genes. Proc Natl Acad Sci USA 70:3160-3164

Wandersman C and Delepelaire P (1990) TolC, an *E. coli* outer membrane protein required for hemolysin secretion. Proc Natl Acad Sci USA 87:4776-4780

Wayne R, Neilands JB (1975) Evidence for common binding sites for ferrichrome compounds and bacteriophage Ø 80 in the cell envelope of *Escherichia coli*. J Bacteriol 121:459-503

Webster R (1991) The *tol* gene products and the import of macromolecules into *Escherichia coli*. Mol Microbiol 5:1005-1011

A TRANSFERRIN-BINDING PROTEIN IN Trypanosoma brucei: DOES IT FUNCTION IN IRON UPTAKE?

Peter Overath[1], Dietmar Schell[1], York-Dieter Stierhof[1], Heinz Schwarz[2] and Dagmar Preis[1]

[1]Max-Planck-Institut für Biologie, Abteilung Membranbiochemie Corrensstrasse 38, and

[2]Max-Planck-Institut für Entwicklungsbiologie, Spemannstrasse 35, D7400 Tübingen, Federal Republic of Germany.

The elucidation of the structure and function of receptors involved in surface recognition, transmembrane signalling and transport is one of the most active fields in present day molecular biology. The receptors of viruses, bacteria and mammalian cells are particularly well characterized, while, in contrast, little is known about the receptors of some of the medically and economically important parasites. The complex life cycle of many parasites involving both vertebrate and invertebrate hosts requires their adaptation to vastly different environments. This adaptation entails the synthesis of specific receptors required, for example, for attachment and invasion of host cells or the uptake of specific nutrients provided by the host. This situation can be illustrated with the African trypanosomes, i.e. Trypanosoma brucei and related species, which cause sleeping sickness in man and Nagana in cattle.

T. brucei is a unicellular flagellated protozoan which divides extracellularly in the blood and tissues of mammals. When a tsetse fly takes a blood meal from its host, these bloodstream forms are transferred to its midgut where they differentiate to procyclic forms. After further differentiation steps, metacyclic trypanosomes are formed in the salivary glands of the fly, which are eventually injected into the mammal the fly uses for a subsequent blood meal, thus completing

NATO ASI Series, Vol. H 63
Dynamics of Membrane Assembly
Edited by J. A. F. Op den Kamp
© Springer-Verlag Berlin Heidelberg 1992

the life cycle. With the exception of the transition
of the bloodstream forms to the procyclic cells (Brun &
Schönenberger, 1981; Czichos et al, 1986; Overath et
al, 1986) very little is known about the signals that
trigger these various differentiation steps.

The characterization of surface molecules that
function in trans-membrane signalling or nutrient
uptake in bloodstream trypanosomes is a particular
challenge because these cells are covered by a dense
glycoprotein coat (Vickerman & Luckins, 1969), which
protects against lytic serum components. For an
individual trypanosome clone, it consists of about 10^7
molecules of a single glycoprotein, the membrane-form
of the variant surface glycoprotein (mfVSG, Cross,
1975; Ferguson et al., 1988). mfVSG extends about
130 Å from the cell surface, and is anchored in the
plasma membrane by a glycosyl-phosphatidylinositol
(GPI) residue attached to the C-terminal amino acid
(Fig.1). It covers not only the cell body and the
flagellum but also the surface of the flagellar pocket,
an invagination of the plasma membrane at the arising
flagellum (Fig.2). Each trypanosome clone can give
rise to variants in which gene rearrangements lead to
the expression of structurally and immunologically
distinct VSGs, a phenomenon designated antigenic
variation (reviewed by Cross, 1990).

mfVSG is the only well characterized surface
molecule of bloodstream forms and is the dominant
antigen for the humoral immune response against the
parasite. Judging by certain growth requirements of
trypanosomes, there must exist other cell surface
molecules, most likely to be invariant proteins. One
class is exemplified by a specific carrier, which
catalyses the uptake of glucose, the main energy source
of bloodstream forms (Gruenberg et al., 1978; Game et
al., 1986; Eisenthal et al., 1989; Ter Kuile &
Opperdoes, 1991). Such carriers are generally integral
membrane proteins organized in multiple lipid bilayer-
spanning α-helices, which are connected on both faces
by polypeptide loops. The VSG-coat shields these
proteins against macromolecules, i.e. specific
antibodies, but it would be permeable to low-molecular
weight solutes such as glucose. The situation must
clearly be different for a second class of proteins,
i.e. receptors, which have to interact with
macromolecular ligands. While solute carriers may be
distributed over the body of the cell, receptors are
likely to be confined to the membrane lining the
flagellar pocket, the only region of the parasite where
exocytosis and endocytosis have been demonstrated both
by electron microscopy, and by the uptake of several
high molecular weight components (reviewed by Balber,
1990).

The best evidence for a receptor in the flagellar
pocket is based on the work of Coppens et al. (1987,
1988). Bloodstream trypanosomes require low density
lipoprotein (LDL) for growth (Black & Vandeweerd, 1989)
which is taken up by receptor-mediated endocytosis in

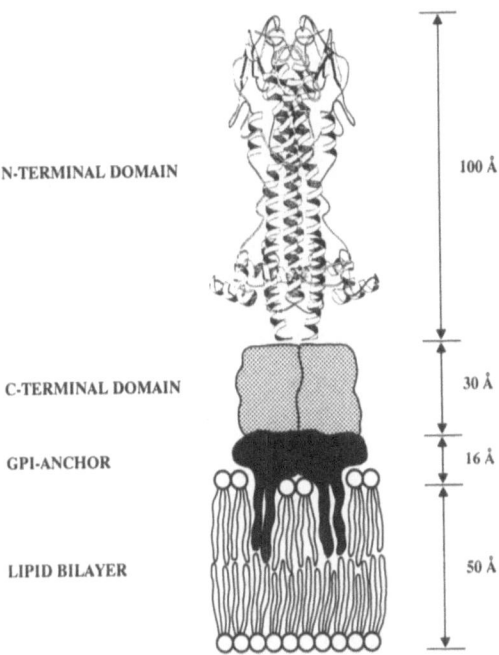

N-TERMINAL DOMAIN   100 Å

C-TERMINAL DOMAIN   30 Å

GPI-ANCHOR   16 Å

LIPID BILAYER   50 Å

Fig.1. Organization of the membrane form of the variant
       surface glycoprotein in bloodstream
       trypanosomes. Each monomer of the dimeric
       molecule consists of an N-terminal domain (see
       Freymann et al., 1990) a C-terminal domain of
       unknown structure and the GPI-anchor (Homans et
       al., 1989), which is inserted into the lipid
       bilayer by a dimyristoylphosphatidylinositol
       residue. The elongated structure of the N-
       terminal domain is characterized by two α-
       helices/monomer.

the flagellar pocket (Coppens et al., 1987, 1988).
Binding experiments with cells suggest the presence of
52 000 low-affinity binding sites (K_D = 250 nM) and
about 1800 high-affinity binding sites/cell (K_D = 5.7
nM). The yield of a candidate LDL receptor (app.
M_r 86 000) purified by affinity chromatography likewise
suggests the presence of about 50 000 LDL-binding
proteins/cell. While 1800 receptor molecules can be
accommodated in the flagellar pocket membrane (area
about 1 μm^2, cf. Coppens et al., 1987), 50 000
molecules would cover this surface entirely.
Therefore, the relationship between the two classes of
LDL binding sites observed in cells, the 86 kd protein
and the topological restraints imposed by the available
membrane area of the flagellar pocket, remains to be
clarified.

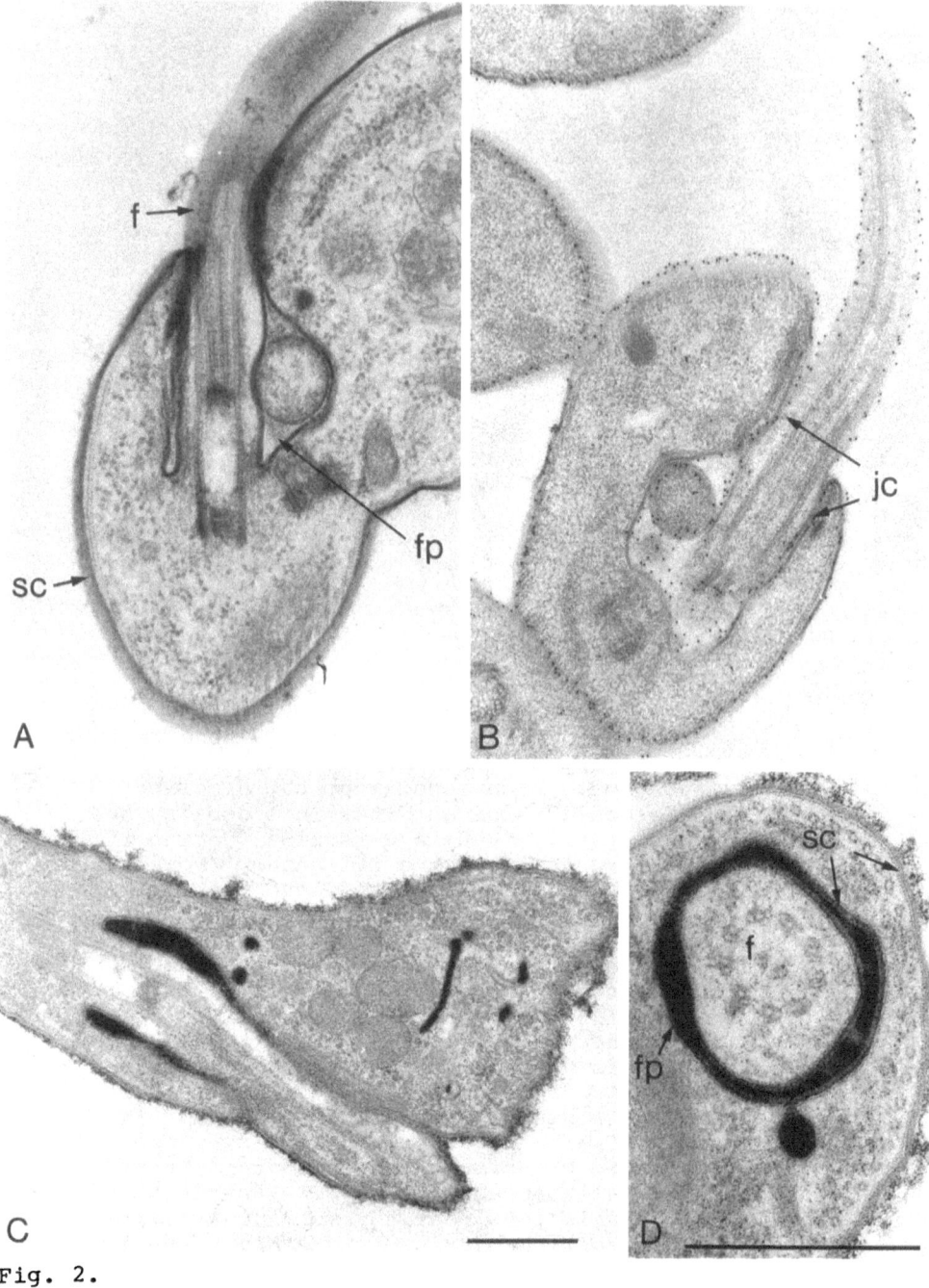

Fig. 2.

For legend to Fig.2 see opposite page.

Our work is directed towards the characterization of a transferrin receptor in trypanosomes. Although bloodstream forms are deficient in cytochromes they contain 1.4×10^6 atoms of iron per cell (Schell et al., 1991a), which may be bound as non-heme iron to a mitochondrial plant-like alternative oxidase (Grant & Sargent, 1960; Clarkson et al, 1989;). Since bovine serum looses its growth-promoting properties after selective depletion of transferrin, it was concluded that transferrin is a growth factor for bloodstream forms (Fig.3; Schell et al, 1991a). The presence of a receptor is implied by experiments showing the endocytosis of transferrin bound to colloidal gold in the flagellar pocket (Webster & Grab, 1988; Webster, 1989) and the rapid uptake of radioactively labeled transferrin (Coppens et al., 1987).

The putative receptor, a transferrin-binding protein (TFBP), was isolated by affinity chromatography (Schell et al., 1991b). The protein has an apparent molecular weight of 42 kd and is modified at the C-terminus by a GPI-anchor, which is sensitive to the T. brucei GPI-specific phospholipase C. About 1 μg of protein was obtained from 10^{10} T. brucei cells (~60 mg total protein); this corresponds to 1000 molecules/cell. The TFBP binds very tightly to human diferric transferrin. Reagents that dissociated the human receptor from transferrin (0.5 M NaSCN, Rudolph & Regoeczi, 1987; Turkewitz et al., 1988) were insufficient for the release of the trypanosomal protein, which was finally eluted under very stringent conditions (50 mM glycine, pH 2). The purified trypanosomal protein retained its ability to bind to transferrin. Finally, TFBP could be isolated from several variant clones expressing different VSGs but it could not be detected in procyclic forms.

Unexpectedly, the transferrin-binding protein was found to be encoded by a gene in the expression site of the VSG gene (Kooter et al., 1988; Pays et al., 1989; Hobbs & Boothroyd, 1990). An expression site with the open reading frames upstream of the VSG gene (ESAGs for

Fig.2. The flagellar pocket of bloodstream forms.
(A) Section of a specimen fixed with glutaraldehyde, tannic acid and OsO_4 and embedded in Epon; sc, surface coat; f, flagellum; fp, flagellar pocket. (B) Lowicryl K4M-embedded trypanosomes on-section-labeled with rabbit anti-VSG antibody and protein A-15 nm gold; jc, junctional complex. (C and D) Visualization of the FP by cytochemical staining of acid phosphatase activity using β-glycero-phosphate as a substrate and Ce^{+++} as a phosphate capture reagent (Robinson & Karnovsky, 1983; Fahimi et al., 1988, Langreth & Balber, 1975; Schell et al., 1990). Phosphatase is also located in intracellular vesicles which can fuse with the FP membrane (D) or in tubular structures. bar = 0.5 μm.

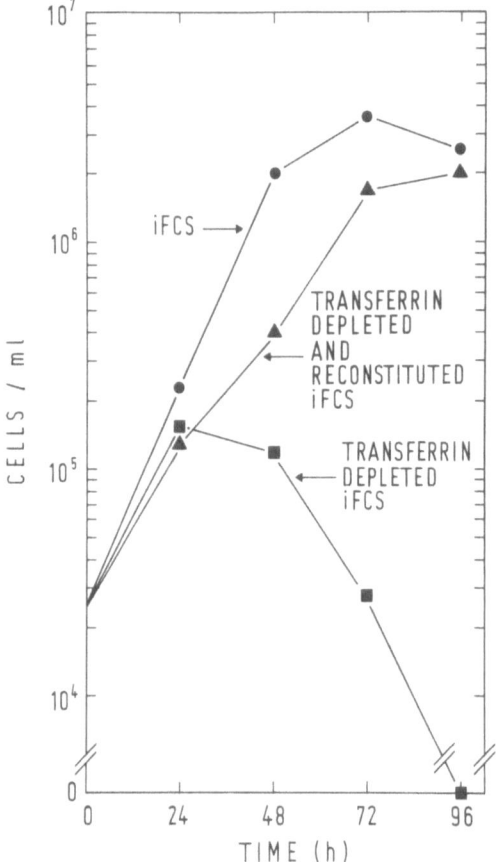

Fig.3. Transferrin as a growth factor for trypanosomes.
Bloodstream forms (clone TC221) were grown in
vitro in medium containing 15% inactivated fetal
bovine serum (iFCS), iFCS depleted of and
reconstituted with transferrin (1 mg/ml) or
transferrin-depleted iFCS. In the transferrin-
depleted medium the cells grow for only 24 h,
thereafter the parasites degenerate and lyse as
indicated by the decrease in cell count (see
Schell et al., 1991a, for further details).

expression-site associated genes) is depicted in Fig.4
together with some of the properties of the proteins
predicted from the DNA sequence. An expression site is
transcribed from an upstream promoter into a large
polycistronic mRNA, which is then spliced to the
individual messages corresponding to the VSG gene and
the ESAGs. The abundance of the mature ESAG mRNAs
varies widely; the VSG mRNA is the most abundant
transcript in bloodstream forms. The absence of TFBP
in procyclic cells is expected considering that in
these forms all transcripts from the expression site

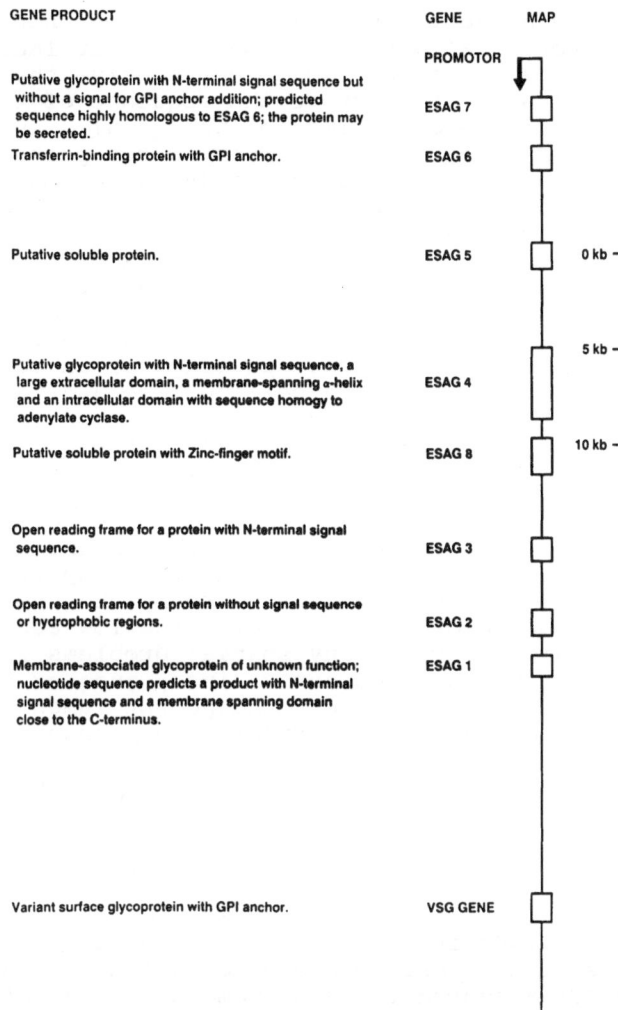

GENE PRODUCT   GENE   MAP

PROMOTOR

Putative glycoprotein with N-terminal signal sequence but without a signal for GPI anchor addition; predicted sequence highly homologous to ESAG 6; the protein may be secreted. — ESAG 7

Transferrin-binding protein with GPI anchor. — ESAG 6

Putative soluble protein. — ESAG 5 — 0 kb

— 5 kb

Putative glycoprotein with N-terminal signal sequence, a large extracellular domain, a membrane-spanning α-helix and an intracellular domain with sequence homogy to adenylate cyclase. — ESAG 4

Putative soluble protein with Zinc-finger motif. — ESAG 8 — 10 kb

Open reading frame for a protein with N-terminal signal sequence. — ESAG 3

Open reading frame for a protein without signal sequence or hydrophobic regions. — ESAG 2

Membrane-associated glycoprotein of unknown function; nucleotide sequence predicts a product with N-terminal signal sequence and a membrane spanning domain close to the C-terminus. — ESAG 1

Variant surface glycoprotein with GPI anchor. — VSG GENE

Fig.4. Genes and gene products of a VSG expression site. For further information see: Alexandre *et al*., 1988; 1990; Cully *et al*., 1985, 1986; Hobbs and Boothroyd, 1990; Pays *et al*., 1989; Revelard *et al*., 1990; Rolin *et al*., 1990; Smiley *et al*., 1990; Son *et al*., 1989; Zomerdijk *et al*., 1990.

are down-regulated to essentially undetectable levels (reviewed by Borst, 1986; Pays & Steinert, 1988; Cross, 1990).

The TFBP is encoded by ESAG 6 (Pays *et al*., 1989), which predicts a protein of 401 amino acids with an N-terminal signal sequence and a stretch of hydrophobic amino acids at the C-terminus characteristic for proteins modified post-translationally by a GPI-

membrane anchor. In addition, the sequence contains
three potential glycosylation sites, at least one of
which appears to be used. Importantly, the sequence of
TFBP bears no homology to the sequence of the human
transferrin receptor (McClelland et al., 1984;
Schneider et al., 1984). This latter protein is a
disulfide-linked homodimer of 95 kd monomers composed
of extracellular, membrane-spanning and cytoplasmic
domains (reviewed by Dautry-Varsat & Lodish, 1984; May
& Cuatrecasas, 1985; Thorstenson & Romslo, 1990)[1].
    In mammalian cells, the complex of transferrin and
receptor is internalized from the cell surface in
coated vesicles which fuse with endosomes. After
release of the iron, the apo-transferrin-receptor
complex recycles to the cell surface and then
dissociates. On this background, there are several
arguments for or against the TFBP of trypanosomes
functioning as a receptor in the uptake of transferrin:
With few exceptions (Fouchier et al., 1988; LeBel &
Beattie, 1988)) GPI-anchored proteins are surface
proteins. Therefore, after synthesis and modification
by a GPI-anchor, TFBP is probably transported to the
flagellar pocket of the parasite by the same route as
mfVSG (Duszenko et al., 1988). The low abundance of
TFBP would not raise any spatial problems in the
flagellar pocket membrane. Since, judging by
immunoelectronmicroscopy, the pocket lumen contains
high concentrations of transferrin presumably trapped
in a matrix of undefined composition (Y.-D. Stierhof
and D. Schell, unpublished results, compare Langreth &
Balber, 1975), TFBP could be rapidly loaded with
transferrin directly after exocytosis and internalized
again by endocytosis. This way, the spread of TFBP
over the cell surface, which is typical for other GPI-
anchored proteins such as mfVSG and procyclin (Bülow et
al., 1988; Roditi et al., 1989), would be prohibited.
Against this scenario are observations demonstrating
that GPI-anchored proteins remain at the cell surface
because they lack a cytoplasmic membrane domain that
allows interaction with clathrin-coated vesicles[2]
(Bülow et al., 1989; Lemansky et al., 1990).
Therefore, with the exception of the rather unique
folate uptake system described in mammalian cells
(Rothberg et al. 1990), GPI-anchored proteins are
generally not considered to function as receptors for
the uptake of high or low molecular weight ligands.
Finally, because TFBP appears to bind transferrin more
tightly than the mammalian receptor, the intracellular

---

[1]The sequence of ESAG 7 (Fig.4) is highly homologous to
that of the adjacent ESAG 6. ESAG 7 predicts a protein
of 340 amino acids with a signal sequence but without a
GPI-anchor addition signal. It is possible that this
protein is secreted by trypanosomes.
[2]A trypanosomal equivalent to coated vesicles from
mammalian cells is described in Shapiro & Webster
(1989).

fate of the complex in trypanosomes may be significantly different from that in mammalian cells.

As a basically extracellular, but nevertheless secluded compartment, the flagellar pocket raises several rather unique questions. Does the junctional complex (jc, Fig.2B) at the entrance to the pocket control the flux of membrane-associated molecules between the pocket membrane and the plasma membrane enclosing the body of the cell and the flagellum? What are the rates at which macromolecules like transferrin and LDL diffuse into the pocket, bind to receptors and enter the cell by endocytosis? Future experiments must show whether the transferrin-TFBP system is a suitable model for answering these questions.

The isolation of TFBP, its postulated location in the flagellar pocket and the observation that it elicits a humoral immune response during chronic infections in rabbits (Hobbs & Boothroyd, 1990) revives the discussion of whether the characterization of invariant surface proteins of trypanosomes will open new avenues for vaccination against sleeping sickness (Borst, 1991). The following mechanisms for the interferring with parasite survival or growth may be considered. Firstly, anti-receptor antibodies may inhibit ligand binding, thereby preventing essential nutrient uptake. Secondly, it should be recalled that bloodstream forms activate the classical complement pathway after binding of antibodies to surface-exposed VSG epitopes while the alternative pathway can only be triggered after cleavage of VSG by trypsin (Mosser & Roberts, 1982; Ferrante & Allison, 1983; Shapiro & Pearson, 1986). Therefore, deliberate immunization with invariant surface proteins may lead to antibody binding at the cell surface or in the flagellar pocket and to complement activation. In particular, invariant surface antigens may become accessible while the parasite exposes two possibly improperly packed VSGs directly after a switch in antigen expression. Theoretical modeling of chronic infections suggests that such a mechanism is only operative for certain VSG combinations (Agur et al., 1989). Thirdly, the binding of antibodies in the concealed compartment of the flagellar pocket will most certainly be unable to induce phagocytosis by macrophages which is the major mechanism for the elimination of parasites after surface-binding of anti-VSG antibodies.

In summary, although our work on invariant surface proteins (Schell et al., 1990; 1991b) is motivated by the idea that the flagellar pocket is the Achilles heel of the parasite and, therefore, "is worth an all out attack" (Borst, 1991), we should anticipate that trypanosomes may outsmart us once again.

Acknowledgements: We thank Dr. M. Chaudhri for the correction of the manuscript and the Fond der Chemischen Industrie for their support.

References

Agur Z, Abiri D, Van der Ploeg LHT (1989) Ordered
    appearance of antigenic variants of African
    trypanosomes explained in a mathematical model
    based on a stochastic switch process and immune-
    selection against putative switch intermediates.
    Proc Nat Acad Sci (USA) 86: 9626-9630

Alexandre S, Guyaux M, Murphy NB, Coquelet H, Pays A,
    Steinert M, Pays E (1988) Putative genes of a
    variant-specific antigen gene- transcription unit
    in Trypanosoma brucei. Mol Cell Biol 8: 2367-2378

Alexandre S, Paindavoine P, Tebabi P, Pays A, Halleux
    S, Steinert M, Pays E (1990) Differential
    expression of a family of putative
    adenylate/guanylate cyclase genes in Trypanosoma
    brucei. Mol Biochem Parasitol 43: 279-288

Balber AE (1990) The pellicle and the membrane of the
    flagellum, flagellar adhesion zone, and flagellar
    pocket - functionally discrete surface domains of
    the blood-stream form of African trypanosomes.
    Crit Rev Immunol 10: 177-201

Black S, Vandeweerd V (1989) Serum-lipoproteins are
    required for multiplication of Trypanosoma brucei
    brucei under axenic culture conditions. Mol
    Biochem Parasitol 37: 65-72

Borst P (1986) Discontinuous transcription and
    antigenic variation in trypanosomes. Annu Rev
    Biochem 55: 701-732.

Borst P (1991) Transferrin receptor, antigenic
    variation and the prospect of a trypansome
    vaccine. Trends Genet in press

Brun R, Schönenberger M (1981) Stimulating effect of
    citrate and cis-aconitate on the transformation of
    Trypanosoma brucei bloodstream forms to procyclic
    forms in vitro. Z. Parasitenkunde 66: 17-24

Bülow R, Nonnengässer C, Overath P (1989) Release of
    the variant surface glycoprotein during
    differentiation of blood-stream to procyclic forms
    of Trypanosoma brucei. Mol Biochem Parasitol 32:
    85-92

Bülow R, Overath P, Davoust J (1988) Rapid lateral
    diffusion of the variant surface glycoprotein in
    the coat of Trypanosoma brucei. Biochem 27: 2384-
    2388

Clarkson AB, Bienen EJ, Pollakis G, Grady RW (1989)
    Respiration of blood-stream forms of the parasite
    Trypanosoma brucei brucei is dependent on a plant-

like alternative oxidase. J Biol Chem 264: 17770-17776

Coppens I, Baudhuin P, Opperdoes FR, Courtoy PJ (1988) Receptors for the host low-density lipoproteins on the hemoflagellate Trypanosoma brucei - purification and involvement in the growth of the parasite. Proc Natl Acad Sci (USA) 85: 6753-6757

Coppens I, Opperdoes FR, Courtoy PJ, Baudhuin P (1987) Receptor-mediated endocytosis in the blood-stream form of Trypanosoma brucei. J Protozool 34: 465-473

Cully DF, Gibbs CP, Cross GAM (1986) Identification of proteins encoded by variant surface glycoprotein expression site-associated genes in Trypanosoma brucei. Mol Biochem Parasitol 21: 189-197.

Cully SF, Ip HS, Cross, GAM (1985) Coordinate transcription of variant surface glycoprotein genes and an expression site associated gene family in Trypanosoma brucei. Cell 42: 173-182

Cross GAM (1975) Identification, purification and properties of clone-specific glycoprotein antigens constituting the surface coat of Trypanosoma brucei. Parasitol 71: 393-417

Cross GAM (1990) Cellular and genetic aspects of antigenic variation in trypanosomes. Annu Rev Immunol 8: 83-110

Czichos J, Nonnengaesser C, Overath P (1986) Trypanosoma brucei: cis-Aconitate and temperature reduction as triggers of synchronous transformation of bloodstream to procyclic trypomastigotes in vitro. Exp Parasitol 62: 283-291

Dautry-Varsat A, Lodish HF (1984) How receptors bring proteins and particles into cells. Sci Am 250(5): 52-58

Duszenko M, Ivanov IE, Ferguson MAJ, Plesken H, Cross GAM (1988) Intracellular transport of a variant surface glycoprotein in Trypanosoma brucei. J Cell Biol 106: 77-86

Eisenthal R, Game S, Holman GD (1989) Specificity and kinetics of hexose-transport in Trypanosoma brucei. Biochim Biophys Acta 985: 81-89

Fahimi HD, Angermüller S, Baumgart E, Völkl A (1988) Recent developments in ultrastructural cytochemistry of phosphatases and oxidases using the cerium technique. In: Eurem 88, Proceedings of the 9th European Congress on Electron Microscopy

(Dickinson HG and Goodhew PJ, eds), Inst Phys Conf Ser, Bristol, No 93, Vol 3 pp 503-507

Ferrante A, Allison AC (1983) Natural agglutinins to African trypanosomes. Paras Immunol 5: 539-546

Ferguson MAJ, Homans SW, Dwek RA, Rademacher TW (1988) Glycosyl-phosphatidylinositol moiety that anchors *Trypanosoma brucei* variant surface glycoprotein to the membrane. Science 239: 753-759

Fouchier F, Bastiani P, Baltz T, Aunis D, Rougon G (1988) Glycosylphosphatidylinositol is involved in the membrane attachment of proteins in granules of chromaffin cells. Biochem J 256: 103-108

Freymann D, Down J, Carrington M, Roditi I, Turner M, Wiley D (1990) 2.9 Å resolution structure of the N-terminal domain of a variant surface glycoprotein from *Trypanosoma brucei*. J Mol Biol 216: 141-160

Game S, Holman G, Eisenthal R (1986) Sugar transport in *Trypanosoma brucei*: a suitable kinetic probe. FEBS Lett 194: 126-130

Grant PT, Sargent JR (1960) Properties of the L-α-glycerophosphate oxidase and its role in the respiration of *Trypanosoma rhodensiense*. Biochem J 76: 229-237

Gruenberg J, Sharma PR, Deshusses J (1978) D-Glucose transport in *Trypanosoma brucei*. Eur J Biochem 89: 461-469.

Hobbs MR, Boothroyd JC (1990) An expression-site-associated gene family of trypanosomes is expressed *in vivo* and shows homology to a variant surface glycoprotein gene. Mol Biochem Parasitol 43: 1-16

Homans SW, Edge CJ, Ferguson MAJ, Dwek RA, Rademacher TW (1989) Solution structure of the glycosylphosphatidylinositol membrane anchor glycan of *Trypanosoma brucei* variant surface glycoprotein. Biochem 28: 2881-2887

Kooter JM, Winter AJ, DeOliveira C, Wagter R, Borst P (1988) Boundaries of telomere conversion in *Trypanosoma brucei*. Gene 69: 1-11

Langreth SG, Balber AE (1975) Protein uptake and digestion in bloodstream and culture forms of *Trypanosoma brucei*. J Protozool 22: 40-53.

LeBel D, Beattie M (1988) The major protein of pancreatic zymogen granule membranes (GP-2) is anchored via covalent bonds to phosphatidylinositol. Biochem Biophys Res Commun 154: 818-823

Lemansky P, Fatemi SH, Gorican B, Meyale S, Rossero R, Tartakoff AM (1990) Dynamics and longevity of the glycolipid-anchored membrane protein, Thy-1. J Cell Biol 110: 1525-1531

May WS, Jr, Cuatrecasas P (1985) Transferrin receptor: Its biological significance. J Membrane Biol 88: 205-215

McClelland A, Kühn LC, Ruddle FH (1984) The human transferrin receptor gene: genomic organization, and the complete primary structure of the receptor deduced from a cDNA sequence. Cell 39: 267-274

Mosser DM, Roberts JF (1982) Trypansoma brucei: Recognition in vitro of two developmental forms by murine macrophages. Exp Parasitol 54: 310-316

Overath P, Czichos J, Haas C (1986) The effect of citrate/cis-aconitate on oxidative metabolism during transformation of Trypanosoma brucei. Eur. J. Biochem. 160: 175-182

Pays E, Tebabi P, Pays A, Coquelet H, Revelard P, Salmon D, Steinert M (1989) The genes and transcripts of an antigen gene expression site from T. brucei. Cell 57: 835-845

Pays E, Steinert M (1988) Control of antigen gene-expression in African trypanosomes. Annu Rev Genet 22: 107-126

Revelard P, Lips S, Pays E (1990) A gene from the VSG expression site of Trypanosoma brucei encodes a protein with both leucine-rich repeats and a putative zink finger. Nucl Acids Res 18: 7299-7303

Robinson JM, Karnovsky MJ (1983) Ultrastructural localization of several phosphatases with cerium. J Histochem Cytochem 31: 1197-1208

Roditi I, Schwarz H, Pearson TW, Beecroft RP, Liu MK, Richardson JP, Bühring HJ, Pleiss J, Bülow R, Williams RO, Overath P (1989) Procyclin gene expression and loss of the variant surface glycoprotein during differentiation of Trypanosoma brucei. J Cell Biol 108: 737-746

Rolin S, Halleux S, Vansande J, Dumont J, Pays E, Steinert M (1990) Stage-specific adenylate-cyclase activity in Trypanosoma brucei. Exp Parasitol 71: 350-352

Rothberg KG, Ying Y, Kolhouse JF, Kamen BA, Anderson RGW (1990) The glycophospholipid-linked folate receptor internalizes folate without entering the clathrin-coated pit endocytic pathway. J Cell Biol 110: 637-649

Rudolph JR, Regoeczi E (1987) Isolation of the rat transferrin receptor by affinity chromatography. J Chromatogr 396: 369-373

Schell D, Borowy NK, Overath P (1991a) Transferrin is a growth factor for the bloodstream form of Trypanosoma brucei. Parasitol Res in press

Schell D, Evers R, Preis D, Ziegelbauer K, Kiefer H, Lottspeich F, Cornelissen AWCA, Overath P (1991b) A transferrin-binding protein of Trypanosoma brucei is encoded by one of the genes in the variant surface glycoprotein gene expression site. EMBO J 10: 1061-1066

Schell D, Stierhof YD, Overath P (1990) Purification and characterization of a tartrate-sensitive acid-phosphatase of Trypanosoma brucei. FEBS Lett 271: 67-70

Schneider C, Owen MJ, Banville D, Williams JG (1984) Primary structure of human transferrin receptor deduced from the mRNA sequence. Nature (Lond) 311: 675-678

Shapiro SZ, Webster P (1989) Coated vesicles from the protozoan parasite Trypanosoma brucei: Purification and characterization. J Protozool 36: 344-349

Shapiro SZ, Pearson TW (1986) African trypanosomiasis: Antigens and host-parasite interactions. In: Parasite antigens, toward new strategies for vaccines (Pearson TW, ed) Marcel Dekker, New York, pp 215-274

Smiley BL, Stadnyk AW, Myler PJ, Stuart K (1990) The trypanosome leucine repeat gene in the variant surface glycoprotein expression site encodes a putative metal-binding domain and a region resembling protein-binding domains of yeast, Drosophila, and mammalian proteins. Mol Cell Biol 10: 6436-6444

Son HJ, Cook GA, Hall T, Donelson JE (1989) Expression site associated genes of Trypanosoma brucei rhodesiense. Mol Biochem Parasitol 33: 59-66

Ter Kuile BH, Opperdoes FR (1991) Glucose uptake by Trypanosoma brucei. Rate-limiting steps in glycolysis and regulation of the glycolytic flux. J Biol Chem 266: 857-862

Thorstenson K, Romslo I (1990) The role of transferrin
     in the mechanism of cellular iron uptake. Biochem
     J 271: 1-10

Turkewitz AP, Amatruda JF, Borhani D, Harrison SC,
     Schwartz AL (1988) A high yield purification of
     the human transferrin receptor and properties of
     its major extracellular fragment. J Biol Chem 263:
     8318-8325

Vickerman K, Luckins AG (1969) Localization of variable
     antigen in the surface coat of Trypanosoma brucei
     using ferritin-conjugated antibody. Nature (Lond.)
     224:1125-1127

Webster P (1989) Endocytosis by African trypanosomes.1.
     3-Dimensional structure of the endocytic
     organelles in Trypanosoma brucei and T.
     congolense. Eur J Cell Biol 49: 295-302

Webster P, Grab DJ (1988) Intracellular colocalization
     of variant surface glycoprotein and transferrin-
     gold in Trypanosoma brucei. J Cell Biol 106: 279-
     288

Zomerdijk JCBM, Quellette M, Ten Asbroek ALMA, Kieft R,
     Bommer AMM, Clayton CE, Borst P (1990) The
     promoter for a variant surface glycoprotein gene-
     expression site in Trypanosoma brucei. EMBO J 9:
     2791-2801

ENVELOPED VIRUSES AS MODELS FOR MEMBRANE ASSEMBLY

Milton J. Schlesinger
Department of Molecular Microbiology, Box 8230
Washington University School of Medicine
St. Louis, MO, 63110, USA

Viruses are small, highly organized nucleoprotein particles that require entry into and exit from a cell in order to propagate. Almost every kind of organism in the biosphere- from the bacteria to man- are infectable by viruses that are often lethal to their host. Virus particles are also highly diverse in structure and composition: their genomes may be either DNA or RNA which can range in size from as little as 2,000 to over 300,000 nucleotides and this nucleic acid can be packaged in particles that include relative simple, highly symmetrical icosahedral structures as well as large, complex pleomorphic shapes. Many virus nucleo-proteins are surrounded by a lipid bilayer which they acquire from their host cell. These viruses are called "enveloped" and they include a number of important human pathogens such as influenza virus and the AIDS virus, HIV. The virus envelope serves to both protect the genome and to facilitate entrance and exit of the virus during its replication in the eukaryotic cell.

Virus infection of a cell initiates with penetration by the particle or its nucleic acid across the cell membrane and concludes with release of newly replicated viruses (as many as 10^4 new particles per cell) by either active secretion from cell membranes or by total disruption of the cell. In this chapter I summarize some of the interactions between animal viruses and host cell membranes during virus replication but the primary emphasis will be on the mechanism of virus

NATO ASI Series, Vol. H 63
Dynamics of Membrane Assembly
Edited by J. A. F. Op den Kamp
© Springer-Verlag Berlin Heidelberg 1992

assembly and release that occurs at the infected cell's surface membrane. I shall also focus on a single kind of virus that has been studied in my laboratory for the past 20 years. This virus is called Sindbis and together with the related Semliki Forest virus are the prototypic viruses for the family called Togaviridae (Schlesinger and Schlesinger, 1985), a group of enveloped viruses that have as their genome a single strand of (+)RNA of 11,700 nucleotides which is infectious when isolated as pure RNA. The entire sequence of the RNA is known and, importantly, it has been cloned as a cDNA in a plasmid construct that permits in vitro transcription of infectious RNA (Rice et al, 1987). The genome is packaged in a protein core, the nucleocapsid, surrounded by a lipid bilayer and spikes composed of virus-specific transmembranal glycoproteins (Fig.1).

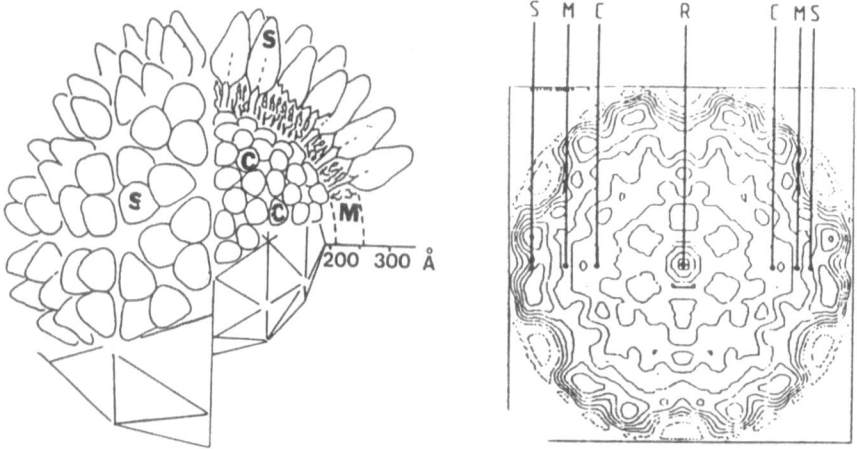

Figure 1. Postulated structures of the Sindbis virus particle. (A) From Harrison, 1983, based on low-angle X-ray scattering. (B) From Fuller, 1987, based on analysis by cryo-electron microscopy of vitrified particles. S, spikes; M, membrane; C, core; R, RNA. Note the crenulated membrane in B. Both the core and intact enveloped particles are arranged in icosahedral symmetry, but they have different lattice structures.

The core is composed of 180 copies of a single polypeptide, the capsid protein, and the externally oriented spikes are heterodimers of two transmembranal glycoproteins, E1 and E2, arrayed as 80 trimers. The composition of the lipid bilayer is asymmetric (Table 1) and reflects that of the host cell's plasma membrane (Hirschberg and Robbins 1974; van Meer et al 1981; Allan and Quinn 1989).

Table 1. Distribution of Phospholipid and Sphingomyelin in Semliki Forest Virus Grown in Baby Hamster Kidney Cells

Membrane Leaflet	PA	PE	PS	PI	PC	SM
Inner	3	22	17	1	9	1
Outer	0	6	0	0	13	26

Numbers are moles/100 moles of total phospholipid PA,phosphatidate;PE phosphatidylethanolamine; PS, phosphatidylserine; PC, phosphatidylcholine; PI, phosphatidylinositol; SM,sphingomyelin.(from Allan and Quinn 1989).

Sindbis virus replication can be divided into discrete events (Strauss and Strauss 1986) that involve several host cell membranes (Fig. 2). The first of these occurs shortly after binding and uptake of virus via the cellular "receptor-mediated endocytosis pathway" which involves clathrin-coated pits and acidic endosomes. In the latter, the virus glycoproteins are conformationally altered and trigger a fusion between virus and endosome membrane that leads to the transfer of the virus core from vesicle to cell cytoplasm. Membrane fusion by virus glycoproteins is a property common to all enveloped viruses but the pH values that lead to fusion differ: some virus membranes fuse at normal physiological pH with plasma membranes whereas other require the more acidic pH found in endosomes (White 1990).

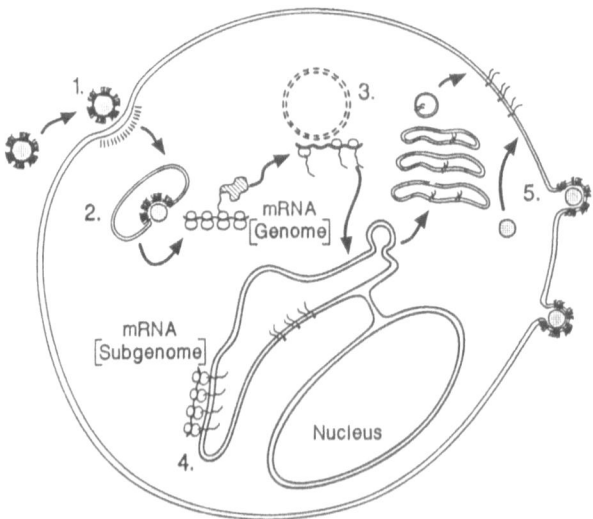

Figure 2. Replication cycle of Sindbis virus with membranal sites noted for the different virus-membrane interactions: 1 and 5, plasma membrane; 2, endosome; 3, endosome (?); 4, rough endoplasmic reticulum.

In the cytoplasm, the core structure collapses and "early events" commence. For Sindbis virus, these include translation of the genomic RNA to form enzymes that replicate and transcribe the viral RNA. There is indirect evidence that these RNA synthetic events occur on membranes associated with endosomes or lysosomes (Froshauer et al 1988). The next stage of virus replication is the formation of a subgenomic mRNA species that encodes all of the virus' structural components. This mRNA is translated as a polyprotein with cotranslational proteolytic processing that produces first the capsid protein and then the two glycoproteins, E1 and E2 plus a small hydrophobic polypeptide, noted 6K. Host cell signalase in the rough endoplasmic reticulum (RER) is the putative processing enzyme for the glycoproteins and 6kD protein. Both glycoproteins have signal sequences (El uses a sequence in the 6K protein) that direct insertion of the protein into the RER membrane and stop-transfer sequences that designate the transmembranal domains of the proteins. During insertion of the polypeptide into the RER lumen, oligosaccharides are added and these are essential for proper folding of the protein (Gibson et al 1979; Schlesinger et al 1985). After completion of nascent polypeptide translation, the two glycoproteins and the 6K protein move through the cell's secretory vesicles to the plasma membrane. Early in transport- during transit from RER to Golgi (Bonatti et al 1989)- all three proteins are acylated with palmitic acid on cysteines that reside in the portions of the polypeptides oriented to the cytoplasmic face of the lipid bilayer(see below). This modification is critical for effective virus particle secretion. Shortly before arrival at the plasma membrane, the E2 glycoprotein is cleaved by a host cell tryptic-like protease present in a late trans-Golgi vesicle.

The final stages of Sindbis virus replication occur at the cell plasma membrane where the nucleocapsid binds to the cytoplasmic domain of virus glycoprotein, E2. In the current model, initial binding allows for a nucleation event which leads to additional protein-protein interactions between virus glycoprotein and nucleocapsid that, in turn, wraps the the lipid bilayer around the core (Fig. 3). In this process host cell membranal proteins are excluded from the virus particle. The final step is a fusion of the lipid bilayer that seals the virus membrane and releases the particle from the cell. Membrane curvature during assembly is severe and may proceed by sharp bends at discrete points.

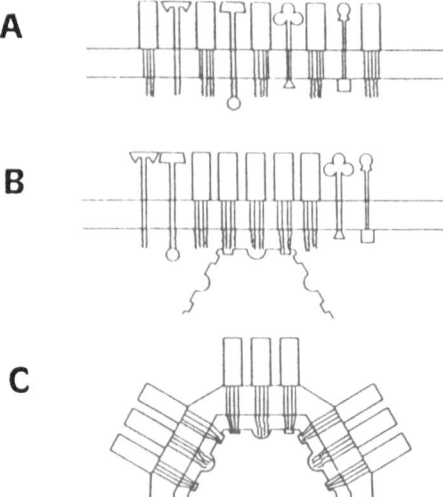

Figure 3. Postulated nucleation and envelopment of virus nucleocapsid by membranes containing virus-specific glycoproteins. (A) Cell plasma membrane with host and virus glycoproteins;(B) Nucleation of assembly (C) Membrane bending. From Fuller 1987.

Much of our current information about enveloped virus assembly and budding has come from electron microscopy, which can provide dramatic visual evidence for this phenomena. Electron micrographs depicting budding of a wide variety of different kinds of RNA viruses are found in a review that also offers detailed interpretations of how virus nucleocapsids interact with lipid bilayers (Dubois-Dalcq et al, 1984). A "classical" view of a virus acquiring a lipid bilayer is shown in electron micrographs of thin sections of infected cells (Fig.4A), but a technique utilizing carbon replicas of virus-infected tissue culture cells that had been rapidly frozen, deep etched and freeze-fractured provides a more detailed picture (Fig.4B and C). Here, one sees the Sindbis virus nucleocapsids bound to the inner surface of the membrane (Panel C) and penetrating the bilayer as the glycoproteins collect in patches on the outer surface (Panel B).

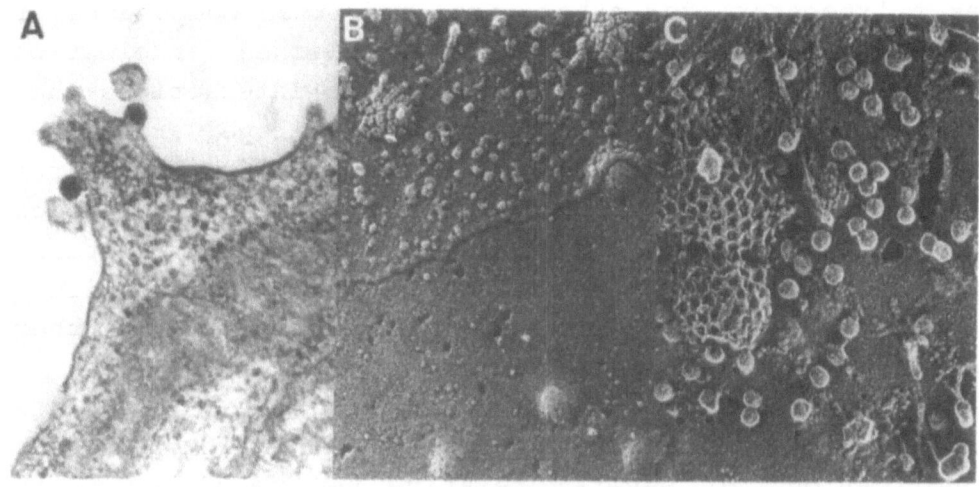

Figure 4. Electron micrographs of budding Sindbis virus. Panels A, stained preparation of a thin section of infected cells (courtesty M. Aach Levy, Washington U.); B and C, replicas of infected cells quick-frozen, deep etched and freeze-fractured (courtesy J.Heuser, Washington U.). Note the hexagonal arrays of the clathrin-coated vesicles in C.

Although togaviruses as well as the myxo-, paramyxo-, rhabdo and retroviruses assemble at the plasma membrane, other viruses use different intracellular membranes. For example, the herpes viruses assemble at the nuclear membrane, the coronaviruses assemble at the RER membrane and the flaviviruses assemble at the Golgi membranes. At least one non-enveloped virus, the rotavirus, utilizes the RER membrane for assembly and secretion from the infected cell.

Despite this diversity of organelles and of virus structures, certain general kinds of activities occur during virus-membrane assembly. In all cases, a membrane-embedded protein encoded by a virus gene acts as a "receptor" for other viral-specified components or "ligands". The initial binding establishes a nucleation site for additional protein-protein interactions that drive the envelopment of the cellular bilayer around the virus core resulting in the extrusion of the virus from the intracellular space (cf Fig.3 and Dubois-Dalcq et al 1984). Normally, the extruded particle retains the lipid envelope, but in the case of the rotavirus, noted above, the membrane is removed during a further maturation process of the particle.

Current research aimed at defining molecular events in togavirus assembly and budding have involved the application of several relative new technologies in biochemistry. One of these was the development of "network antibodies" in which an initial antibody was raised against a peptide corresponding to the E2 glycoprotein's cytoplasmic domain (Fig. 5). Antibodies, called anti-idiotypes, were then raised against this initial antibody in order to find a complementary structure reactive to the initial peptide. Such anti-idiotypes were found to bind to the virus nucleocapsid (Vaux et al 1988), thus confirming the postulated recognition between nucleocapsid and glycoprotein.

A second methodology, noted "reverse genetics", involves site-directed mutagenesis and consists of substituting amino acids at various positions in the cytoplasmic portion of the E2 glycoprotein. This approach is applicable to the Sindbis and Semliki Forest viruses because both viral genomes have been cloned into cDNAs that are transcribable into infectious RNAs.

Thus, by selecting appropriate restriction-enzyme
derived fragments of these clones and replicating them
in the single-stranded M13 bacteriophage system
supplemented with an oligonucleotide that contains a
"mutated" codon, it is possible to prepare all possible
amino acid substitutions at all possible sites desired.
In studies carried out thus far (Gaedigk-Nitschko et al
1990; Gaedigk-Nitschko and Schlesinger 1991), mutations
were made at 5 positions in E2; three were
substitutions at cysteines (two cysteines were altered
in one mutant and one in another), one at a conserved
tyrosine and another at a serine near the carboxyl
terminus of E2 (Fig. 5).

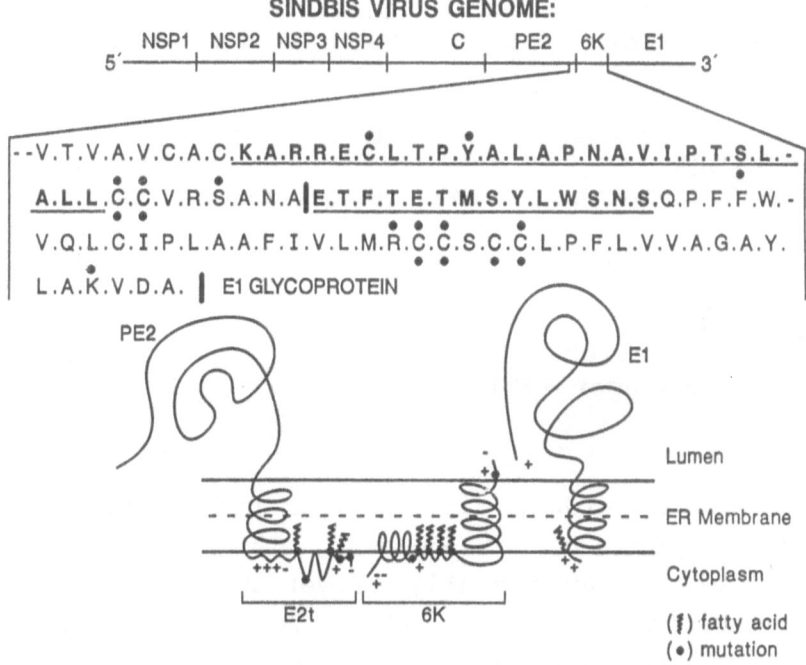

Figure 5. A proposed topology for Sindbis virus
glycoproteins E1 and E2 and the 6K protein with sites
shown where site-directed mutations were constructed.

Three distinct phenotypes were found for these four
mutants and their properties relative to the wild type
Sindbis virus are summarized in Table 2 together with
another set of mutants obtained by site-mutagenesis of
the 6K protein.

Table 2. Summary of Properties of E2 and 6K Mutants

Mutant	Growth		Particle	Others
	CEF	C7-10		
E2				
C395S	Normal	Slow	Multi-cored	less palmitate
Y399F	Normal	Slow	Multi-cored	excess capsid membrane-bound
C415S				
C416A	None	–	None	no progeny virus
S419C	Slow	Slow	Multi-cored	altered E2-6K cleavage
6K				
R34A	Normal	Slow	Multi-cored	
C35S,C36A				
C38S,C39A	Normal	Slow	Multi-cored	almost no palmitate
K52A	Slow	Slow	Not detected	altered 6K-E1 cleavage;

Data in Gaedigk-Nitschko et al 1990;Gaedigk-Nitschko
and Schlesinger 1991.

Despite individual differences, the mutants had
several properties in common. They all secreted
particles at a significantly slower rate than the wild
type from vertebrate cells and all made infectious virus
slower than the wild type in insect cells. Even more
interesting was our observation that the particles
secreted from both types of host cells were much more
stable to thermal inactivation at 56° C, displaying

multi-hit kinetics that indicated the presence of multi-cored virus particles.

Each cysteine-substituted mutant had lower levels of acylated palmitic acid, suggesting this amino acid was a site for fatty acylation. The mutation near the carboxy-terminal of E2 interferes with signal protease processing between E2 and 6K and affects maturation of these proteins as well as their appropriate orientation in the bilayer.

The mutations in the 6K protein show that this small hydrophobic, membranal-bound protein is essential for effective virus assembly. This protein is highly enriched in cysteines (10% of its amino acids) and most are palmitoylated in a manner suggesting a lipid-type structure with a polypeptide head group. The precise function of 6K is unknown; however, small amounts can be detected in purified virions (Gaedigk-Nitschko and Schlesinger 1990).

An intriguing property of these mutants is the differential effect of the mutation on host cell growth, i.e. mutant virus formation is much slower in insect cells than in vertebrate cells. One explanation for this could be the profound difference in membrane composition between these two different species leading to a much lower viscosity of the insect cell membrane. Sindbis virus grown in insect cells reflect this lower viscosity (Table 3). Perhaps nucleation and budding are more difficult to achieve in these cells.

Table 3. Properties of Membranes in Sindbis
virus grown in mosquito or BHK cells.

Property	BHK cell	Mosquito cell
Cholest. Phos.Lipid	0.83	0.61
Amino-Lipid tot.Phos.Lipid	0.32	0.68
Anisotropy(25C)	0.256	0.182
Microviscosity	7.08	2.62

Data from Moore NF et al 1976.

The slow growth on insect cells by the mutants has
been exploited in attempts to isolate second-site
revertants. Two of these have been obtained thus far:
one for the initial tyrosine to phenylalanine
substitution and one for the initial cysteine to serine
substitution in E2. The revertant of the tyrosine-
>phenylalanine mutation makes very tiny plaques and
shuts off host-cell protein synthesis much faster than
the wild type cell. The revertant of the cysteine-
>serine mutation appears to make normal, single-cored
particles. We postulate that some second site
revertants should have mutations in the domains of the
capsid protein that bind to the glycoprotein sequences
and these new mutations should allow for a mapping of
the capsid "receptor" structure. Revertants of the
mutants altered in fatty acylation should help in
defining the role of fatty acids in the assembly and
budding process. It will be necessary to analyze the
genotypes of these revertants to insure that the

original mutation is still intact and to identify the new mutation in the revertant.

A third approach to determine how virus proteins interact with membranes utilizes small peptides designed to specifically inhibit protein-protein interactions. With regard to Sindbis virus, we discovered that peptides whose sequences match regions in the E2 glycoprotein carboxyl terminus could block budding of virus particles (Schlesinger, 1989). Peptides were effective when added to virus infected cells during a one cycle growth curve in chicken embryo fibroblast cells. Under the conditions tested, there were minimal effects on other factors in virus replication such as formation and maturation of intracellular structural proteins and formation of virus nucleocapsids. The smallest and most effective peptide had the sequence, L.T.P.Y.A.L.A, which is identical to the E2 positions 396-402 (cf Fig.5) that is highly conserved among togaviruses. Modification of this peptide by an octanoyl group at the amino terminus and amidation at the carboxy terminus increased the potency by about 10 fold. Consistent with these results are recent data that show a binding between cytoplasmic domains of the p62/E2 glycoproteins of Semliki Forest virus and its nucleocapsid (Metsikko and Garoff 1990).

The experimental approaches and data summarized here for studying assembly and membrane-protein interactions for Sindbis and Semliki Forest viruses will surely be applied to other enveloped viruses. And it is clear from the examples cited above that enveloped viruses offer superb experimental systems for studying protein-lipid interactions and membrane assembly.

REFERENCES

Allan D, Quinn P (1989) Membrane phospholipid asymmetry in Semliki Forest virus grown in BHK cells. Biochim Biophys Acta 987:199-204

Bonatti S, Migliaccio G, Simon K (1989) Palmitoylation of viral membrane glycoproteins takes place after exit from the endoplasmic reticulum. J Biol Chem 264:12590-12595

Dubois-Dalcq M, Holmes KV, Rentier B (1984) Assembly of enveloped RNA viruses. Springer, Berlin Heidelberg New York

Froshauer S, Kartenbeck J, Helenius A (1988) Alphavirus replicase is located on the cytoplasmic surface of endosomes and lysosomes. J Cell Biol 107:2075-2086

Fuller S (1987) The T=4 envelope of Sindbis virus is organized by interactions with a complementary T=3 capsid. Cell 48:923-934

Gaedigk-Nitschko K, Schlesinger MJ (1990) The Sindbis virus 6K protein can be detected in virions and is acylated with fatty acid. Virology 175:274-281

Gaedigk-Nitschko K, Schlesinger MJ (1991) Site-directed mutations in Sindbis virus E2 glycoprotein's cytoplasmic domain and the 6K protein lead to similar defects in virus assembly and budding. Virology 183: in press

Gaedigk-Nitschko K, Ding M, Schlesinger MJ (1990) Site-directed mutations in the Sindbis virus 6K protein reveal sites for fatty acylation and the underacylated protein affects virus release and virion structure. Virology 175:282-291

Garoff H, Simons K (1974) Location of the spike glycoprotein in the Semliki Forest virus membrane. Proc Natl Acad Sci USA 71:3988-3992

Gibson R, Schlesinger S, Kornfeld S (1979) The nonglycosylated glycoprotein of vesicular stomatitis virus is temperature sensitive and undergoes intracellular aggregation at elevated temperatures. J Biol Chem 254:3600-3607

Harrison SC, David A, Jumblatt J, Darnell J (1971) Lipid and protein organization in Sindbis virus. J Mol Biol 60:523-528

Hisrchberg CB, Robbins PW (1974) The glycolipids and phospholipids of Sindbis virus and their relation to the lipids of the host cell plasma membrane. Virology 61:602-608

Metsikko K, Garoff H (1990) Oligomers of the cytoplasmic domain of the p62/E2 membrane protein of Semliki Forest virus bind to the nucleocapsid in vitro. J Virol 64:4678-4683

Moore NF, Barenholz Y, Wagner RR (1976) Microviscosity of togavirus membranes studied by fluorescence depolarization: influence of envelope proteins and the host cell. J Virol 19:126-135

Renkonen O, Kaarainen L, Simons K, Gahmberg CG (1971) The lipid class composition of Semlike Forest virus and of plasma membranes of the host cells. Virology 46:318-326.

Rice CM, Levis R, Strauss JH, Huang HV (1987) Production of infectious RNA transcripts from Sindbis virus cDNA clones: Mapping of lethal mutations, rescue of a temperature-sensitive marker, and in vitro mutagenesis to generate defined mutants. J Virol 61:3809-3819

Schlesinger M (1989). Inhibition of Sindbis virus assembly by peptides that mimic the cytoplasmic domain of the virus glycoprotein. In: Krausslich H-G, Orozolan S, Wimmer E (eds) Viral Proteinases as targets for chemotherapy, Curr Commun in Mol Biol Cold Spring Harbor, New York p 73

Schlesinger S, Koyama AH, Malfer C, Gee SL, Schlesinger MJ (1985) The effects of inhibitors of glycosidase I on the formation of Sindbis virus. Virus Res 2:139-149

Schlesinger S, Schlesinger MJ (eds) (1986) The togaviridae and flaviviridae. Plenum New York London

Strauss EG, and Strauss JH (1986). Structure and replication of the alphavirus genome. In: Schlesinger S, Schlesinger MJ (eds) The Togaviridae and Flaviviridae Plenum, New York p35

van Meer G, Simons K, Op den Kamp JAF, van Deenen LLM (1981) Phospholipid asymmetry in Semliki Forest virus grown on baby hamster kidney (BHK-21) cells. Biochemistry 20:1974-1981

Vaux DJT, Helenius A, Mellman I (1988) Spike-nucleocapsid interaction in Semliki Forest virus reconstructed using network antibodies. Nature 336:36-42

White JM (1990) Viral and cellular membrane fusion proteins. Annu Rev Physiol 52:675-697

THE BIOGENESIS OF VACCINIA VIRUS

Gareth Griffiths and Beate Sodeik
European Molecular Biology Laboratory
Postfach 10.2209
W-6900 Heidelberg
Germany

THE USE OF VIRAL MEMBRANE PROTEINS AS TOOLS TO FOLLOW BIOSYNTHETIC MEMBRANE TRAFFIC

Our current understanding of the processes involved in the biosynthetic route of membrane proteins, from their site of synthesis, the rough ER to the plasma membrane owes a great deal to the use of model viral membrane proteins as tools. The three most popular proteins in this respect have been the G protein of vesicular stomatitis virus (VSV), the haemagglutinin (HA) of influenza virus and the spike proteins of Semliki Forest virus (SFV) (see Simons and Warren, 1984).

There are three reasons why these viral membrane proteins have been so successful for molecular, biochemical as well as morphological studies of membrane traffic in general.

1. Within a few hours of the initial infection all protein synthesis of the host cell is effectively switched off. Radioactive amino acids will thus only be incorporated into a relatively small number of viral proteins (into one or two spanning membrane glycoproteins in the case of the three viruses mentioned above).

2. The infected cell produces extremely high levels of each of the viral proteins. In a baby hamster kidney cell (BHK) infected with SFV we calculated that a cell makes on average $\approx 1.2 \times 10^5$ molecules of each protein per minute (Quinn et al.,

NATO ASI Series, Vol. H 63
Dynamics of Membrane Assembly
Edited by J. A. F. Op den Kamp
© Springer-Verlag Berlin Heidelberg 1992

1984).   This high level of expression obviously facilitates
detection by both biochemical and morphological means.

3.   The viruses are simple structures consisting of 1 RNA molecule and
multiple copies of the nucleocapsid protein(s) and the 1-2 spanning
membrane glycoproteins.   In the case of the membrane proteins, for
example, the virus is totally dependent on the host's machinery for
the synthesis, glycosylation and targeting to their final cellular
destination.

Two other points need to be emphasized before we go to the more
complex situation with vaccinia virus.   First, the budding event is due
to an interaction between a cytoplasmically synthesized nucleocapsid
and the cytoplasmic tail of one of the viral spanning membrane
glycoproteins (Fuller, 1987; Vaux et al., 1990).   Second, some
information contained within the structure of that spanning membrane
protein determines the final destination of that protein within the cell
and consequently, the site where viral budding occurs.   In the case of
VSV, SFV and influenza the proteins have information that allows them
to go out to the plasma membrane, where budding occurs.   Other viruses
may bud into intracellular sites.   Corona virus, for example, buds
predominantly into the intermediate compartment between the ER and
Golgi (Tooze et al. 1984; 1988).   Recent transfection experiments have
shown that the E1 membrane glycoprotein, which is believed to
determine the site of budding of this virus, by itself has the
information which retains it specifically in the intermediate
compartment (Machamer et al., 1990).   Similarly, the Bunya viruses
such as Punta Tora virus bud into the Golgi complex.   Again, a complex
of the two glycoproteins of this virus by itself has the information for
retention in the Golgi complex during intracellular transport (Matsuoka
et al. 1988; Chen et al. 1991a, 1991b).

The second point to consider is that a threshold of concentration of the
viral membrane protein appears to be required for the budding event.   In
the case of SFV, which buds at the plasma membrane, the requirement
for a concentration threshold prevents the capsids from interacting
with the cytoplasmic tail of the spike proteins in the ER and in the
Golgi complex.   When, however, transport of these spikes through the

Golgi is blocked with monensin these proteins accumulated in one (medial) compartment of this organelle to such an extent that aberrant budding into this compartment can now be observed (Griffiths et al., 1984; Quinn et al., 1984).

THE BIOGENESIS OF VACCINIA VIRUS: ROLE OF THE INTERMEDIATE COMPARTMENT BETWEEN THE ROUGH ENDOPLASMIC RETICULUM AND THE GOLGI COMPLEX

Vaccinia is among the most complex of animal viruses with a genome size of 191 x 10^6 bp that codes for about 263 potential genes (Goebel et al., 1990). The virus is surrounded by up to four different membranes which are sequentially acquired during the process of morphogenesis.

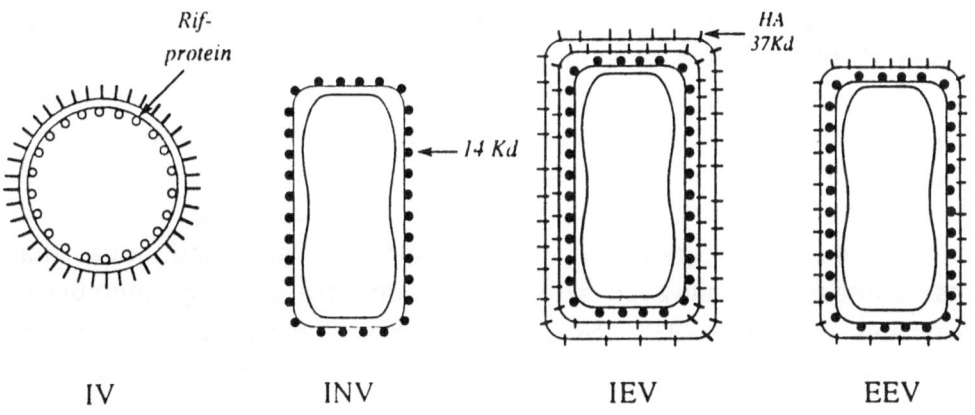

Figure 1:

Schematic model of the four morphologically different vaccinia forms.

(1) **IV** = spherical Immature Virus with 2 membranes. The cytoplasmic site of the inner membrane can be labelled with anti-62kDa. Note that we are unable to label the INV and later forms with this antibody. We assume this is due to technical reasons, such as steric hindrance.

368

(2) **INV** = brick-shaped **I**ntracellular **N**aked **V**irus with 2 membranes. The outer membrane can be labelled with anti-14kDa.
(3) **IEV** = **I**ntracellular **E**nveloped **V**irus with 4 membranes. The two outermost membranes can be labelled with both, anti-37kDa and anti-HA.
(4) **EEV** = **E**xtracellular **E**nveloped **V**irus with 3 membranes. The outermost membrane can be labelled with both anti-37kDa and anti-HA.

Closely related to the causative agent of smallpox, vaccinia is of great interest for a number of reasons. First, since it is a DNA virus that replicates in the cytoplasm, it has become a powerful model to study DNA replication and transcription (Moss, 1990a, b, 1991). Second, because of the ease with which foreign genes can be inserted into the viral genome and, via infection, expressed in several different tissue culture cells, it has become a popular expression system. Despite the enormous amount of literature on vaccinia virus, very little is known about the cytoplasmic events leading to its assembly and release by infected cells. In fact, a surprising dogma has developed with respect to the acquisition of the first of the viral membranes, namely that it originates from new membrane synthesis by virally-encoded enzymes (Dales and Pogo, 1981). This model is problematic for two reasons. First, if true, it would be the first example of de novo synthesis of membrane rather than the more commonly accepted idea that membranes are always synthesized on pre-existing membrane templates (Palade, 1983). Second, upon critical examination the actual published evidence in support of de novo synthesis rests solely on weak electron microscopic data from the early 1970's.

By analogy to the budding of other membrane viruses such as Semliki Forest virus (SFV) and vesicular stomatitis virus (VSV), we started the vaccinia virus project with the working hypothesis that the virus acquires its membranes from the bilayers of a pre-existing host cell compartment(s). The hypothesis further postulates that the cell compartment(s) would acquire a key, newly synthesized viral membrane protein(s). Thus, as for other animal membrane viruses, these hypothetical proteins would determine the site of viral budding. We have now followed vaccinia virus infection in a variety of tissue culture cells, primarily BHK, Hela and RK_{13}. Subsequently, a whole range of established markers for cellular compartments of the exocytic and endocytic pathway were matched by immunocytochemistry at the light- and electron microscopic levels with a spectrum of antibodies against characterized viral proteins that were kindly provided by many groups. The antibodies included one against a 62 kDa cytoplasmic protein (from Drs. B. Moss and R. Doms, NIH, Bethesda) that is presumably involved in the interaction of the capsid with the first budding membrane and appears to interact with the antibiotic rifampicin, which selectively and reversibly blocks the first "budding" event (Moss, 1969). This protein was localized to the cytoplasmic site of the first viral membrane (Fig. 2C). Another antibody, against a 14 kDa peripheral membrane protein (from Dr. M. Esteban, SUNY, New York), was localized to the second viral membrane (Fig. 2D), while an antibody against the 85 kDa vaccinia virus haemagglutinin (from Dr. H. Shida, Kyoto University) localized to the third and fourth membranes (the so-called envelope) (see Fig. 1 for a summary of these localizations).

Overall, our data suggest that the first two membranes of vaccinia originate from the intermediate compartment that is isolated between the rough endoplasmic reticulum and the Golgi complex (also referred to as "salvage" compartment) (see Pelham, 1989). The main evidence for this model rests on co-localization of the viral DNA and 62 kDa protein with three different markers of the intermediate compartment. First, the 53 kDa protein of Schweizer et al., 1988 (in collaboration with Dr. Hans-Peter Hauri, Biocenter, Basel). Second, the corona virus E1 expressed using the vaccinia virus (in collaboration with Dr. Carolyn Machamer, John Hopkins University, Baltimore) and third, the recently characterized small GTP-binding protein rab2 that had been previously localized to the intermediate compartment (antibodies provided by Dr. M. Zerial, EMBL, Heidelberg; Chavrier et al, 1990) (see Figs. 2A and 2B).

Our current model for the biogenesis of the vaccinia is that the nucleocapsid becomes engulfed by a cisternal (two membraned) structure that originates from, and is continuous with, the intermediate compartment. Following this envelopment or "budding" the 14 kDa protein binds peripherally to some, as yet unidentified, vaccinia

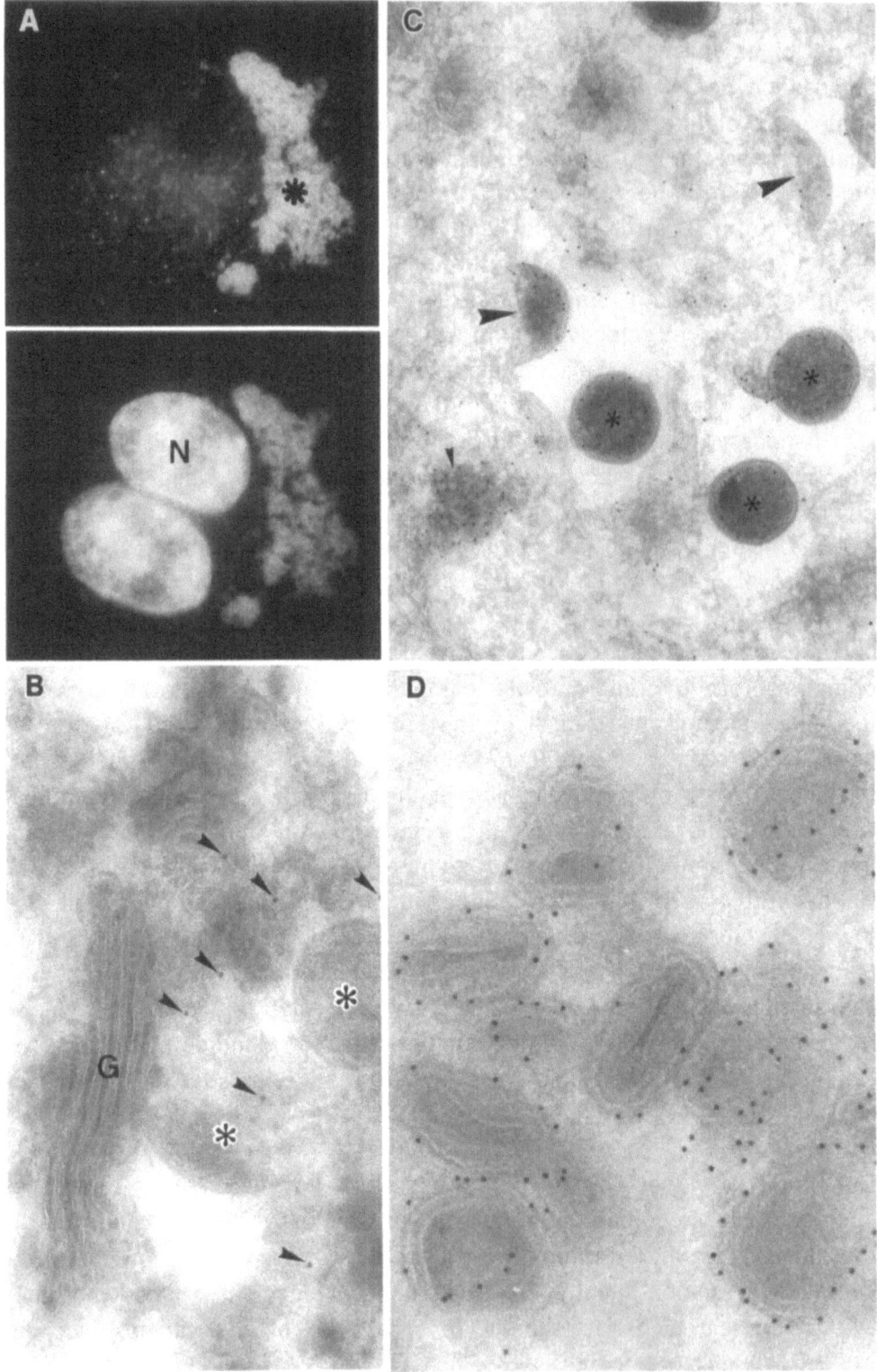

Figure 2:

(A) Double fluorescence labelling of BHK cells 9 h post infection with vaccinia virus. The upper panel shows the localization of rab2 protein and the lower panel shows the corresponding DNA localization using the Hoechst dye which labells both the nuclei (N) and the viral "factory" (asterisk). After vaccinia infection rab2 localizes almost exclusively with the viral factories.

(B) Cryosection of a BHK cell infected with vaccinia virus for 8h in the presence of rifampicin. Subsequently, the drug was chased out for 20 mins in order to enrich for immature virions. The section was labelled with rabbit anit-peptide antibodies against rab2 followed by protein A gold. The label (small arrowheads) is associated with membranes which are close to viral budding structures (asterisks) as well as close to the Golgi apparatus (G).

(C) Cryosection of a Hela cell 8h post infection with vaccinia virus. The section was immunolabelled with antibodies against the 62kDa rifampicin-sensitive vaccinia protein. The labelling is found in the cytosolic regions corresponding to the viral factories (small arrowhead) and at the cytoplasmic site of the first viral budding membranes (large arrowhead). The peripheral regions of fully budded, so-called immature virions are also labelled (asterisks).

(D) Cryosection of a Hela cell grown in suspension, 19h post infection with vaccinia. The section was labelled with rabbit anti-peptide antibodies against the 14kDa vaccinia protein. The label is associated only with the outer membrane of the intracellular mature virions. Immature virions were never labelled.

membrane protein. Using a recombinant vaccinia virus which expressesthe 14 kDa under an inducible promotor it has recently been shown that the 14 kDa is not necessary for the formation of infectious intracellular virions but for the acquisition of the third and fourth enveloping membranes (Rodriguez and Smith, 1990). The precise intracellular origin of these membranes is not yet clear.

We have now used a large number of markers for the intermediate compartment. In addition to the 53 kDa, the corona E1 and rab2 similar results have been obtained for example with antibodies against mouse ypt-1 another GTP-binding protein (in collaboration with

Dr. D. Gallwitz, Göttingen). Our current working model is summarized in Fig. 3.

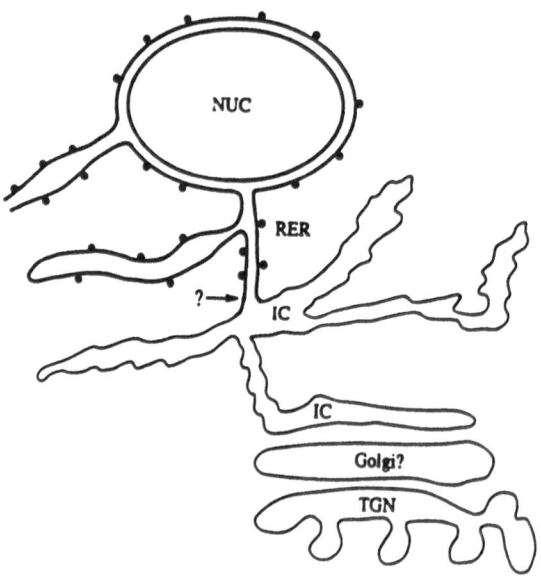

Figure 3:

Schematic model of the exocytic pathway. This model proposes that the intermediate compartment (IC) is continuous with the rough ER (proximally) and with a cisternal structure on the cis side of the Golgi complex (distally). An important point of dispute is whether the RER and IC are really continuous (this model) or discontinuous (?). In the latter case an additional vesicular transport step would be required. The idea we favour is that the RER/IC have different functional domains that are in direct continuity. The Golgi complex is shown as a single compartment; the ? indicates that the precise number of compartments is still open. The available evidence in the literature now argues that the TGN is structurally and functionally distinct from the more proximal Golgi compartments.

Collectively, these data argue that the intermediate compartment is in direct continuity between the rough ER and the first Golgi cisterna. This is essentially a structural model put forward by others, in particular Lindsey and Ellisman (1985a, b) and differs from the more commonly accepted idea of a "salvage" compartment that is discontinuous from both the rough ER and the Golgi complex (Pelham, 1989; Klausner, 1989). We emphasize, however, that the organization of the intermediate compartment is extremely complex and further data are required to define its precise boundaries and to unequivocally distinguish between these two models.

REFERENCES:

Chavrier P, Parton RG, Hauri HP, Simons K and Zerial M (1990) Localization of low molecular weight GTP-binding proteins to exocytic and endocytic compartments. Cell 62: 317-329

Chen SY, Matsuoka Y, Compans RW (1991a) Assembly and polarized release of Punta Toro virus and effects of brefeldin A. J Vir 65:1427-1439

Chen SY, Matsuoka Y, Compans RW (1991b) Golgi complex localization of the Punta Toro virus G2 protein requires its association with the G1 protein. Vir 183: 351-365

Dales S and Pogo BGT (1981) Biology of poxviruses. In: Kingsbury DW and Hausen HZ (eds) Virology Monographs. Springer-Verlag, Vienna New York

Fuller SD (1987) The T-4 envelope of Sindbis virus is organized by interaction with a complementary T-3 capsid. Cell 48: 923-934

Goebel SJ, Johnson GP, Perkus ME, Davis SW, Winslow JP and Paoletti E (1990) The complete DNA sequence of vaccinia virus. Vir 179: 247-266

Griffiths G, Quinn P, Matthieu-Costello O and Hoppeler H (1984) Density of newly synthesized plasma membrane proteins in intracellular membranes. I. Stereological studies. J Cell Biol 96: 835-850

Klausner RD (1989) Sorting and traffic in the central vacuolar system. Cell 57: 703-706

Lindsey JD, Ellisman MH (1985a) The neuronal endomembrane system I. Direct links between rough endoplasmic reticulum and the cis element of the Golgi apparatus. J Neurosci 5: 3111-3123

Lindsey JD, Ellisman MH (1985b) The neuronal endomembrane system II. the multiple forms of the Golgi apparatus cis element. J Neurosci 5: 3124-3134

Machamer CE, Mentone SA, Rose JK and Farquhar MG (1990) The E1 glycoprotein of an avian coronavirus is targeted to the cis Golgi complex. Proc Natl Acad Sci USA 87: 6944-6948

Matsuoka Y, Ihara T, Bishop DHL, Compans RW (1988) Intracellular accumulation of Punta Toro virus glycoproteins expressed from cloned cDNA. Vir 167, 251-260

Moss B (1990a) Poxviridae and their replication. In: Fields BN and Knipe DM (eds) Virology. Raven Press Ltd, New York

Moss B (1990b) Replication of poxviruses. In: Fields BN and Knipe DM (eds) Virology. Raven Press Ltd, New York, p 2079-2111

Moss B (1991) Vaccinia virus: A tool for research and vaccine development. Science 252: 1662-1667

Palade GE (1983) Membrane biogenesis: An overview. Meth in Enzymology 96: XXIX-LV

Pelham HRB (1989) Control of protein exit from the endoplasmic reticulum. Ann Rev Cell Biol 5: 1-23

Quinn P, Griffiths G and Warren G (1984) Density of newly synthesized plasma membrane proteins in intracellular membranes. J Cell Biol 98: 2142-2147

Schweizer A, Fransen JAM, Baechi T, Ginsel L and Hauri HP (1988) Identification, by a monoclonal antibody, of a 53-kD protein associated with a tubulo-vesicular compartment at the cis-side of the Golgi apparatus. J Cell Biol. 107: 1643-1653

Simons K and Warren G (1984) Semliki Forest virus: A probe for membrane traffic in the animal cell. Advances in Protein Chemistry 36: 79-125

Tooze J, Tooze S and Warren G (1984) Replication of coronavirus MHV-A59 in sac(-) cells: determination of the first budding of progeny virions. Eur J Cell Biol 33: 281-293

Tooze SA, Tooze J and Warren G (1988) Site of addition of N-acetyl-galactosamine to the E1 glycoprotein of mouse hepatitis virus A59. J Cell Biol 106: 1475-1487

Vaux D, Tooze J and Fuller S (1990) Identification by anti-idiotype antibodies of an intracellular membrane protein that recognizes a mammalian endoplasmic retention signal. Nat 345: 495-502

A VIRAL PROTON CHANNEL

Andreas Schlegel & Christoph Kempf
Institute of Biochemistry
University of Berne
Freiestrasse 3
CH-3012 Bern
Switzerland

The fusion of biological membranes is an important natural process generating, maintaining and destroying life. It can be found in both pathogenic and nonpathogenic processes. Viruses, as pathogens, have to overcome a major barrier, namely the host cell membrane, in order to enter the target organism. Different ways have evolved to perform the entry. Enveloped viruses for example which acquire a lipid bilayer around their genome during virus maturation deliver their nucleic acid by the fusion of their envelope with a cellular membrane. Some of them, for example the causative agent of AIDS (HIV), bind to cellular receptors on the plasma membrane and enter the cell by fusion at neutral pH. Other enveloped viruses such as Influenza virus, Semliki Forest virus (SFV) or Vesicular Stomatitis virus (VSV) are taken up by the host cells via endocytosis and the fusion of cellular and viral membrane is triggered by the acidic pH ($<$ 6.2 for SFV) within the endosome.

SFV, a member of the family Togaviridae, is composed of the nucleocapisd, made out of the capsid protein (C-protein) and a single stranded (+) RNA, which is surrounded by an envelope. The envelope contains lipids derived from the host cell membrane and the three viral proteins E_1, E_2 and E_3, E_1 and E_2 representing integral membrane proteins. They form heterotrimeric aggregates, $E_1E_2E_3$, which are clustered as trimers $(E_1E_2E_3)_3$, called spikes.

Fusion of viral and endosomal membrane leads to the release of the nucleocapsid and subsequently to infection of the cell. Replication, transcription and translation of the viral RNA result in the production of viral structural proteins. The

envelope proteins are transported to the plasma membrane and new nucleocapsids are assembled within the cytoplasm of the host cell. In the last step of virus maturation nucleocapsids bind to the cytoplasmic tails of the E_2 envelope proteins and are subsequently surrounded by the host cell plasma membrane containing the viral spike proteins. This process results in the budding of progeny virions. At this stage of infection the fusogenic viral proteins are expressed on the host cell plasma membrane. If at this stage of infection, the cells are exposed to a mildly acidic, extracellular pH, analogous to the pH in the endosome, cell cell fusion can be induced (for reviews see Garoff et al 1982; White et al 1983). This virus-induced fusion is also referred to as fusion from within (FFWI). SFV-induced FFWI is a multistep process and some of the associated dynamic events occuring at the plasma membrane such as i) a low pH induced conformational change of a plasma membrane protein, most probably the viral envelope protein E_1 (Koblet et al 1985; Omar and Koblet 1989), ii) the influx of protons into the cells that undergo cell-cell fusion (Kempf et al 1987 and 1988a), iii) a hyperpolarization of the plasma membrane potential (Kempf et al 1988b) and iv) Na^+/K^+-fluxes across the plasma membrane (Kempf et al 1988b) have been described.
There are several lines of evidence suggesting that the conformational change of a viral envelope protein represents a crucial step which initiates membrane fusion.
1. Viruses which enter cells via endocytosis contain fusogenic proteins that have the general property to undergo low pH induced conformational changes. Low pH treatment of SFV irreversibly leads to a complete trypsin resistance of the E_1 protein (Kielian and Helenius 1985) and to an increase of the hydrophobic properties of the spike proteins (Omar and Koblet 1988). Hemagglutinin (HA) which is the fusogenic protein of Influenza virus irreversibly undergoes a low pH induced conformational change leading to the exposure of the fusion peptide that is supposed to interact with the target membrane (for reviews see Hoekstra and Kok 1989; Hoekstra 1990; White 1990). The G protein of VSV undergoes a low pH induced

conformational change as well which, in contrast to the SFV E_1 and Influenza HA, is a reversible process.

2. The conformational change of a plasma membrane protein, probably the E_1 viral protein, precedes the fusion of SFV infected cells (Omar and Koblet 1989). The conformational change is a very fast process which is complete for SFV within 10 seconds after acidification (Kielian and Helenius 1985) or after 15 seconds for Influenza HA (Hoekstra 1990), whereas cell-cell fusion takes minutes to be completed.

3. SFV-induced cell-cell FFWI was considered to exclusively occur at mildly acidic pH. However, we have recently shown that such cell fusion can also be triggred by a transient acidification of the cytoplasm of infected cells at an extracellular, neutral pH (Kempf et al 1990). Results were obtained by utilising NH_4Cl pulses combined with covalent modification of cell surface proteins. Briefly, exposure of SFV-infected cells to NH_4Cl at neutral pH and subsequent removal of the NH_4Cl by repetitive washings leads to a transient acidification of the cytoplasm and polykaryon formation. This result allowed two interpretations: i) the intracellular acidfication of SFV-infected cells triggered directly the fusion process without affecting the extracellular matrix or ii) the re-establishment of the intracellular pH by proton extrusion led to a temporary acidification of the extracellular space in close proximity to the cell surface. The latter possibility could be excluded by performing the experiments at either various buffer capacities or in presence of cations which can compete with protons for binding to proteins and therefore prevent conformational changes or by applying protein modifying chemicals such as dithiothreitol or butanedione which block SFV-induced fusion only after a conformational change had occurred. None of these treatments was capable of hampering SFV-induced FFWI triggered by an acid load. Thus, these results implied that the proton influx observed upon acidification of the extracellular space is required for the fusion process. Our findings allowed a reinterpretation of the general consensus on how SFV-induced

FFWI takes place. So far it was assumed that the low extracellular pH led to a conformational change of the viral spike protein E_1 and exposure of a hydrophobic sequence which interacts with the opposing membrane. This in turn would result in lipid mixing culminating in membrane fusion. In view of our results, however, we postulated that acidification of the extracellular environment leads to a conformational change of the spike in such a way that part of the protein folds back into the anchoring membrane to form a proton channel, resulting in proton influx. A transient drop of the intracellular pH might activate additional factors which are required for SFV-induced FFWI. An acid load leading to fusion would short cut this event. The postulated interaction with the opposing membrane would occur via a different part ("fusion peptide") of the protein.

4. Recently we have shown that the low pH induced conformational change of SFV spike proteins leads to the opening of a proton channel (Schlegel et al manuscript submitted). With this observation we were able to link the conformational change of viral proteins to the proton influx reported for FFWI. We utilised isolated virions to show that SFV envelope proteins can mediate proton translocation, making use of the property of isolated nucleocapsids to shrink upon low pH exposure. This contraction is irreversible and results in a decrease of the particle diameter and a higher sedimentation rate (Soederlund et al 1972). The nucleocapsids were therefore used as pH sensitive markers in order to determine a proton influx into the interior of SFV particles. As a control we used shaved spikeless SFV particles, which were obtained by proteolytic digestion and were shown to be completely devoid of protruding spike proteins. A schematic representation of the experiments performed with intact and spikeless SFV particles is given in Fig. 1. Intact and spikeless SFV was exposed to low pH (5.8), neutralized and uncoated with detergent in order to release the nucleocapsids. The contraction of the nucleocapsids was monitored by sedimentation analysis on linear sucrose gradients and by

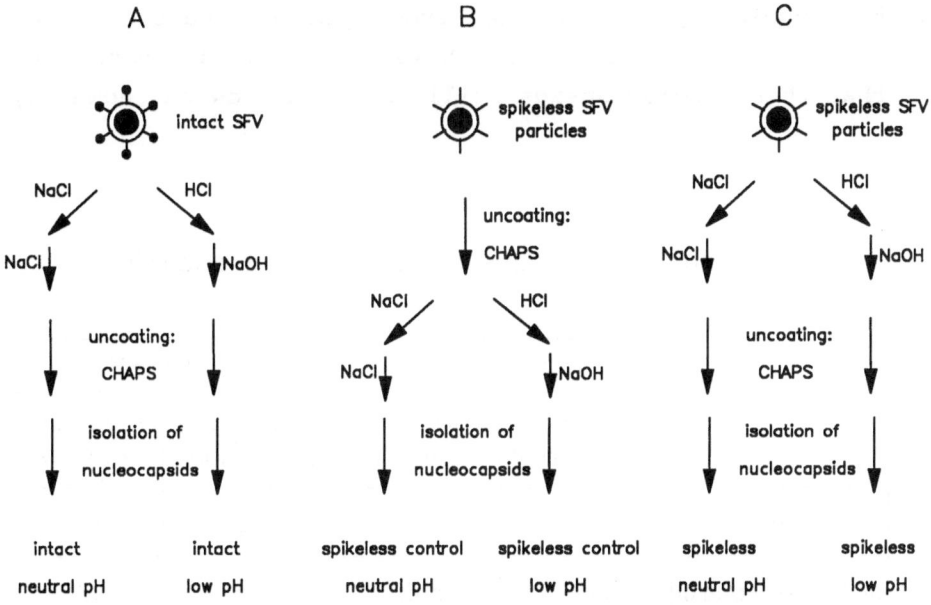

Figure 1. Schematic representation of the experiments: SFV and spikeless SFV were either exposed to pH 7.4 or 5.8, respectively. After restoring the pH of the acid samples to 7.4 all probes were uncoated with detergent and the nucleocapsids analysed by sucrose density centrifugation and electron microscopy (A, C). Spikeless particles were uncoated with detergent and subsequently exposed to pH 5.8 or 7.4 before analysis, respectively (B).

electron microscopy. The data obtained are summarized in table 1. The nucleocapsids derived from low pH treated intact SFV had a significantly higher S value and a much smaller particle diameter. These results clearly showed that exposure of intact SFV to low pH causes a contraction of the nucleocapsids similar to that found with isolated nucleocapsids (Soederlund et al 1972). This can only be explained by a proton influx into the virion. This proton influx could be either mediated by the viral spike proteins or due to a leakiness of the lipid envelope. To eliminate the latter possibility we have used spikeless SFV particles obtained by proteolytic digestion with bromelain. Control experiments (Fig. 1) performed to test whether nucleocapsids isolated from spikeless particles retain the contractability upon low pH exposure, demonstrated (table 1) that these nucleocapsids still shrink at low pH. However,

TABLE 1

Summarized results of nucleocapsid analysis

	neutral pH		low pH	
	S−value (sedimentation)	particle diameter in nm (EM)	S−value (sedimentation)	particle diameter in nm (EM)
intact	153	52	162	40
spikeless control	142	n.d.	146	n.d.
spikeless	142.4	44	142.7	45

The preparations are named according to figure 1

these experiments also revealed that the S values of nucleocapsids obtained from spikeless SFV particles are lower than the corresponding ones from intact SFV. This shift may be caused by the digestion procedure which seems to affect the general sedimentation behaviour of the cores. Nevertheless, the data clearly showed that these nucleocapsids could still be used as sensitive indicators to detect a proton translocation into the interior of spikeless particles. Thus we exposed spikeless SFV particles to low pH and compared the nucleocapsid obtained after uncoating with detergent with nucleocapids of untreated spikeless particles. These experiment revealed that the S values and the diameters of both populations of nucleocapsids were identical. From this we concluded that it was not a general leakiness of the envelope but rather the ectodomain of the spike proteins which was responsible for the proton translocation across the membrane of intact SFV.

Summarizing all these results one can say that both the fusion of SFV virions with the endosomal membrane and SFV-induced FFWI require a conformational change of a viral spike protein which mediates proton translocation across the viral or cellular membrane. Taking into account the observation that SFV particles which contain only the E_1 envelope protein and the transmembrane anchor of E_2 are capable of inducing membrane fusion (Omar and Koblet 1988) it is tempting to speculate that it has to be the E_1 protein.

At this point the question arises whether the backfolding hypothesis (Kempf et al 1990) still can be used as a valid model or whether regulatory functions of the E_1 protein have to be postulated which lead to the opening of a channel formed by another protein. As discussed elsewhere (Schlegel et al manuscript submitted) there are several theoretical considerations supporting the hypothesis that the E_1 protein itself might act as proton channel: i) it contains a hydrophobic stretch of amino acids close to the C-terminal, membrane anchoring domain and computer predictions denote this peptide segment as a potentially membrane associated helix, ii) helical wheel analysis of this peptide revealed that it may form an amphiphilic α-helix exposing five out of six Ser residues in close vicinity to one another iii) experiments with synthetic, α-helical membrane spanning peptides that contain Ser residues in a similar way have shown that such peptides act as ion channels (DeGrado and Lear 1990; Lear et al 1988). Together with the finding that the spike proteins are arranged as trimers (Vogel et al 1986) these considerations lend further support to the backfolding model in such a way that the proton channel may be formed by these amphiphilic, Ser containing α-helices (Fig 2). The hypothesis that the E_1 protein may regulate the opening of a proton channel formed by another protein deserves also some attention. It was recently postulated that the M2-protein of influenza virus - a minor viral membrane protein - may function as a channel (Sugrue and Hay 1991). On the other hand, Sindbis as well as SFV contain a minor envelope protein, too, namely the 6K protein (Gaedigk-Nitschko and Schlesinger 1990). Side directed mutations in the Sindbis virus 6K protein have shown that this protein is essential for efficient virus assembly and budding (Gaedigk-Nitschko et al 1990; Gaedigk-Nitschko and Schlesinger 1991). It remains to be elucidated

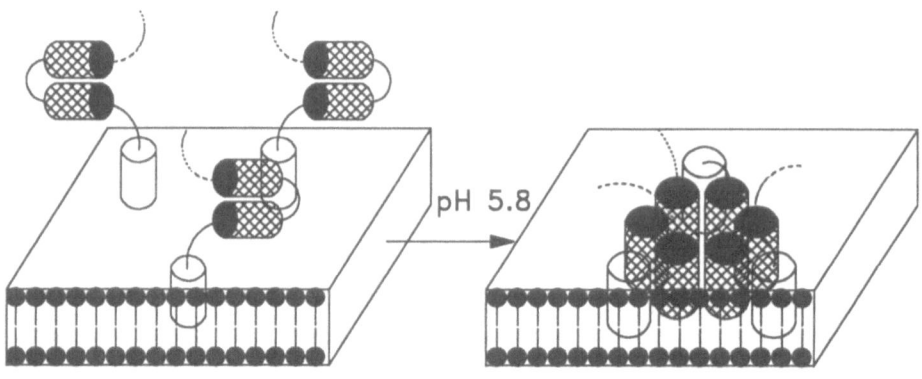

Figure 2. Model of low pH induced backfolding of SFV E_1 proteins leading to the formation of a proton channel.

whether the 6K protein, in analogy to the ascribed, putative function of the influenza M2-protein, has a similar function, namely to form a proton channel which would be regulated by the ectodomain of the E_1 protein.

In order to test whether the E_1 protein acts itself as proton channel or whether it functions as a regulatory protein we focused on the backfolding model as a working hypothesis.

The following criteria which can be experimentally tested should be fulfilled: i) if the E_1 protein folds back into the anchoring membrane additional membrane spanning domains must result, ii) complete proteolytical shaving should lead to the generation of additional particle bound protein fragments, iii) it should be possible to isolate these fragments together with the shaved particles, iv) if the low pH induced backfolding is irreversible then low pH exposure, neutralization and digestion at pH 7.2 should produce the same fragments.

We therefore exposed SFV virions to pH 5.6 and 6.0, neutralized the former sample and kept the latter at the indicated acidic pH in order to digest them subsequently at pH 7.2 (preparation 5.6/7.2) and pH 6.0 (preparation 6.0/6.0), respectively. The control preparation was kept at pH 7.2 during the whole experiment and digested at pH 7.2 (preparation 7.2/7.2). Shaved particles were purified and concentrated by centrifugation and prepared directly for gel electrophoresis. The results are depicted in Fig. 3. Both

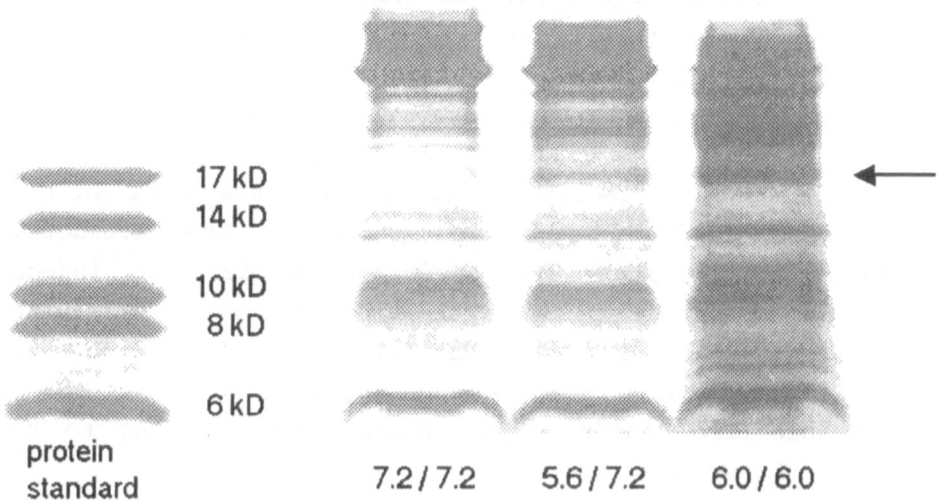

Figure 3. Video picture of a SDS gelelectrophoresis of spikeless particles isolated after digestion of low pH treated SFV virions. The virions were exposed to pH 7.2, 5.6 and 6.0 and subsequently digested at pH 7.2, 7.2 and 6.0 (preparation 7.2/7.2, 5.6/7.2 and 6.0/6.0), respectively.

samples which have been exposed to low pH, preparation 5.6/7.2 and 6.0/6.0, contained an additional peptide with an apparent molecular weight of 17 kD which was not present in control preparations (7.2/7.2). This 17 kD peptide must be the result of an irreversible low pH induced conformational change because it can be found in both acidic preparations (5.6/7.2 and 6.0/6.0). This result substantiates the backfolding hypothesis but still requires additional prove.

Currently neither of the two hypothesis - the backfolding model or the hypothesis that the E1 protein may regulate the opening of a proton channel formed by the 6K protein - can be excluded. Nevertheless, it can be concluded that a low pH induced conformational change of a viral spike protein leads to the opening of a proton channel. This process may not only be of biological relevance for the SFV-induced FFWI (Kempf et al 1990) but may also play a crucial role in the virus-endosome fusion and the subsequent uncoating of viral RNA. It was recently proposed for rubella virus, another member of the family Togaviridae, that acidification of the capsid might be crucial for the uncoating process (Mauracher et al 1991). In their study they showed that the rubella virus nucleocapsid undergoes a low pH induced conformational change, which leads to an alteration from having hydrophilic to having hydrophobic properties. This solubility shift may be required for efficient uncoating. Together with the data presented in this paper these findings lend support to the attractive hypothesis that within the endosome a low pH induced conformational change of a viral spike protein takes place resulting in i) a proton influx into the virion, ii) the fusion of viral and endosomal membrane and iii) the low pH induced solubility shift of the nucleocapsid leading to the uncoating of the viral RNA.

REFERENCES

DeGrado WF, Lear JD (1990) Conformationally constrained α-helical peptide models for protein ion channels. Biopolymers 29: 205-213

Gaedigk-Nitschko K, Schlesinger MJ (1990) The Sindbis virus 6 K protein can be detected in virions and is acylated with fatty acids. Virology 175: 274-281

Gaedigk-Nitschko K, Ding M, Aach Levy M, Schlesinger MJ (1990) Site-directed mutations in the Sindbis virus 6 K protein reveal sites for fatty acylation and the underacylated protein affects virus release and virion structure. Virology 175: 282-291

Gaedigk-Nitschko K, Schlesinger MJ (1991) Side-directed mutations in Sindbis virus E2 glycoprotein's cytoplasmic domain and the 6 K protein lead to similar defects in virus assembly and budding. Virology 183: in press

Garoff H, Kondor-Koch C, Riedel H (1982) Structure and assembly of alpha viruses. Curr Top Microbiol Immunol 99: 1-50

Hoekstra D, Kok JW (1989) Entry mechanisms of enveloped viruses. Implications for fusion of intracellular membranes. Biosci Rep 9: 273-305

Hoekstra D (1990) Membrane fusion of enveloped viruses: especially a matter of proteins. J Bioenerg Biomembrane 22: 121-155

Kempf C, Michel MR, Kohler U, Koblet H (1987) Can viral envelope proteins act as or induce proton channels? Biosci Rep 7: 761-769

Kempf C, Michel MR, Kohler U, Koblet H (1988a) Exposure of Semliki Forest virus-infected baby hamster kidney cells to low pH leads to a proton influx and a rapid depletion of intracellular ATP which in turn prevents cell-cell fusion. Arch Virol 99: 111-115

Kempf C, Michel MR, Kohler U, Koblet H, Oetliker H (1988b) Dynamic changes in plasma membrane properties of Semliki Forest virus infected cells related to cell fusion. Biosci Rep 8: 241-254

Kempf C, Michel MR, Omar A, Jentsch P, Morell A (1990) Semliki Forest virus induced cell-cell fusion at neutral extracellular pH. Biosci Rep 10: 363-374

Kielian M, Helenius A (1985) pH induced alterations in the fusogenic spike protein of Semliki Forest virus. J Cell Biol 101: 2284-2291

Koblet H, Kempf C, Kohler U, Omar A (1985) Conformational changes at pH 6 on the cell surface of Semliki Forest virus-infected Aedes albopictus cells. Virology 143: 334-336

Koblet H, Omar A, Kohler U, Kempf C (1988) Investigations of cell-cell fusion in Semliki Forest virus (SFV) infected C6/36 (mosquito) cells. In "Invertebrate and fish tissue culture" (Kuroda Y, Kurstak E, Maramorosch K, eds). Japan Scientific Societies Press Tokyo/Springer Berlin pp 140-143

Lear JD, Wasserman ZR, DeGrado WF (1988) Synthetic amphiphilic peptide models for protein ion channels. Science 240: 1177-1181

Mauracher CA, Gillam S, Shukin R, Tingle AJ (1991) pH-dependent solubility shift of rubella virus capsid protein. Virology 181: 773-777

Omar A, Koblet H (1988) Semliki Forest virus particles containing only the E1 envelope glycoprotein are infectious and can induce cell-cell fusion. Virology 166: 17-23

Omar A, Koblet H (1989) The use of sulfite to study the mechanism of membrane fusion induced by E1 of Semliki Forest virus. Virology 168: 177-179

Soederlund H, Kaeaeriaeinen L, von Bonsdorff CH, Weckstroem P (1972) Properties of Semliki Forest virus nucleocapsid. Virology 47: 753-760

Sugrue RJ, Hay AJ (1991) Structural characteristics of the M2 protein of influenza A viruses: Evidence that it forms a tetrameric channel. Virology 180: 617-624

Vogel HH, Provencher SW, von Bonsdorff CH, Adrian M, Dubochet J (1986) Envelope structure of Semliki Forest virus reconstructed from cryoelectron micrographs. Nature 320: 533-535

White J, Kielian M, Helenius A (1983) Membrane fusion proteins of enveloped animal viruses. Quart Rev Biophys 16: 151-195

White JM (1990) Viral and cellular membrane fusion proteins. Annu Rev Physiol 52: 675-697

Acknowledgemnts
This work was supported in part by the Swiss National Science Foundation (Grant 31-25732.88 to CK)

SUBJECT INDEX

NATO ASI Series H

NATO ASI Series H

NATO ASI Series H

NATO ASI Series H